LIFE'S DEVICES

STEVEN VOGEL

Life's Devices

THE PHYSICAL WORLD OF
ANIMALS AND PLANTS

ILLUSTRATED BY
ROSEMARY ANNE CALVERT

PRINCETON UNIVERSITY PRESS
PRINCETON, N.J.

Copyright © 1988 by Princeton University Press
Published by Princeton University Press, 41 William Street,
Princeton, New Jersey 08540
In the United Kingdom: Princeton University Press,
Chichester, West Sussex

Library of Congress Cataloging-in-Publication Data
Vogel, Steven, 1940–
Life's devices: the physical world of animals and plants /
Steven Vogel; illustrated by Rosemary Anne Calvert.
p. cm.
Bibliography: p.
Includes index.
ISBN 0-691-08504-8 (alk. paper) ISBN 0-691-02418-9 (pbk.)
1. Biophysics. 2. Biomechanics. I. Title.
QH505.V634 1988
574.19'1—dc19   88-17603

This book has been composed in Linotron Baskerville

Princeton University Press books are printed on acid-free
paper, and meet the guidelines for permanence and
durability of the Committee on Production Guidelines for
Book Longevity of the Council on Library Resources

Printed in the United States of America

9   8   7   6

To my father, Max Vogel,

with great affection and gratitude,
on the occasion of
his seventy-fifth birthday

Contents

CONTENTS

Preface

How might we begin looking at living things? The ecologist, Marston Bates, once made a suggestion, saying "I think I'll start with a rabbit sitting under a raspberry bush and from this gradually go into the mechanics of the situation." This book is about just that—the mechanics of the situation.

The questions of concern here are enormously diverse, ranging from why trees so rarely fall over and the significance of the hull shape of baby sea turtles to the relative scarcity among organisms of right angles, metals, and wheels. Most of them, though, are exceedingly unsophisticated questions, not the sort that would occur only to a specialist at some frontier of research. They're about the ordinary activities of ordinary creatures—questions a person might pose while exploring a coastline or tramping through a forest. For me much of their appeal rests on just that commonplace character. More immediately, they have the advantage of making no special assumptions about what a reader already knows. Thus the people to whom the present book is directed comprise no well-circumscribed and carefully defined group. It's just me, talking about the things I find interesting, and noticing, in the process, that while the sophistication with which I address questions may have increased with age and experience, the questions themselves are as ingenuous as ever.

What draws these particular questions together is the kind of explanations to which they yield. Neither inheritance nor development nor cellular processes nor molecular events will play much of a role here—the present theater is a simpler one, even if the actors speak a language unfamiliar among biologists. For explanations, rationalizations, and syntheses we'll look mainly toward geometry and physical science; I want to get a step or so beyond a "gee whiz" presentation of living things as marvelous contraptions. For a nonbiologist, at least, the approach ought to be a natural one since it invokes notions close to one's everyday experience and intuitive sense of reality. Gravity and elasticity have an immediacy that cells and molecules do not.

Still, decent candor requires that I own up to a slight snag. The bits of physics that play a part here are well-plumbed, well-understood, and not the least bit controversial. They are, furthermore (as a bumper sticker once proclaimed), what makes the world go around. Oddly and unfortunately, they're not dear to the hearts of all of us—for some they're items of positive antipathy. Perhaps some social scientist might study the division of society into those superior sorts for whom matters mechanical are rational and those others for whom they're lurking demons or revealed truth. Education seems to make little difference. By this criterion the auto mechanic has culture while most lawyers remain primitive. The division extends to scientists, with biologists as divided as any, except that most of us were subjected to a few college courses in physics and mathematics, so we can't easily admit either innocence or fear. As James Thurber said about the founding editor of *The New Yorker*, "Ross approached all things mechanical, to reach for a simile, like Henry James approaching Brigitte Bardot. There was awe in it, and embarrassment, and helplessness."

Perhaps the problem is that at one stage or another many of us stepped out of an orderly sequence in quantitative science, and that subsequent material was then presented in unfamiliar terms of reference. I've faced the problem and tried to exclude as few people as possible from the intuitive satisfactions of the present topic. Both trigonometry and logarithms have been expurgated, and the little calculus I need has been reduced to words and graphs. If you don't recall what an equation is or that *ma* implies that some *m* is multiplied by some *a*, there's an appendix that makes the points. And if you don't know the difference between force and work or about the scaling of areas and volumes, you'll be told in the early chapters.

The result, perhaps inevitably, is a substantial dose of background early on and a text in which the ratio of words about physics to words about biology generally decreases chapter by chapter. The reader especially afflicted with physics-anxiety may thus find less joy in the early chapters than further along. Still, the casual reader can take a cavalier attitude toward the first few chapters and yet get some yield from the later ones. Biological background, incidentally, should present few problems. The book is, after all, about biology, which is in any case less hierarchical than are the physical sciences. The crucial matter of the operation of evolution by natural selection is dealt with in the first chapter; other items are easily available in dictionary, encyclopedia, or any ancient textbook.

While I hope that people outside the academic world find *Life's Devices* of interest, it makes no serious attempt to disguise its origins in my activities as a college teacher. It grows out of a course in biomechanics given by Steve Wainwright and me at Duke University and another given with Mike LaBarbera at the Friday Harbor Laboratories of the University of Washington. Its gestation has run parallel to that of a course entitled "Life in a Physical Context" given at Duke for the past few years in the graduate program in Liberal Studies—classes mainly of adult nonscientists. Calibration of the level of presentation has taken advantage of those students, to their occasional discomfort.

Beyond being an incidental artifact of my own trade, the mildly pedagogical character of the book is my deliberate attempt to write something usable as the primary text for a course. Nonprofessional students, undergraduates in particular, seem to find the present subject matter appealing, and teaching it has been a lot of fun. But especially in science, it's most awkward to have a course without a text; the primary or even secondary literature is not accessible to the complete novice. So a book usable for teaching might precipitate a course; without one a course is less likely to happen. The most obvious instructional addendum is the appendix of problems and demonstrations. Others are such things as the index of citations and the designation of general references, both incorporated in the bibliography.

A course built around the present material might have either of two missions. It might be given to undergraduates who do not intend to go further with biology or even with science. It then has the dual objectives of providing an unusual view of a very immediate world and of teaching some rather useful physics in the appealing context of life's diversity. Or it might be offered for people whose major interests are biological. The book should then serve as an entree to more specialized ones such as those by Alexander (1983), Wainwright et al. (1976), McMahon (1984), Currey (1984), Vincent (1982), and the present author (1981). In the latter role, such a course fits an undergraduate (U.S.) curriculum between traditional courses in physiology and ecology. Indeed, one of the points I'm trying to make with this book is that its subject has an intellectual coherence comparable to fields in which we ordinarily teach courses.

I would be less than candid if I omitted mentioning the substantial arbitrary and accidental elements in determining what's here and what's not. I'm writing about an area lacking any strong tradition of what is and what is not in its purview; thus personal taste, my own investigations, and the accidents of scientific association have played a large part. On one hand I've indulged a bias toward topics that I like; on the other I've avoided some that just seemed to get too complex too quickly—animal locomotion is the most glaringly slighted—where the distinction between summarization and oversimplification was elusive. Suspicion of bias—insider trading, sweetheart deals, and the like—is hard to suppress when I notice the heavy representation in the bibliography of people whom I account as personal friends.

While all the illustrations have been drawn specifically for this book, several represent only slight modifications of specific originals. Permission to use the following originals has been given by the various copyright holders and is gratefully acknowledged. Figure 5.3: Cambridge University Press (*Journal of the Marine Biological Association of the U.K.*); Figure 5.4: The Marine Biological Laboratory (*Biological Bulletin*); Figures 5.7, 6.13, 13.1, and 13.4: American Association for the Advancement of Science (*Science*); Figure 7.2: American Society of Mammalogists (*Journal of Mammalogy*); Figure 7.4: Academic Press Inc., Ltd. (*Animal Behaviour*); Figure 9.4: Society for Experimental Biology (*Symposium 34*); Figure 15.1: Sigma Xi, The Scientific Research Society (*American Scientist*). In addition, David R. Maddison permitted use of the drawing from which Figure 6.6 was derived, and Richard B. Emlet allowed use of a photograph upon which Figure 10.5 was based; I thank them both.

The impulse to do this book derives from my proclivity to proselytize for its subject matter and from the fact that my last book caused me less grief and got a better reception than I expected. Several people urged me to write a book, but this isn't the one any of them had in mind. I have, though, received generous help at every stage from planning to proofreading. I cannot recall the names of all the people whose suggestions have been included, much less the names of those whose ideas somehow didn't fit. A list would begin with at least half of my colleagues in the Department of Zoology and all of the students in our biomechanics group ("BLIMP"). Six graduate students kindly took the opportunity to turn the tables and criticize the curmudgeon ("Reciprocity," according to Walt Kelly, "is the spites of life")—all or part of the first draft was read by Hugh Crenshaw, Olaf Ellers, Matthew Healy, Carlton Heine, Anne Moore, and Katherine Weiss. The subsequent draft was scrutinized by Jane Vogel and the members of the 1987 liberal studies class. Professor Emeritus John Gregg has been especially helpful, commenting on both the entire manuscript and an earlier collection of prose with regard to matters logical and linguistic as well as scientific. No problem seemed dauntingly serious, no pitfall too wide to jump over after a dose of the quiet sagacity of Judith May and Diane Grobman at Princeton University Press. Norman Rudnick did more than any author should expect from a copy editor—he deftly removed my foot from my mouth on many matters physical, providing lessons to lessen the innocence of the biologist. Duke University provided the requisite sabbatical semester, and the DUPAC clinic restored the corporeal me to a condition equal to the present undertaking. To all of these I express my gratitude.

LIFE'S DEVICES

CHAPTER 1

Constraints and opportunities

"Throw physic to the dogs: I'll none of it."
Shakespeare, *Macbeth*

BIOLOGY conveys two curiously contrasting messages. In a strictly genetic sense all organisms are unarguably of one family. Our numerous common features, especially at the molecular level, indicate at least a close cousinhood, a common descent from one or a few very similar ancestors. On the other hand, what a gloriously diverse family we are, so rich and varied in size and form! The extreme heterogeneity of life impresses us all—trained biologists or amateur naturalists—with the innovative potency of the evolutionary process. The squirrel cannot be mistaken for the tree it climbs, and neither much resembles its personal menage of microorganisms. The apposition of this overwhelming diversity with the clear case for universal kinship tempts us to assume that nature can truly make anything—that, given sufficient time, all is possible though evolutionary innovation.

Some factors, though, are beyond adjustment by natural selection. Some organisms fly, others do not, but all experience the same acceleration due to gravity at the surface of the earth. Some, but not many, can walk on water, but all face the same value of that liquid's surface tension if they attempt the trick. No amount of practice will enable you to stand in any posture other than one in which your "center of gravity," an abstract consequence of your form, is above your feet. If an object, whether sea horse or saw horse, is enlarged but not changed in shape, the larger version will have less surface area relative to its volume than before. In short, there is an underlying world with which life must contend. Put perhaps more pretentiously, the rules of the physical sciences and the basic properties of practical materials impose powerful constraints on the range of designs available for living systems. The case for the pervasive operation of such constraints has been pointedly put forth in a recent essay by Alexander (1986).[1]

Were these restrictions the physical world's sole impact on life, we

---

[1] Allusions to a person and a year refer to entries in the bibliography at the end of the book. This practice, used in most scientific journals, not only gives some idea of the antiquity of the source, but emphasizes the fact that we who do science and write about it are real people.

might be content to work out a set of limits—quantitative fences that mark the extent of the permissible perambulations of natural design. There is, however, a more positive side, at least from our point of view as observers, investigators, and rummagers for rules. The physics and mathematics relevant to the world of organisms are rich in phenomena and interrelationships that are far from self-evident, and the materials on earth are themselves complex and diverse. Tiny cells with thin walls can withstand far greater pressures than would produce a blowout in any vertebrate artery, yet the materials of cellular and arterial walls have similar properties. The slime a snail crawls on may be alternately solid enough to push against and sufficiently liquid for a localized slide. An ant can lift many times its own weight with muscles not substantially different from our own. (But no Prometheus could exist among ants—as Went, 1968, remarked, the minimum sustainable flame in our atmosphere is large enough to prevent an ant from coming close enough to add fuel.) By capitalizing on such possibilities the evolutionary process appears to our unending fascination as a designer of the greatest subtlety and ingenuity.

This book is about such phenomena—the ways in which the world of organisms bumps up against a nonbiological reality. Its theme is that much of the design of organisms reflects the inescapable properties of the physical world in which life has evolved, with consequences deriving from both constraints and opportunities. In one sense it is a long essay defending that single argument against a vague opponent—the traditional disdain for or disregard of physics by biologists. In fact, the theme will function mainly as a compass in a walk through a miscellany of ideas, rules, and phenomena of both physical and biological origin. We'll consider, though, not the entire range of relevant items of physics, but a limited set of mostly mechanical and largely macroscopic matters. I mean to work through various bits of physics relevant to the design and operation of organisms and to illustrate their pervasive influence wherever I have appropriate examples.

The macroscopic bias should be emphasized. This book in places deals with some rather bizarre phenomena but never gets far from a kind of everyday reality. Explanations, where possible, deliberately ignore the existence of atoms and molecules, waves and rays, and similar bits of deus ex machina. Not that these aren't as real as our grosser selves (or so implies some very strong evidence); rather, in explanations for the general reader, they have an unavoidable air of ecclesiastical revealed truth. More importantly, to incorporate particle physics in a more rigorous view of the immediate world would take far more space and complexity than a single book. After all, can you think of any part of your perceptual

reality that demands the odd assumption that matter is ultimately particulate—that if you could slice cheese sufficiently thin it would no longer
be cheese? Maybe Democritus, commonly credited with the "invention"
of atoms, just made a lucky guess as an accident of his inability to imagine
anything infinitesimally small! Only when we consider the phenomenon
of diffusion (Chapter 8) do we need to recognize atoms and a real world
in which matter cannot be subdivided ad infinitum.

ABOUT SIZE

The largeness of people was implicit in our blithe disposal of molecules. The general topic of size receives undivided attention in Chapter
3, but, in fact, the widespread role of size is one of several secondary
themes throughout the book.

The ease with which we can avoid worrying about atoms reflects the
vast gap in scale between them and us, between the size of atoms or small
molecules and even small organisms. Cells (or unicellular creatures) may
be small, but inhabitants of the atomic realm are *much* smaller. There are,
roughly, as many molecules in a cell as there are cells in the cat observing
me write. (The point is crucial in Schrödinger's 1944 classic essay, "What
is Life." One of his arguments is that well-ordered structures can be built
of individually ill-behaved atoms only if enough atoms are used so that
their actions are statistically dependable.)

But from smallest to largest, we organisms ourselves occupy an extensive size range—from the tiniest bacterium about 0.3 micrometers long
(about a hundred-thousandth of an inch) to a whale about 30 meters long
(100 feet). (Some trees are 100 meters high but are no more massive than
the whale.) The range is about 100,000,000-fold; eight orders of magnitude we call it, counting the zeroes, or factors of ten. An excellent introduction to the truly cosmic subject of size is *Powers of Ten* by Morrison
and Morrison (1982).

Among organisms, humans are near an extreme—we're relatively big
creatures a meter or two long. Only a little over an order of magnitude
separates us from the largest living things, but six to seven orders lie
between us and the smallest. On a scale of orders of magnitude, a "typical" organism would be between a millimeter and a centimeter in
length—roughly an eighth of an inch. The point about size isn't trivial—
the appearance of the physical environment to an organism and the phenomena of immediate relevance to its life depend most strongly on how
big the organism is. You may not need to imagine the world of an atom,
but you'll find challenge enough in trying to get some intuitive sense of
the physical world of small creatures. Incidentally, for all of our fixation

on microscopes, biologists have not usually had much of that intuitive sense to which we'll aspire here.

The relationship between size and reality can be best put with a half-serious example. Consider all animals that live in air, that is, neither in water nor in some solid material. These creatures are much denser than the medium around them and therefore can fall if released from a height. But size enters into any examination of this business of falling. We can divide organisms according to the consequences of a fall into four categories that depend mainly on size.

In the first category, made up of creatures above roughly 100 kilograms (220 pounds) in mass, injury is possible if the animal falls a distance as short as its own height—tripping is a potential danger to cows, horses, and the like. The fall of an elephant is a matter of the utmost gravity. (We, especially as we get older, run a similar risk even at a lower mass; the upright posture of a human gives us an unusually great height relative to our mass.)

In the second category, comprising animals with masses between about 100 kilograms and 100 grams (4 ounces), falling may be injurious, but the fall must involve a distance greater than the height of the animal. Dogs should avoid cliffs, and cats must climb down trees with deliberation, but squirrels, near the lower limit, can take riskier-looking leaps of faith. Hedgehogs (about 500 to 1000 grams in mass) are also just above the lower limit but, according to Vincent and Owers (1986), cope with falls using a special device—spines that can act as shock absorbers.

In the third category, from 100 grams down to perhaps 100 milligrams (give or take an order of magnitude), no height is great enough to cause substantial injury from a fall—the hazard, if any, is the predator at ground level. Falls may all too often befall nestling birds, but do we ever notice one injured by impact? A few years ago, at the instigation of my skeptical colleague, Knut Schmidt-Nielsen, I dropped two adult mice from the roof of a five-story building onto pavement. Not only were they uninjured (briefly stunned, though), but they adopted a spread-eagle, parachutelike posture and fell stably. It certainly looked as if the neural circuitry of these small rodents was arranged to deal with the circumstance. (The extent to which this posture reduces falling speed might bear looking into.)

The fourth category includes the smallest airborne organisms, for whom falling itself takes on a peculiar meaning. Upon release, the creature (by which I mean either plant or animal—the word "organism" is awkwardly deficient in commonplace synonyms) goes downward only in an uncertain, statistical sense. Air is never still, and if falling speed is comparable or less than the speeds of upward and downward movement of air, then the direction of a fall is no longer dependably earthward. In fact, air is host to quite a diversity of seeds, pollen, spores, and tiny animals, to the great discomfort of those of us with allergies.

On the surface of the earth, gravity (gravitational acceleration, strictly) is everywhere the same. Yet its practical effects are widely divergent, depending mainly upon the size of the organism in question. As Haldane (1928) put what took me far more words, "you can drop a mouse down a thousand-yard mine shaft and, on arriving at the bottom, it gets a slight shock and walks away. A rat is killed, a man is broken, a horse splashes."

PHYSICAL VERSUS BIOLOGICAL SCIENCE

"Interdisciplinary" is a contemporary buzzword. By the usual divisions among fields, the present topic is, if it matters, thoroughly interdisciplinary. The mix does generate a few practical peculiarities, mainly a jumbled lot of antecedents with some resulting oddness in presentation.

Ordinarily we probably make too much of the distinction between biological and physical science, between living and nonliving devices. It certainly isn't a practice sanctified by antiquity. Galileo, whom we regard as a physical scientist, figured out that jumping animals, from fleas on up, should reach about the same maximum height irrespective of their body sizes (Haldane, 1928). (More will be said about jumping in Chapter 14.) A key element in developing the idea of conservation of energy was established by a German physician, Mayer, in 1841 from observations on the oxidation of blood, and the basic law for laminar flow of fluids in pipes was determined about the same time by a French physician, Poiseuille.

Physics and biology, with separate histories for the past few centuries, have developed their necessarily specialized terminologies in different and virtually opposite ways. Biology goes in for horrendous words of classical derivation, from *Strongylocentrotus droehbachiensis* (a sea urchin whose roe is accounted a delicacy by some) to anterior zygopophysis (a minor protuberance on a vertebra). Each word has been defined more precisely than your workaday household noun in order to reduce misunderstanding and terminological controversy. That the jargon tends to exclude the uninitiated and those without youthfully spongelike memories is not (for better or worse) given much consideration.

By contrast, physics (and engineering) eschews Greco-Latin obfuscation and pretension; in doing so, it creates an equally serious difficulty. The most ordinary, garden-variety words are given precise definitions that unavoidably differ from their commonplace meanings. It takes *work* to pull something upward but not to hold it suspended. Stress and strain

7

are entirely distinct, the former commonly causing the latter. Mass is not the same as weight, even if they are functionally equivalent on terra firma. Both physical and biological practices will plague the reader, but the former tends to be more subtly subversive—a bit of biological jargon is jarring when you don't know its meaning, but an ordinary word with a special definition for scientific use easily passes unnoticed.

The next chapter will be largely given to the task of establishing a necessary physical base, with a fair dose of the associated terminology. Biological terminology will enter piecemeal—for present purposes physics does a better job of providing a logical framework.

One term from physics needs special attention at the start: *energy*, which gets the most cavalier treatment by press and politicians. We ought to be able simply to define it with care and proceed from there. While it *does* have a precise meaning in the physical sciences, the trouble is that the meaning doesn't lend itself to expression in mere words. Basic dictionaries and textbooks are little help—they define energy as the capacity for doing work, unblushingly evading the issue! Feynman (et al. 1963), comes right out with the unusually candid admission (no company man was he, whether teaching physics or serving on the commission probing the shuttle explosion), "It is important to realize that in physics today, we have no knowledge of what energy *is*. We do not have a picture that energy comes in little blobs of a definite amount."

In practice the idea of energy explains so much—the law of conservation of energy is sometimes considered the greatest generalization of physics. Ultimately that's the advantage of energy. For us it is more of a difficulty—it's just too easy to hide behind a word with no ready definition and thereby to avoid some crucial explanations. So the word and the concept will be only a parenthetical presence until the final chapter.

EVOLUTION AND NATURAL DESIGN

The words "evolution" and "design" have already surfaced; I find it hard to avoid either in any general discussion. Used together, they represent a subtle contradiction, one that ought to be resolved before we go further. If the process of evolution is incapable of anticipation, that is, if it is blindly purposeless, the term "design" is seriously misleading—in common usage, design implies anticipation and purpose. The problem is not just terminological. Why do organisms appear to be well designed if they are not designed at all? Perhaps it's best to begin by reviewing the logical scheme for which "evolution by natural selection" is the quick encapsulation.

First, some observations. Every organism of which we have any knowledge is capable of producing more than one offspring; thus, populations of organisms are always capable of increasing. It takes, though, some minimum quantity of resources for an organism to survive and reproduce, and, in the long run, the resources available to any population are limited. Next, three consequences. One is that a population in a particular area ought to increase to some maximum. A second is that once the maximum is reached, more individuals will be produced than can find adequate resources. The third is that some individuals will not survive to reproduce. Pause here to consider further observations. Individuals in any population vary in ways that affect their success in reproduction, and at least some of this individual variation is passed on to their offspring. Now a final consequence. Features that confer increased relative success in reproduction will appear more often or in exaggerated form in the individuals of the next generation. We say, in short, that these features will have been "naturally selected," that is, by selection only from preexisting, even if latent, variations.

The model, at this level, is one of the least controversial items of modern science—every aspect has been observed and tested, and competing models for the generation of biological diversity (even if logically without flaw) uniformly fail to correspond to reality. Indeed, given geological time and the variation generated by an imperfect hereditary mechanism, it is difficult to see how evolution could be avoided. Remaining argument devolves about details—whether the process is usually steady or episodic, the roles of specific genetic mechanisms (such as sexual recombination), and so forth. The model has no place for anticipatory design, and there is no need (indeed, no evidence) that an environmental challenge can determine the character of the variation upon which natural selection can act.

Selection, quite clearly, operates most directly on individual organisms. The main test, defining its "fitness," is an organism's success in engendering progeny. (Some adjustment has to be made for indirect contributions that aid the reproduction of one's kinfolk, but this is of little present concern.) The selective process knows nothing about species; no clear evidence indicates that any organism ever does anything "for the good of the species." Nor does the process care directly about parts of an organism. Legions of cells die on schedule in the development of an individual; in no way can we speak of such cells as more or less "fit" than any others. Trees commonly shed leaves; the shed leaves were not therefore less fit— the term fitness is inapplicable here since it refers only to the reproductive potential of potentially reproductive individuals, that is, the whole trees.

This book is mainly about organisms, so we will be concerned with a

9

level of biological organization upon which the invisible hand of the selective process should incur fairly immediate consequences. It is the immediacy of operation of that unseen hand that makes organisms appear well designed—as a colleague of mine put it, "The good designs literally eat the bad designs." But it must be emphasized that we mean "design" in a somewhat unusual sense, implying only a functionally competent arrangement of parts resulting from natural selection. In its more common sense, implying anticipation, "design" is a misnomer—it connotes the teleological heresy of goal or purpose. Still, verbal simplicity is obtained by talking teleologically—teeth are for biting and ears for hearing. And the attribution of purpose isn't a bad guide to investigation—biting isn't just an amusing activity incidental to the possession of teeth. If an organism is arranged in a way that seems functionally inappropriate, the most likely explanation (by the test of experience) is that one's view of its functioning is faulty. As the late Frits Went said, "Teleology is a great mistress, but no one you'd like to be seen with in public."

We functional, organismic biologists are sometimes accused of assuming a kind of perfection in the living world—"adaptationism" has become the pejorative term—largely because we find the presumption of a decent fit between organism and habitat a useful working hypothesis. But the designs of nature are certainly imperfect. At the very least, perfection would require an infinite number of generations in an unchanging world, and a fixed world entails not only a stable physical environment but the preposterous notion that no competing species undergoes evolutionary change. Furthermore, we're dealing with an incremental process of trial and error. In such a scheme, major innovation is not a simple matter—features that will ultimately prove useful are most unlikely to persist through stages in which they are deleterious or neutral. So-called hopeful monsters are not in good odor. Many good designs are simply not available on the evolutionary landscape because they involve unbridgeable functional discontinuities. Instead, obviously jury-rigged arrangements occur because they entail milder transitions. In addition, the constraints on what evolution can come up with must be greater in more multifunctional structures. Finally, a fundamentally poorer, but established and thus well-tuned, design may win in competition with one that is basically better but still flawed.

I make these points with some sense of urgency since this book is incorrigibly adaptationist in its outlook and teleological in its verbiage. The limitations of this viewpoint will not insistently be repeated, so the requisite grain of salt should be in the mind of reader as well as author. Incidentally, the ad hoc character of many features of organisms are recounted with grace and wit in some of the essays of Stephen Jay Gould, not just as an argument against extreme adaptationism but as evidence for the blindly mechanical and thus somewhat blundering process of evolution. His collection entitled *The Panda's Thumb* (1980) is particularly appropriate here.

SIMPLIFYING REALITY—MODELS

This book is, in the final analysis, about organisms rather than physical science—the latter merely provides tools to disentangle some aspects of the organization of life. But, beyond using physics to organize the sequence of things, we'll take an approach more common (historically, at least) in the physical sciences. Biologists love their organisms, collectively, singly, sliced, macerated, or homogenized. Abstractions and models are vaguely suspect or reprehensible. As D'Arcy Thompson (1942) put it, biologists are "deeply reluctant to compare the living with the dead, or to explain by geometry or by mechanics the things which have their part in the mystery of life." But we will repeatedly use the "dead" to explain the "living." Explanation requires simplification, and nothing is so un-simple as an organism. And the most immediate sort of simplification is the use of nonliving models, whether physical or (even) mathematical.

Science is, in fact, utterly addicted to models for simplification and generalization. Even a tiny aspect of the world is just too complex to yield to simultaneous and systematic analysis of all of its diverse characteristics. Consider, for a moment, your left thumb—how many facets of this minor appendage might be measured, recorded, and subjected to statistical treatment? Simplification and abstraction have marked all progress in science; one begins very simply and then adds elements of complication as necessary and possible. We'll do just that, introducing some topic and asking very simple questions about it, then repeatedly returning to the same topic with questions that require more sophisticated analyses. Acceleration, for instance, will be discussed with reference to simple jumps, to jumps with air resistance and the trajectories of projectiles, and to the mechanics of the supply and storage of the work of propulsion.

CONTRASTING TWO TECHNOLOGIES

Much of the popularity of science fiction, I think, comes from its common focus on technologies alternative to the one developed on earth through human activity in the late twentieth century. A similar attraction must underly popular support for the search for extraterrestrial intelligence—the possibility of comparing what we've made here with alternative scenarios holds a strong intellectual appeal. But extraterrestrial life,

11

much less intelligence, is elusive and its discovery is only a very remote prospect (the recent recognition of its remoteness was described by Horowitz, 1986). And the stuff of science fiction is both pretty anthropocentric and ultimately fictional.

Such a comparison between our technology and an alternative can nonetheless be made and turns out to be an unavoidable, if perhaps adventitious, aspect of the present book. The alternative technology available for our examination is the one generated here on earth through the operation of natural selection, which has resulted (in the most corporeal sense) in ourselves. The comparison is particularly interesting in that, first, the generating mechanisms are as different as can be—natural selection, strictly, implies no anticipation or calculation, unlike human design. Second, both sorts of technology use the substances available on the surface of the same planet. The contrast between them is another secondary theme, best introduced through a set of comparisons between "natural" (but not entirely unhuman) and "human" (not completely unnatural) technologies.

(1) Surfaces of and within organisms are curved, most commonly cylindrical, but sometimes with spherical or elliptical elements. (The major theme of Wainwright 1988 is the ubiquity of such shapes.) Flat surfaces are less common. By contrast, people make load-bearing flat surfaces in profusion—floors, roofs, walls, even the surfaces of beams. Cylindrical elements—pipes, cans, bicycle frames—are certainly not scarce but don't dominate.

(2) Our technology is rife with right angles—never mind pyramids, it's the 90° angle to which we seem addicted. It appears in almost every door, window, floor tile, box, book pages, many letters of our alphabet, the pockets of my shirt, and on and on. Yet right angles are surprisingly rare among organisms. Tree trunks are generally at right angles to the ground or horizon, but other examples are not easy to find.

(3) We use a few pliant materials–plastic hinges, elastic bands, rubber pads, and so forth; but relative to the abundance of our stiff stuff, soft and stretchy substances are unusual. We manage to live with the awkward tendency of stiff materials to fracture. We even fabricate them in curious geometries to take advantage of their limited deformability—coiled springs of steel spring to mind. Nature is typically pliant—skin, muscle, viscera, even fresh wood (dry timber is several times stiffer). Stiff material does occur—teeth, clam shells, big bones—but less commonly.

(4) Our preferred structural materials are most often made of single components above the molecular level, and the values of their properties are the same (isotropic) whatever the direction of measurement—we mostly use metals and ceramics. Nature's materials are composites, com-

binations of two or more components, almost always arranged so that the materials' mechanical behavior depends on the direction in which they're loaded. We do make such anisotropic composites—we combine oriented glass fibers and glue to make fiberglass—but their use is limited. (And "composite material" seems always to be preceded in the popular press by "advanced"!)

(5) Substantial pieces of metal, either pure or alloyed, never occur in nature, even though metallic atoms are crucial to the biochemistry of all organisms, and tiny chunks are basic in magnetic sense organs. Ours is an overwhelmingly metallic technology, and we capitalize on the impressive mechanical advantages and diversity of properties available in metals.

(6) Both gases and liquids resist being squeezed and thus can be used as structural materials; air and water are the cheapest and most available of substances. Occasionally we use air as a compression-resisting material—in blimps, inflatable buildings, door closers, and so forth—but I can't think of a clear case where nature employs air in such a manner. Conversely, nature makes elaborate and extensive use of water as such a compression-resisting material in sea anemones, penises, squid tentacles, worms, sharks, and elsewhere; but we use it in only a few devices such as fire hoses that collapse when not being used.

(7) Life may tolerate a reasonable range of ambient temperatures, but organisms are basically isothermal machines rather than heat engines and do their business without depending on large internal differences in temperature. Heat conduction, therefore, is not a major issue in organisms—handy, since we aren't built of the wonderfully conductive metals. But our functional parts (cells and so forth) are often very small, and a formally analogous process, "molecular diffusion" (Chapter 8), is always crucial. Human technology makes impressively elaborate use of heat conduction but less of diffusion.

One can continue such a list, although the items get more obscure and complex. You'll notice that I haven't given more than a hint of an explanation of the differences between the two technologies for any of these examples. The notion of evolution can provide some basis for the distinctions. Beyond that, explanations and rationalizations will come later, at least for those items about which something reasonable can be said.

A variety of variables

"Time is nature's way of keeping everything from
happening all at once."

Attributed to Woody Allen

THIS will be a chapter about tools—terminological, conceptual, and logical—a set of tools from the physical folk for us to use in analyzing biological situations. If terms were all that mattered, a glossary would do; as it happens, we face somewhat worse complexity. The relevant items are simply too intricately interrelated to make much sense except as components of sequences and hierarchies. It's a big dose of abstractions, but you can easily return to the present pages whenever necessary, so complete mastery isn't crucial.

MEASURING QUANTITIES

Even when we're trying to decide some essentially qualitative question, we almost always use quantitative reasoning and methods (e.g., measurement and calculation). There's nothing really strange about the process— houses come in a range of prices; you want a house; you use quantities such as price and monthly payment to determine a qualitative issue: Should you buy a particular one? Multifaceted quantitative data may be required—which is preferable in some application, a suspension bridge or an arch bridge? The only novelty here is that this sort of reasoning is so ubiquitous. An outsider may view science as an endless pursuit of data, but for all our addiction to numbers they're just a means, not an end in themselves. To pursue an earlier example, it's important to know the value of the acceleration due to gravity at the earth's surface because with it we can judge how far you may fall without serious prospect of damage upon impact.

Measurement and quantification also have a more subtle aspect. For some purposes, very good data are mandatory. If you want to predict eclipses, a very accurate measure of gravitational forces is crucial, much more accurate than for worrying about the hazards of falling. How much accuracy is sufficient depends entirely on the question one is trying to answer. The process is the same as that used by the designer of a product to be made on an assembly line—it's necessary to specify just how inaccurately formed each component may be and still fit satisfactorily into the final product. A good design is one that is tolerant of sloppily made components, and a robust hypothesis can often be tested with crude data. The common notion that scientists are as accurate as possible represents a serious misunderstanding. The trick is to figure out beforehand what level of accuracy is required and then to waste no effort doing substantially better. I think I've never needed to work to closer than a part in a thousand when doing science, but the numbers on my tax return pretend to better than a part in a million.

Indeed, we'll have occasion to talk in completely quantitative terms with a *lack* of accuracy quite beyond everyday experience. For some issues, order-of-magnitude accuracy is entirely sufficient; in fact, just that crudity characterized the discussion of size in the last chapter. Given the variety of shapes in which organisms come, great accuracy in specifying size would have amounted (in that instance) to a tacit delusion.

Quantification implies mathematics. So without some mathematics, it would be quite awkward to explain almost any of the things this book is about and, worse, quite impossible to provide any sense of scale or proportion. But the requirements for mathematical formality are not really great—I have found it possible to do end runs around trigonometry and logarithms and to do what calculus I need in words and graphs. The irreducible minimum, though, is algebra. If the formalities of algebra have left your immediate recall, a few pages of reminders are provided in Appendix 1.

And measurement implies units. It would be handiest to have only a single set of units, built up from something other than historical accident. The metric system, one of the lasting results of the French revolution, is popularly considered such a system. (I've always viewed the scheme as insufficiently revolutionary—they should have gone further and changed the counting system from a decimal, or base of ten, to a duodecimal, or base of twelve, rather than merely decimating the units.)

Recently, a more rigid version of the metric system has been urged on the scientific community—the so-called Système Internationale, SI for short. It involves a few unfamiliar units that will be taken up as needed—kilograms, watts, and meters may be ordinary, but newtons, joules, and pascals are less so. Its main attraction isn't just that it is the way science is *supposed* to be done but that it provides us with a most convenient internal consistency and circumvents all sorts of constants and conversion factors. The main features of SI (commonly and redundantly called the "SI system") are the following:

(1) Fundamental units: kilograms for mass, meters for distance, and seconds for time. The prefix "kilo" on "gram" is a minor bit of awkwardness tolerated to keep the basic units intuitively familiar without rejecting

the original metric names. It means that a milligram, a thousandth of a gram, is a millionth of the fundamental unit. A biologist, used to small scale work, would surely have set the gram as the basic unit of mass.

(2) Prefixes for multiples and fractions of fundamental units in steps of three orders of magnitude (a factor of $10 \times 10 \times 10 = 1000$): millimeters are okay, but centimeters are officially suppressed (we'll occasionally use centimeters anyway on account of their handy size and familiarity). For our needs six prefixes will suffice:

giga-	(G)	a billion (U.S.) times	$10^9 \times$
mega-	(M)	a million times	$10^6 \times$
kilo-	(k)	a thousand times	$10^3 \times$
milli-	(m)	a thousandth of	$10^{-3} \times$
micro-	(μ)	a millionth of	$10^{-6} \times$
nano-	(n)	a billionth (U.S.) of	$10^{-9} \times$

It's easy if you're a proper classicist (which I am not)—the enlarging prefixes come from Greek (giga = giant; mega = large; kilo = thousand) while the diminishing prefixes come mainly from Latin (milli = thousand; micro = small; nano = dwarf).

(3) Rules for combining units: the main one for us is that prefixes in complex units can attach only to numerators, not denominators. Thus, a kilogram per cubic meter is fine, but a microgram per cubic millimeter (logically the same thing) is unlawful. The results are sometimes amusing—the breaking strength of a spider's thread is properly given in meganewtons of force (a meganewton is about the weight of a 100-metric-ton mass) per *square meter* of cross section (a *very* thick strand).

DIMENSIONS AND UNITS

To begin with, explaining even our limited bit of biology involves lots of things such as force, mass, density, pressure, and acceleration—"physical quantities" or "physical variables" we'll call them. What these have in common is that they are quantitative attributes of physical objects or situations. To put them into a systematic scheme, we require some explicit way to express, not just the quantities, but the relationships among them. The simplest approach is to show that some things called the "dimensions" of the quantities are just combinations of a very few basic ones. (Note here the peculiarity of specialized jargon—the term "dimension" is being used in quite a different sense from the familiar length, width, and height.)

The actual choice of which dimensions to declare basic is a bit arbitrary. A common practice is the use of the following three (eventually, a fourth, temperature, will sneak in):

Mass	given the symbol	M
Length	given the symbol	L
Time	given the symbol	T

Each dimension is a property of the physical quantity with the same name—but not necessarily a property unique to that quantity. Thus the quantity "mass" refers to an amount of matter; the dimension "mass" is more general and is a property of anything that can be measured on the same scale (or with the same units) as mass. For instance, pressure, stress, and stiffness share the same dimensions. This curious and abstract distinction between dimensions and quantities raises little practical trouble. Nor does definition of the physical quantities corresponding to these basic dimensions—we'll just assume common knowledge of what is meant by mass (quantity of matter), length (the distance between two points on a specified path), and time (the rate of conversion of future to past). Quite obviously each can be measured in a proper quantitative fashion; almost as obviously you'll recognize that each has a real zero point (not like the arbitrary zero of the ordinary temperature scales) and that none can be explained in terms of the other two.

Note that neither physical quantities nor dimensions are synonymous with units. The dimensions of a quantity constitute the scale on which we measure it; the units are the rather more specific matter of how we mark or subdivide that scale. Notice, though, that the names of units (such as "meters per second") often reveal the underlying dimensions (such as "length over time")—unit names are a great help to a fallible memory trying to remember the dimensions of some variable!

The business of figuring out the basic dimensions for a physical quantity is neither mysterious nor difficult. "Speed" is a measure of how long (time) it takes to go a certain distance (length). Or length per unit time. Or length/time. Or, in dimensions, L/T. Or, as we'll express the matter henceforth,

speed has dimensions of $(length)^1 (time)^{-1}$ or LT^{-1}.

Volume, to give another example, can be viewed as the product of a length, a width, and a height, three length dimensions, so

volume has dimensions of $(length)^3$ or L^3.

Formulas for volumes may contain other items, but they must be pure numbers quite devoid of any dimension. Thus, the formula for the volume of a sphere (Appendix 1) is the product of a fraction (four-thirds), the ratio of the circumference to the diameter of a circle (pi, or 3.1416), and the radius to the third power—"four-thirds" and "pi" are dimensionless, while "radius" has the dimension of length. Also, as we see here, a given length may appear more than once in a formula. Still, any expression for volume *always* has three length dimensions, even if a single physical variable (radius here) has to provide all three.

A physical quantity may be dimensionless; it gets that way by being defined as the ratio of a variable having a certain dimension to another variable with the same dimension. For example, "strain" (yet again a special meaning of an ordinary word) is the distance a material is stretched (by a stress) divided by its original length, the ratio of one length to another. Since length divided by length leaves no resulting dimension at all (recall pi), strain is quite nakedly dimensionless and, of course, unitless as well. You might say, if it helps your intuition, that strain is measured in meters per meter, but that's a sophomoric sort of sophistry. We'll encounter other dimensionless variables and find that they're surprisingly ordinary. Their main oddity is really a nice convenience: they come out the same no matter what units are used to obtain them! Thus, for strain, one gets the same numerical result whether original length is in inches and extension is also in inches or whether the two are both in cubits or fathoms. A strain of 0.01 just means that something has been stretched 1% beyond its original length.

It's worthwhile being careful to keep the dimensions distinct from the physical quantities, and it's even more important to maintain a careful distinction between the quantities and their units. Don't *ever* indicate units within a formula; when units must be specified, mention them in a separate bit of text. In an equation such as $F = mg$ there shouldn't be even the merest hint of the abbreviation for a milligram. The biological literature has too many instances of violations of this rule, which not only produce confusion but subject us to well-deserved ridicule. And don't mix systems of units in using a formula—first convert each datum to a consistent system, such as SI, appropriate to the formula.

Physical quantities

I once wrote a polemical paper arguing that the physical characteristics of an animal's world are relevant to its behavior (Vogel 1981a); in the process I listed some physical quantities that seemed to be at least occasionally important ("bioportentous parameters" in the paper). The list was compiled from contributions solicited as I took lunch at a different table each day at a marine laboratory. After collecting about thirty items

I began to imagine that people were avoiding me; with no real end in prospect, I arbitrarily ended the exercise. We'll not worry about quite so many quantities in this book, and we certainly don't need to begin with all of those that we will eventually encounter. But it will be useful to step through a set of explanations or evasions, moving from the simpler and more intuitively real to the more derived and abstract. Table 2.1 summarizes the prose and gives the combination of basic dimensions for each physical quantity; you might find it useful to dog-ear its page.

Mass, length, time

Mass, length, and time have already appeared and we'll not attempt further definition. Note, though, that lowercase italic "*m*," "*l*," and "*t*" will designate these quantities, with the uppercase roman letters reserved for the corresponding dimensions.

Area

An area is a length in one direction multiplied by a length at a right angle to the first and often multiplied by some dimensionless factor (perhaps pi) as well. So it's always proportional to length times length, or length squared. We'll use the symbol S (think of "surface") whether we're talking about surface area, cross-sectional area, projected or shadow area, or any other area. (Occasionally it's useful to invent nominal areas—

TABLE 2.1. A VARIETY OF VARIABLES

Quantity	Symbol	Origin	Dimensions	SI Units
Mass	m	(fundamental)	M	kilogram (kg)
Length	l	(fundamental)	L	meter
Time	t	(fundamental)	T	second
Area	S	∝* to length squared	L^2	square meter
Volume	V	∝ to length cubed	L^3	cubic meter
Velocity	v	length/time	LT^{-1}	meter/second
Acceleration	a	speed change/time	LT^{-2}	meter/second^2
Force	F	mass × acceleration	MLT^{-2}	newton
Momentum	mv	mass × velocity	MLT^{-1}	kg meter/second
Stress	σ	force/area	$ML^{-1}T^{-2}$	⌈ newton/meter^2
Pressure	p	force/area	$ML^{-1}T^{-2}$	{ or ⌊ pascal
Work	W	force × length	ML^2T^{-2}	joule
Power	P	work/time	ML^2T^{-3}	watt
Density	ρ	mass/volume	ML^{-3}	kg/meter^3

* ∝ is the symbol for "is proportional to."

quantities that have the dimensions of an area and thus behave like areas in formulas. For example, volume raised to the two-thirds power is a nominal area of some use for comparing biological items of diverse shapes.)

Volume

As mentioned earlier, volume is the product of three lengths (and some constant that depends on the shape at hand). Each length with which a volume is specified is at right angles to the other two—this puts us clearly in a three-dimensional world and solid geometry. Thus, volume is proportional to length cubed (third power); we'll use the symbol V.

Speed and velocity

Speed is the ratio of a distance (length) to the time it takes to go that distance; to it we'll assign the symbol v, using the lowercase to distinguish it from volume. The v really stands for "velocity," a special sort of speed which properly includes some specification of the direction as well as the rapidity of travel, and about which more will be said shortly.

Acceleration

If speed is viewed as the rate at which your location changes, then acceleration can be visualized as the rate at which your speed changes. In short, it is the ratio of a difference in speeds (which retains the character and dimensions of a speed) to the time that elapses as the speed alters from one to the other. It includes the notion of deceleration—when the speed happens to be decreasing, and the acceleration is negative. If you run into a wall, the magnitude of your acceleration (negative) may be inconveniently great—your speed changes a lot over a short period of time. Increasing the time period through controlled collapse of your vehicle can usefully reduce the absolute value of the acceleration and the corporeal consequences. We use the symbol a and recognize its dimensions as length divided by time (speed) and divided again by time, i.e., length divided by the square of time.

One particular value of acceleration is especially relevant here, and allusions to it were scattered through the first chapter—it is the downward acceleration of an object under the sole influence of the gravitational force at the surface of the earth. This value is about 9.8 meters per second squared (32 feet per second squared); it's usually referred to as g. In our perceptual world that's an unpleasantly high value, corresponding (to use the benchmark of my youth) to a time for steady acceleration from 0 to 60 miles per hour of 2.7 seconds.

Force

Here things begin to get more complex. It's more difficult to set a large mass in motion than a small one and to speed up a given mass in a shorter period of time—to give something a large acceleration. In these situations "difficulty" is essentially synonymous with "force." (Similarly, a large mass slowed down by impact can give something a substantial push, a clear disadvantage of being kicked or punched.) Two factors are involved, mass and acceleration, and force, with symbol F, is defined as the product of the two—perhaps the most useful of all the equations we'll encounter is this one, simply put as $F = ma$. A special name has been coined for the SI unit of force, the "newton" (N) (after Sir Isaac). It's the same as a kilogram-meter per second squared (from mass times acceleration).

But force can also be exerted without any noticeable motion being involved, as when you stand on the floor or attempt to squeeze an incompressible object such as a brick. So, while force is defined in terms of acceleration, the former can exist without the latter—the squeeze on the brick is measured as the force which would, if applied to one side of a movable mass, give it a certain acceleration. Acceleration, it turns out, happens only as a result of a *net*, or *unbalanced*, force. Objects stay at rest or in steady motion (constant velocity) unless acted upon by an unbalanced force. When you squeeze the brick it responds by pushing outward against you with equal force, so the force is balanced. The reality of this countervailing force will be no small matter when we talk about the behavior of nonrigid materials.

What about the force on the floor? Here gravity enters—an object is attracted to the earth with a force proportional to its mass times the mass of the earth and inversely proportional to the square of its distance from the earth's center. Since the earth's mass is constant, and the distance to the earth's center (about 4000 miles) is practically the same for all objects near the surface, the proportionality of the force to the mass of the object is all that normally matters. This force is exposed when objects left to themselves (dropped, for instance) accelerate downward—$F = ma$, where a becomes g, the acceleration due to gravity. Even if the object isn't free to fall, though, the force is still there, and it's still proportional to the mass of the object; it is countered by another force that denies the object its freedom, so there is no net force to cause acceleration.

A bathroom scale measures the unhappy consequences—the force, whether measured in pounds, stones, or newtons, is what we ordinarily call "weight." Again, an object weighs an amount equal to its mass times the acceleration it would experience if falling freely. On the earth, where

21

that acceleration is about 9.8 meters per second squared, as mentioned before, $f = mg$ for one kilogram becomes $1 \times 9.8 = 9.8$ newtons. Thus, a mass of one kilogram has a weight (here on earth) of just 9.8 newtons. The relationship, so tidy in SI units, is messy when it comes to pounds. The weight (force of gravity) of a mass of one pound is also called a pound, so one doesn't know unless it is made explicit whether a "pound" is a mass or a weight. A force equal to a pound weight will still give an acceleration g to a pound mass, but the proportionality constant is swallowed by the terminology. In this book, using SI units, kilograms will always specify mass (with pounds of mass given parenthetically) and newtons will specify force.

In all of our personal experience, weight is proportional to mass with the same constant of proportionality (the gravitational acceleration on earth). So it's difficult to get a personal sense of the fundamental difference between the two quantities. On the moon, the same 1 kilogram mass would weigh only 1.6 newtons, or 0.37 pounds (weight); the consequences are revealing. Consider the gait of an astronaut on the moon. He hops upward quite well—the same muscular force makes his mass go further upwards when the downward acceleration is less. He hops forward as impressively, but we notice that it takes quite a forward tilt to get going—the same mass must be accelerated, but the concomitant push backward on the ground requires weight, and the latter is in short supply. Even odder is the business of cornering—to change the direction of motion requires as much force as on earth. Again, it's a matter of accelerating an unchanged mass. But the force must be exerted against the ground, and the scarcity of weight means that an astronaut must lean far over to come around a turn. What fascinates me is just how splendidly well the human neuromuscular machine copes with this personal and evolutionary novelty! To repeat, on the moon an object has less weight than on earth but is not one bit less massive.

And we use beam balances and spring scales more or less interchangeably even though one indicates weight and the other mass. It's amusing to contemplate the consequences of a small change in the earth's gravity. One class of device would require recalibration; the other would work quite as before. The reader can figure out which is which.

Momentum

Momentum is another composite variable, the product of mass and speed (really velocity). It turns out to be especially useful, as first recognized by Descartes who referred to it as the "quantity of motion"—a 10 kilogram object going 1 meter per second has as much "motion" as a 1 kilogram object going 10 meters per second. No special symbol or SI unit attaches to momentum—we merely use the symbol mv and the unit kilo-gram-meter per second. We'll have a lot to say about momentum later.

Stress

If you exert ten newtons of force pushing the broad side of a knife against a carrot, the carrot will be none the worse; if you use the same force and the sharp edge, the carrot will be cleaved. Clearly, the area over which the force is applied is of interest—more often than not what matters is not force per se but force divided by the area against which it is applied, which is what is meant by "stress" in the physical as opposed to the psychological sense. The symbol used is the Greek lowercase sigma (σ); the SI unit is designated the newton per square meter or, alterna-tively, the pascal (Pa) after Blaise Pascal (1623–1662). A stress inevitably causes a strain (mentioned earlier); the two are most emphatically not the same thing—the former is how much you aggress upon something, the latter how much it is deformed as a result.

Pressure

Related in a slightly confusing way to stress is pressure; both are forces divided by areas. We'll find the following distinction convenient. Stress is a unidirectional phenomenon—you press (should I say stress?) down-ward with the knife. Pressure is omnidirectional—you pump air into a tire and the tire inflates in any direction it can get away with. We'll use a lowercase p for pressure; the SI unit is again the newton per square me-ter or (more often for pressure) the pascal. It's a very small unit—atmos-pheric pressure is about 101,000 pascals (Pa), and my (nonhypertensive) arterial blood pressure ranges from about 14,000 (systolic) to 11,000 (di-astolic) Pa.

Besides pascals there are quite a few other units of pressure. Some raise no difficulties—pounds per square inch (14.7 of them equal an at-mosphere), for instance. Others are distinctly odd—inches of mercury or millimeters of water are units of height with some peculiar specification of a material. They refer to the results of applying a pressure to a column of liquid (Figure 2.1)—the more pressure, the higher the column that can be supported. Neither the area of the exposed surface of liquid nor the diameter of the column nor the empty volume above the column make any difference in the height that a given pressure will support. To get these latter into respectable forces over areas, one multiplies the given height by the gravitational acceleration on earth and then by the density (see below) of the indicated material. Thus 29.9 inches of mercury times 0.0254 meters per inch (converting inches to meters) times 9.8 meters per second squared (acceleration) times 13,600 kilograms per cubic me-

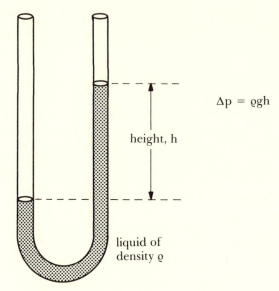

$\Delta p = \varrho g h$

height, h

liquid of
density ϱ

FIGURE 2.1. A simple U-tube manometer for
measuring a pressure difference. If there's another liquid above the depressed portion of
manometer liquid (shaded), as when working
with pressures in water, then one uses the difference in their densities in the formula. If, as is
usually the case, the device is used to measure
pressures in a gas, the slight density of the gas
can be ignored.

ter (density of mercury) gives 101,000 Pa—the first number is the figure
the (U.S.) weather reporter gives for typical barometric (atmospheric)
pressure at sea level.

One other detail about pressure. We live with a background pressure
of about an atmosphere. Its value has to be considered only in rare devices such as very early steam engines—in these a stream of water con-
densed steam in a cylinder, thus creating a partial vacuum, and the outside pressure of an atmosphere then pushed the piston inward. Or in the
inspiratory movements of the breathing scheme of animals such as we—
you force your rib cage outward and upward and your diaphragm downward, and air enters your lungs, pushed inward by the atmosphere of
pressure outside (101,000 Pa) working against a slightly lower pressure
(minimally about 90,000 Pa) inside (Figure 2.2). For most purposes,
though, the background pressure is presumed and ignored, and what is

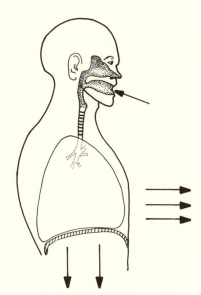

FIGURE 2.2. A person having an inspiration. Air enters the mouth because the diaphragm flattens, moving downward, and the rib cage moves forward and upward.

cited is pressure above atmospheric, called "gauge pressure"—that's what a pressure gauge for tires or blood indicates.

Work

Applying a force to something needn't accomplish much at all in a strictly physical sense. We accomplish something when work is done; the work done is defined as the product of the force applied to an object times the distance the object moves while the force is acting and in the direction of application of the force. Thus, a chain on a chandelier may exert an upward force on the chandelier without pause for a century or two, but it does no work, requires no motor, and eats no food. By contrast, the carriage of an ascending elevator must be worked on by a motor or descending counterweights. For work we'll use the symbol W; the SI unit is the joule (after James Joule, 1818–1889), sometimes called the newton-meter.

We tend to confuse force and work, probably because our muscles have to degrade food to produce force even without doing work—holding a weight or even standing erect is fatiguing. Why, once we've lifted something, we can't just lock it in place isn't at all clear. After all, you do work to raise a flag, but you wouldn't think of holding the rope all day. Odder still, the problem isn't a peculiarity of all muscles; those that hold the halves of a clam's shell together for long periods have largely solved the

25

problem with a so-called catch mechanism. (One wonders about Atlas, but the ancient myth makers say nothing of any affinity for clams.)

A textbook of high school physical science to which my son was subjected showed a man pushing a desk sideways across the floor. The desk weighed 100 pounds (force) and was pushed 10 feet (distance); the legend, a more literal term than usual, declared the work to be 1000 foot-pounds. Don't make the same mistake—weight is always a downward force while proper floors are horizontal. The man does no work directly against the weight of the desk and, if the floor is slick or wheels provided, he may do almost no work even against the frictional resistance to sliding. To put it another way, the weight of the desk gives us little idea of its resistance to sideways motion, the resistance the force of the man "works" against. Of course, the weight does influence the friction between desk and floor, so it plays some role in the resistance to motion.

It always takes energy to do work; it needn't take energy to apply a force. As noted in Chapter 1, energy is, in fact, defined as the capacity to do work. But no more about that queerest of concepts for quite a while— seventeenth and eighteenth century mechanics did quite well without it and so will we.

Power

Power is simply the time rate of doing work, or work done divided by time taken. A more powerful motor can lift a weight faster or else lift a heavier weight at the same rate—the physical quantity corresponds reasonably well with our everyday usage. We'll use the symbol P; the SI unit is the watt (after James Watt, 1736–1819) and as an electrical unit is familiar to all (except those who keep light bulbs in all the sockets to prevent electricity from leaking into the house). A watt is, of course, a joule per second. If you multiply power by time you get energy or work, so a watt-second is a joule—a trivial point except for the predilection of electric companies to bill us for "kilowatt-hours" (1 kilowatt-hour = 3.6 megajoules), in short, for energy or work. "Metabolic rate," usually given in kilocalories per hour (or per some other unit of time) is, of course, just power. A calorie is 4.2 joules; a kilocalorie (or Calorie) per hour is 1.16 watts. The other common household (or, better, barnyard) unit of power is the horsepower—746 watts.

Density

Density is mass divided by volume—the ratio of how much substance is present to how much of the world it takes up. It's a useful "material property," that is, all samples of the same substance have the same density under the same conditions. Thus, it is a characteristic of a substance rather than of a particular object—water, for instance, has a density of about 1000 kilograms per cubic meter. Most organisms have about this value, but metals are mostly denser (Table 2.2), and gases are all much less dense (about 1.2 kilograms per cubic meter for air). When we say something is very heavy we mean ambiguously that it is either very massive or very dense. Fortunately, the distinction can usually be made from the context of the remark. As a symbol for density, we'll use the Greek letter rho (ρ).

Newton's Laws

Newton's three great laws of motion were introduced above without specific notice—they bear repeating with emphasis. The first law is that matter of unbalanced forces:

A body stays at rest or in uniform motion in a straight line unless a force is applied to it.

The second law is equivalent to our definition of force, using the equation, $F = ma$.

Acceleration is proportional to the applied force and in the same direction as the force.

The third law is in the assertions that the brick fights the squeeze and that the floor pushes up against your weight.

TABLE 2.2. DENSITIES (IN KILOGRAMS PER CUBIC METER) OF A VARIETY OF MATERIALS

Timber, white pine	370
Wood, fresh red oak	600
Gasoline	680
Fats and many oils	915 to 945
Seawater	1026
Spider silk	1260
Antler, red deer	1860
Femur, cow	2060
Tympanic bulla, fin whale	2470
Shell, mollusk	2700
Aluminum	2700
Tooth enamel, human	2900
Iron	7870
Mercury	13,550

When one body exerts a force on another, the second always exerts a force on the first; the two forces are equal in magnitude, opposite in direction, and act along the same line.

MOTION

With the introduction of speed and acceleration, we've tacitly brought in the notion of motion, something about as ordinary as mass but in practice a little more complicated. To begin with, there is a distinction between *speed* and *acceleration*—the former is a measure of the amount of motion, the latter of its unsteadiness. The difference involves the general issue of rates and gradients, which will occupy a large part of Chapter 4. For now, the main point is a reminder that the two are fundamentally different, however we confuse them in everyday affairs. Another important division is between types of motion, whose description again uses terms that deviate from their everyday meanings.

Translational motion

Translation here is any change in an object's position that meets a particular requirement—the object must not change its orientation; that is, a line drawn on the object must remain pointed in the same direction. Change in position implies some frame of reference since the change can be described only with reference to the position of something else. For our purposes, the frame of reference will usually be the surface of the earth. The arbitrary nature of the frame of reference needs emphasis; if you're uncomfortable with it, so were the ancients and so are, apparently, some contemporary biologists. I once worked on the flight of small insects and made a tediously large number of measurements on fruit flies "flying" in a (small) wind tunnel—the air moved while the flies stayed put. I recall having to stammer out in several talks ad hoc justifications of the shift of frame of reference—I never found a generally satisfactory explanation of something I found intuitively obvious and usually resorted to analogies. Try observing a dandelion seed (fruit, really) moving in a gentle wind—one imagines the fluff on top, having more drag, will be tilted downwind relative to the seed beneath. But the flier is fully vertical! In reality it knows nothing about the wind—it's traveling at wind speed, so the wind doesn't exist in its frame of reference, quite detached from terra firma. The stillness of the air around a drifting hot-air balloon is remarked upon by almost every first-time passenger. Yet one still encounters references to the problem of the ruffled feathers of a bird in a tail wind.

It's important to recognize also that, by this present definition, trans-

28

lational motion may be along a curve or even in a circle. The pedals and your feet *translate in a circle* when you ride a bicycle, and the seats of a ferris wheel translate in a circle around its axis of rotation because the pedals, feet, and seats remain horizontal (parallel to the line of the horizon) throughout. More peculiarly, a small element of bathwater translates in a circle as it moves around the micromaelstrom toward the drain. Both these situations—circular translation of a solid element and circulatory motion of a fluid in a vortex—turn out to have biological relevance, the first on account of the scarcity of true wheel-and-axle arrangements in nature and the second in the explanation of fluid-dynamic lift.

Rotational motion

But the rotor of the ferris wheel and both wheels and cranks of the bicycle do rotate, that is, they change their *orientation* (angle) with respect to a line of fixed direction, the essential feature of rotational motion.

In practice these two sorts of motion may be combined—the wheel of a moving bicycle does both at once—rotates and advances. A particularly interesting case is the motion of the moon around the earth. The same side of the moon always faces us, so it must rotate once around an internal axis every time it translates in a full circuit around the earth. Incidentally, our sensory machinery makes the proper distinction between rotation and translation—the former makes you dizzy while the latter does not. The cultural evolution of dances may reflect this fundamental distinction!

FORCES AS VECTORS

Force, velocity, acceleration, and momentum are what are called "vector" quantities as opposed to a mere "scalar" quantity such as mass. The difference is that a vector has a direction as well as a magnitude. (For a force, a third specification, called "line of action" is often important—it was mentioned in the statement of Newton's third law of motion and will reappear shortly.)

In a one-dimensional world, a force can be exerted in only two directions, both along the same line. We arbitrarily designate one direction of force as positive and the other as negative, add the forces algebraically, and thereby get the resultant (net) force. If the net force *is not* zero, the object on which the force acts accelerates (Newton's second law)—the simplest example is an object falling at the urging of gravity just after release, before its increasing speed has incurred much air resistance. If the net force *is* zero, the object upon which the forces act does not accelerate (Newton's first law). An example is an object that has fallen in air

29

at the urging of gravity for a considerable distance and picked up enough speed so that upward force of drag due to air resistance (drag is always opposite to the direction of motion with respect to the air) just reaches the downward force of gravity. With no unbalanced force the object thereafter falls at a steady speed. (If there's a wind, the object may simultaneously move laterally with respect to the ground, but nothing else alters.) For smallish items of low density, that is, for biologically interesting bodies, such a steady descent constitutes most of a fall.

In a two-dimensional world, combining forces gets a little more elaborate. Two or more forces acting in the same or opposite directions add algebraically in the ordinary way. Two forces at an angle to each other will act on an object in a direction between those of the forces and with a resultant force less than the ordinary sum of the two forces. We can make a useful diagram of the situation by representing the directions of the forces by arrows and making the length of each arrow proportional to the magnitude of the force (Figure 2.3a).

A simple graphical technique, vector summation, combines two or more forces to determine the resultant force. Leave one arrow attached to the original point and then move (translate) the others without stretching or rotating them. Fix the tail of one of the loose forces to the head of the attached one, fix the tail of another loose force to the head of the newly fixed one, and continue in this way until all forces are represented. If the head of the final force gets just back to the original point, the net force is zero, and we say that the forces are "balanced." If the final force has its head somewhere else, one merely draws a line (vector) from the original point to the head of the final force, with an arrow pointing to-

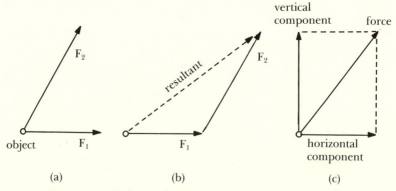

FIGURE 2.3. In (a) and (b) a pair of force vectors acting on an object are combined (summed) to give a resultant (dotted). In (c) a vector is resolved into two components at right angles to each other; summing these latter would restore the original vector.

ward the head. This line depicts the magnitude and direction of the net force (Figure 2.3b). This graphical process permits an end run around the trigonometry. If you prefer something more concrete than force vectors, think of two people pulling a heavy sled, each by a separate rope and in a slightly different direction, and ask about the net effect. Or think of a migrating bird heading in one direction with a wind blowing in another and ask where and how fast the bird is going—velocity vectors work just the same way as force vectors. And subtraction of vectors works the same way as summation—one merely takes the vector to be subtracted, puts the arrowhead on its other end, and then adds this reversed vector instead.

Just as two plus three make five, five can be dismembered into two and three. Similarly, a vector can be dismembered ("resolved") into two components. One particular resolution of a vector will be useful when we consider the aerodynamic forces, lift and drag. A vector in any direction can be resolved into a pair of components each at a right angle to the other. Thus, any velocity can be viewed as the combination of one component velocity on a north-south axis and another on an east-west axis. Or a force directed obliquely upward can be resolved into a vertical component and a horizontal component. Resolution takes only a pair of dotted lines on a graph, as shown in Figure 2.3c, or a quick application of the Pythagorean theorem (Appendix 1).

Dealing with vectors in three dimensions is, in principle at least, no different. It's awkward, though, with two-dimensional paper, so we'll stick to a two-dimensional slice of reality and leave the third to the imagination.

So far, we've used forces only to get objects into translational motion. What makes things rotate? If you push on the axis of a wheel you merely make it translate; if you apply a force to the rim it begins to spin. (The decently skeptical will argue that pushing on the axle of a bicycle—or a rolling pin—will cause rotation. It does so because the rim touches the ground. Taking a bicycle or rolling pin as a translationally fixed frame of reference, the ground or dough is moving and applying a force by contact and friction with the rim.) *Just how effective is the force in inducing rotation depends on how far from the axis of rotation is its line of action.* So the effectiveness of a force, what we call the *moment of the force* (which has nothing whatsoever to do with time), is the product of the magnitude of the force and the perpendicular distance from the line of action of the force to the axis of rotation, F times D in Figure 2.4. Note that a moment has the same dimensions as work, but the two are fundamentally different—for work, force and distance run in the same direction, while for a moment the two are perpendicular.

Equipped with the notion of a moment of force, you can make sense

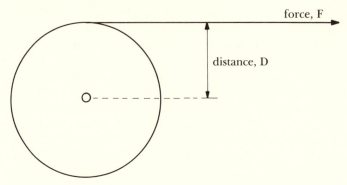

FIGURE 2.4. A wheel may be set into rotation by a force on its rim; the effectiveness of the force is given by its "moment," the product of the force and the perpendicular (or shortest) distance from its line of action to the axis of rotation.

of lots of practical devices. Levers, for instance, are practical applications of nothing more than these moments. To extract a nail with a pry-bar you have to exert a moment in one rotational direction that just exceeds (at the start) and then equals (for steady motion) the moment in the opposite direction caused by the resistance of the nail to extraction (Figure 2.5). A pair of pliers, wire snips, scissors, or shears operates in a similar way—you can understand why some have long handles and short business ends and others the opposite, and why you can turn a nut more effectively with pliers if you hold the pliers near the ends of the handles.

Another large class of devices, from crankshafts to bicycles and including just about all of our muscles, tendons, and joints, can be illustrated by a so-called windlass (Figure 2.6). With such a device a person applying

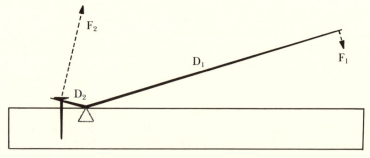

FIGURE 2.5. A pry-bar extracts a nail. Applying a modest force, F_1, at the end of the handle results in a substantially greater force, F_2, lifting the nail since the lever arm, D_1, of F_1 is much larger than the lever arm, D_2, of F_2.

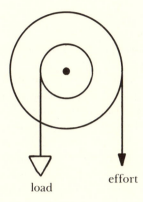

FIGURE 2.6. A windlass. A small down-ward force (effort) results in a large up-ward force on the load since the moment arm of the former exceeds that of the latter. A winch with a long handle and a narrow drum works the same way.

load   effort

a small force can generate a very large force. It isn't a matter of some-thing for nothing—to make the large force move something a short dis-tance the small force must move a long distance. An extreme version of the windlass was used (with teams of horses) to pull out the thousands of tree stumps involved in excavation of the Erie Canal between the drain-ages of the Hudson River and the Great Lakes in the 1820s (Tarkov 1986). In the arrangements of our muscles, the load and effort are or-dinarily reversed—muscle generates lots of force but doesn't shorten all that much, so, as in Figure 2.7, the load is at the end of the long arm (which may be literally an arm).

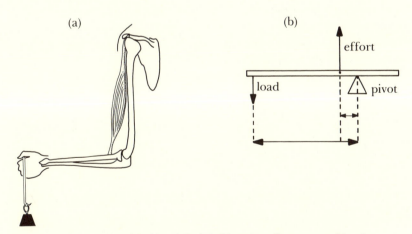

FIGURE 2.7. (a) a hand-held weight is held up by a force exerted by the biceps muscle of the upper arm; the lower end of the biceps is attached to the forearm. (b) The lever arrangement by which a limited shortening of the biceps, attached close to the axis (pivot) of rotation (the elbow), causes a wide sweep of the hand, much farther from the axis.

CENTER OF GRAVITY

A beam balance or a seesaw is a device whose operation is based on the equality or near-equality of a pair of moments, one clockwise and the other counterclockwise. A small weight may balance a large one if their distances from the support (fulcrum) are large and small, respectively. We can view such arrangements from the opposite perspective as well. Consider a beam with fixed weights in which only the support point is adjustable—one can slide the beam over the support to find the balance point; it's called the "center of gravity" of the beam. That's the point (or axis) at which the entire downward force of the beam (including the weights) acts as if concentrated—we can express the weight of the beam by a single downward vector through that point.

If the beam is weightless aside from the weights hanging from it, then the location of the center of gravity is easy to calculate—it's the unique place that makes the moments equal and opposite. If the beam alone, of nonuniform thickness, *is* the weight in question, then finding the center of gravity gets a little messier. One has to treat it as lots of little elements, each with its tiny weight and individual distance from the unknown balance point. Or one can let gravity do the test. Merely hang the object so it is free to rotate; it will reorient so the point from which it is hung is at or above the center of gravity (Figure 2.8). Doing the test twice, with different hanging points, will usually locate the specific center of gravity in two dimensions; doing it three times is necessary for an object irregular in all three dimensions.

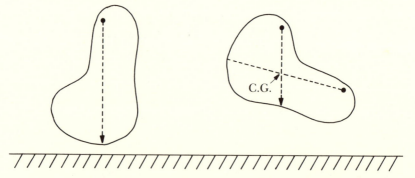

FIGURE 2.8. Locating the center of gravity of a two-dimensionally irregular object. The object is first freely hung from an arbitrarily located point and a vertical line is drawn through that point. The object is then hung from another point and a second vertical line is drawn. The center of gravity is located at the intersection of the two lines.

This odd abstraction, center of gravity, is crucial to what is often called gravitational stability. A person just can't stand not having the body's center of gravity above an area that includes the feet and any floor between them. A submerged body (not immobilized by contact with bottom mud) will ordinarily roll and orient itself so its center of gravity lies exactly beneath an even more abstract center of gravity—that of the water that would have been present had the body not taken up the space.

CONSERVATION LAWS

Perhaps the most elegant and powerful statements in physics are the conservation laws. Each states that in any process taking place in a properly isolated system something remains quite dependably unchanged. That is, amidst all of the confusing alterations of states of matter, chemical arrangements, velocities, shapes, and so forth, some item will add up to the same total amount at any time as at any other time. Two such laws will be of very practical use to us as we proceed.

Conservation of mass

It may seem obvious to us, but it was far from clear to the ancients that mass does not change in any normal process. Most of the difficulty concerned chemical changes, and the confusion came mostly from ignorance about the nature and mass of gases.

Conservation of mass may not make you sit up and take notice, but the notion has an important application in fluid mechanics where it's called the "principle of continuity." Consider a functionally incompressible fluid—water, for instance—flowing through a rigid pipe. Clearly, the rate at which mass flows past any point along the pipe must be the same as for any other point—nothing can accumulate anywhere since the water is incompressible and the pipe rigid. (Neither description is *perfectly* true, but both may be *very* good approximations.) Equivalently, the volumetric rate at which water enters one end of the pipe must be the same as that at which water leaves the other end. That's the basis for the use of a nozzle—constrict the exit, and, if the volume leaving per unit time is unchanged, then the water has to leave at a higher speed. In some cases the pipe needn't have walls—water in a column falling downward from a vertical spigot accelerates as it falls; since its speed increases, the cross-sectional area of the column decreases. The mass of water passing each horizontal plane in a unit of time must be the same or the conservation of mass would be violated.

It turns out that the principle works for gases in quite as simple a form—unless the speed of flow is higher than ever happens in living systems, gases are not compressed by flowing, and the relationship between mass and volume doesn't depend on speed. (For cases where compression is appreciable, density is no longer constant—the conservation law still holds, but its application is more complicated.) This principle of continuity is basic to the understanding of all circulatory, respiratory, and similar fluid transport systems in organisms. Since flow in the tiny capillaries of the body or in the small tracheoles of the lungs is vastly slower than flow in the aorta or bronchi, the total cross-sectional area of capillaries or tracheoles *must* exceed that of the aorta or bronchi—the details will get attention in Chapter 8.

A numerical example will illustrate the way the principle of continuity can be used. Let us ask what, very roughly, is the number of individual capillaries open and functional in a person at rest, certainly a tough thing to get by a direct count. The principle helps by requiring that all the blood leaving the left side of the heart in a given time must pass through the capillaries of the body in the same time. We know the output of the heart from its stroke volume and rate of beating—1×10^{-4} cubic meters per second. (I've put everything in SI units—the numbers may look strangely small, but the calculations become very easy.) This number must equal the cross-sectional area of a typical capillary times the speed of flow through a capillary times the number of capillaries. It's easy to measure the area of a capillary and the speed of blood flow through it—functioning capillaries and moving blood cells are visible with a good microscope in thin tissues such as the extended tongue or the interdigital webbing of a frog, or the mesenteries of a rabbit. The area is 3×10^{-11} square meters and the flow rate is 5×10^{-4} meters per second. So the number of capillaries must be about 7,000,000,000.

The principle of continuity is a handy tool in all sorts of situations. After a storm, a creek rises. Does the speed of flow in the creek increase in direct proportion to the total water it carries? Certainly not—if this were true the creek wouldn't rise. Incidentally, the piers of the old London Bridge (built in 1209, it lasted until 1832) were protected with "cutwaters" (essentially tiny islands) so wide that almost half of the cross-sectional area of the Thames was blocked. As a result, the speed of flow nearly doubled at the bridge, giving rise to a substantial hazard and to the expression, "shooting the bridge," for a boat passing under.

Conservation of momentum

The main reason it is handy to consider the composite quantity, momentum (mass times velocity), is that recognition of the concept of momentum makes it possible to state another and even more powerful conservation law. Mass is a scalar quantity, while velocity is a vector, so their product has a direction, that of the velocity. We can thus recognize that an object traveling in one direction will have positive momentum while one traveling in the opposite direction will have negative momentum. If two blobs of clay of equal mass come together at equal and opposite speeds and stick, the resulting mass will be at rest. The sum of their momenta before the collision was zero; it remains so afterward.

Consider the hose and nozzle again. Constrict the nozzle and the hose pushes back on you as the water exits. What has happened is that the speed of the water leaving the hose has increased from its speed when it was traveling along the hose, but the mass of water leaving in any period of time has not decreased (the principle of continuity forbids it). So the *mv* of water leaving the nozzle is greater than that of water flowing within the hose, and the *mv* difference appears in the reverse direction. If the movement of the hose is not damped (by holding it), things can get rather well dampened. If you stand and hold the hose, we assert that the counteracting *mv* appears as a change in the rotation of the earth transmitted through your feet. The earth has a largish mass and doesn't complain. In fact, the water pushes against air and earth and thus returns the momentum.

Other cases are less frivolous. Physics books make much of rocketry (expel enough gas backward and you'll move forward) and the recoil of cannon; in each case the conservation law permits nice quantitative predictions about the performance of the devices. A fish swims forward by imparting rearward momentum to the water—a liquid or gas has mass and thus when moving has momentum. Conversely, water passing an attached object slows down—its momentum decreases. The object experiences a force, drag, which pushes it, increasing the momentum of the object plus the substratum. If you can measure the rate of decrease in momentum of the water flowing over the object, it turns out that you can calculate the drag without even touching the object. The measurement isn't especially difficult—since the water is incompressible, the principle of continuity permits you to figure out the momenta upstream and downstream solely from measurements (quite a few, though) of velocity. The trick is useful when dealing with shy or uncooperative organisms.

CHAPTER 3

Size and shape

"Elephant and hippopotamus have grown clumsy
    as well as big, and the elk is of necessity less graceful
than the gazelle."

D'Arcy W. Thompson

GULLIVER, according to Swift, was twelve times as tall as a Lilliputian, and thus 12^3, or 1728, times as heavy. Consequently he was served 1728 times as much food as a Lilliputian would consume and received 108 (Lilliputian) gallons instead of a half-pint of wine. Swift apparently saw no inconsistency in the extrapolation from body weight to food and drink and reported no adverse effects of the drinking.

More recently, and less excusably, some psychologists tested the effects of LSD on an elephant. They took a dose known to have a minimal effect on a cat and multiplied it by the ratio of the weight of an elephant to that of a cat. Upon receiving the injection, the elephant began having violent convulsions and promptly died. West et al. (1962) concluded that elephants are remarkably sensitive to LSD.

Max Kleiber (author of the essay "An Old Professor of Animal Husbandry Ruminates") calculated that if a mouse consumed food at the same rate, relative to its body mass, as does a cow, the mouse would need fur 20 cm thick to stay warm. To accommodate the fur coat, it would, (un)naturally, need either stilts or very long legs. By contrast, if a cow used fuel at the same rate, again relative to its body mass, as does a mouse, its body temperature would exceed the boiling point of water, and we'd be supplied with precooked steaks (Kleiber, 1961).

Size, as mentioned in Chapter 1, is no small consideration in the design of organisms. We living things come in a wide range of sizes—one can emphasize the point by comparing a bacterium and a whale. Since our intuitive scale of size seems more closely associated with length than with volume, we get little sense of the situation from the bland assertion that 10^8 cubed, or 10^{24}, of the smallest bacteria would fill the volume of a whale. So let's lay those 10^{24} bacteria end to end and note the length of the line. It turns out that the line would encircle the earth 7.6 billion times—still too large to imagine! Alternatively, it would go to the sun and back fully a million times—still hard to envision. Or it would do the round trip to the nearest star (3.3 light years away) about 5 times, quite an astronomical distance, and not a lot of help either.

But a whale and a bacterium, whatever their similarities in overall shape or biochemical processes, are profoundly different sorts of creatures. Let's look at the range of lengths within some more biologically homogeneous sets of organisms (Table 3.1). Not only are the ranges less than 10^8-fold, they are *much* less—the commonest factor is 10^3, a *hundred thousandfold* less than 10^8. And the sets with larger factors, algae and vascular plants, are structurally less uniform than the others. Clearly, organisms are not like stretch socks—nature seems to have found that one size does not fit all, in both Darwinian and ordinary senses of "fit." That peoplelike organisms are roughly a meter rather than a millimeter in length is no mere accident.

LENGTHS, SURFACE AREAS, AND VOLUMES

Without a doubt, nothing is more important in determining how size affects biological design than the relationship between surface area and volume. Contact between an organism and its surroundings is a function of its surface, while its internal processes and structure depend mainly on its volume. And the two do not maintain a simple proportionality—unless you change the shape of a body you cannot simultaneously double both its surface area and its volume.

While perhaps not intuitively obvious, the situation can be easily illustrated with a few quick and facile calculations. We use rules relating changes in length to changes in area and volume all the time, but we rarely recognize the underlying generalities. Consider cubes of different sizes (Figure 3.1). Surface area (S) and volume (V) are related to the length (l) of a side by the formulas:

$$S = 6l^2 \text{ and } V = l^3. \tag{3.1}$$

Now consider a set of spheres and the analogous formulas for surface area and volume in relation to radius (r):

TABLE 3.1. THE RANGE OF SIZES WITHIN GENERAL CATEGORIES OF ORGANISMS

Group	Length Range	Factor
Insects	10^{-4} to 10^{-1}m	1000
Fish	10^{-2} to 10^{+1}m	1000
Mammals	10^{-1} to 10^{+2}m	1000
Vascular plants	10^{-2} to 10^{+2}m	10,000
Algae	10^{-5} to 10^0m	100,000

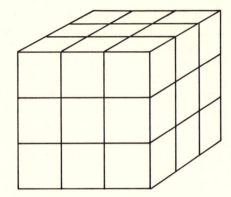

FIGURE 3.1. A pair of cubes. Each edge of the cube on the right is three times as long as an edge of the one on the left. The cube on the right has nine times the surface and no less than twenty-seven times the volume of the one on the left.

$$S = 4\pi r^2 \text{ and } V = 4\pi r^3/3. \tag{3.2}$$

And then imagine a set of circular cylinders (tuna fish cans, roughly) in which radius and height are equal ($r = h$). Their surface areas and volumes are

$$S = 4\pi r^2 \text{ and } V = \pi r^3. \tag{3.3}$$

In each case the formula for area specifies the length of something squared, and the formula for volume requires some length cubed. The situation is general, since areas always have dimensions of L^2 and volumes of L^3, at least for any set of "isometric" objects, that is, objects of varying size but the same shape. It is, of course, not usually true for a collection of nonisometric ("allometric") objects, objects differing in shape.

So what happens if you double *any* length measure of a solid body without changing its shape? You increase both surface and volume, not twofold but fourfold (2 squared) in surface and eightfold (2 cubed) in volume. If you triple any length, you increase surface by 3 squared, or 9-fold, and volume by 3 cubed, or 27-fold. The constants—6, 4π, etc.—in the formulas, being constant, make no difference to the argument.

Put as formal proportionalities then, the rules for sets of isometric (geometrically similar) objects are

Surfaces are proportional to the squares of lengths.

Volumes are proportional to the cubes of lengths.

We began by asking about the relationship between surface and volume as size (length) changes; we can now state that relationship merely by dividing each element of the surface proportionality by the corresponding one in the volume proportionality

Surface/volume is proportional to $l^2/l^3 = 1/l$, or

Surface/volume is inversely proportional to length, or

$$S/V \propto l^{-1}. \tag{3.4}$$

This tells us that a large object will have *less surface relative to its volume* than will a small object of the same shape. From this viewpoint, the main difference between bacterium and whale is that the bacterium has 100,000,000 times as much surface, *relative to its volume*, as does the whale. In short, the whale is big inside with little outside, while the bacterium is big outside with little inside.

In case after case, this reduction in relative surface as a body gets larger has elicited a specific response from the evolutionary process. The essence of the response is not a defiance of geometry, for that's simply not possible. Instead it involves an evasion of a premise, that of isometry, or constancy of shape. A whale and a bacterium are vastly further from geometrical isometry than they appear, one is tempted to say, on the surface. A few "gee whiz" examples emphasize the point.

(1) A tree appears far less rotund than either a whale or a bacterium—that is, it has a lot of very obvious leaf surface. In fact, gas exchange between the photosynthetic cells of leaves and the atmosphere occurs at the walls of tortuous internal passages, so the functional surface area is from ten to over thirty times greater than meets the eye. For an orange tree with 2000 leaves, the outer surface was measured as 200 m²; but the internal surface for gas exchange is thirty times greater—6000 m², or 0.6 hectares (1.5 acres).

(2) Plants absorb water and minerals from the soil, another exchange process that depends on having adequate surface. A figure is available for the surface area of the root hairs of a square meter of lawn planted with Kentucky bluegrass—the ten billion (U.S.) hairs have a surface of 350 m², double that of the floor of my whole house.

(3) Our lungs are the functional interface between us and the atmosphere. The capacity of a pair of lungs is about 6 liters, but this modest volume, divided among 30,000,000 alveoli, is bounded by a surface of 50 to 100 m², about the floor area of a large classroom.

(4) Topologically and physiologically (but not ordinarily) speaking, the cavity of one's entire digestive system is outside the body. Absorption of

food into bloodstream and cells takes place mainly across the wall of the small intestine, of which each of us has about 7 m looped around within our abdomens. This 7-m pipe, though, has projections (villi) on it, which have smaller projections (millivilli?), and the surface cells have a border of microvilli. The result of all the villification is a surface of 2000 m², or 0.5 acres.

(5) Even within an organism there are exchange processes and the attendant proliferation of surface. Our muscles exchange various substances, oxygen and carbon dioxide in particular, with the bloodstream, doing so across the walls of capillaries, with a combined surface area of about 6000 m², the same as the total internal surface of the leaves of the orange tree.

What happens, then, is an increasingly large deviation from sphericity in increasingly large biological systems, whether whole organisms or the internal components of organisms. Such an arrangement is recognizable even at the cellular level—many crucial bits of metabolic machinery operate on surfaces, and cells are full of internal membranous surface. If a stack of surfaces maintains its absolute (*not relative*) internal spacing whatever the size of cell or organism (as with the spacing between sheets in a pile of paper), then this particular allometry will keep the system's surface area proportional to its volume. Increase in size, whether of component cells or of organisms, and whether during the growth of an individual organism or the evolution of some lineage, is typically allometric rather than isometric—the parts increase out of proportion to one another. And thus shape turns out to be a function of size. The details form the subject matter of the topic commonly called "scaling."

THE SCALING OF VARIOUS PHENOMENA

The original example arguing the importance of scale involved falling under the influence of the earth's gravity. By this time it should be clear that many other biologically relevant phenomena also are strongly size-dependent. We now have enough background to consider another example illustrating the pervasiveness of scale and our consequent difficulty in imagining the world of organisms differing in size from ourselves.

Animals attack each other in various ways. Chimps and humans throw projectiles, mainly nicely dense rocks, as one of various aggressive tactics. Smaller animals, even manually dexterous ones do not. Are we simply more clever? Probably not—our size merely preadapts us for throwing things. The momentum of a projectile, a reasonable measure of its destructive capability, is its mass times its velocity. Mass is proportional to length cubed, while achievable velocity is probably proportional to the length of the throwing appendage and thus to the length of the organism. If an animal can throw a projectile of mass proportionate to its own, then the momentum of the projectile will be proportional to the animal's length to the fourth power—quite a drastic relationship. Probably, no small animal can impart enough momentum to a projectile to make it a significant weapon! There are, someone will counter, the nematocysts produced by the stinging cells of jellyfish and other coelenterates. But they aren't projectiles in quite the familiar sense—they require direct contact, and the trigger is pulled by the prey not the predator.

General rules can be suggested through such arguments, rules that ought to be relevant to students of animal behavior, natural history, and ecology. Kicking and hitting, for the same reasons as throwing, are the prerogatives of large creatures—the projectile is simply part of the aggressor. Biting, crushing, and squeezing will work for fairly small as well as very large organisms, and these forms of violence are extremely common in nature. But for still smaller systems, the advantage gradually shifts to defense—with more sharply curved surfaces, relative resistance to deformation improves (Chapter 11 will have more on this phenomenon). It becomes more practical for a predator to burrow into the prey if a mechanical sort of aggression is to be used—roundworms and chiggers really get under one's skin. Small prey, though, have a very serious nonmechanical disability. Their surface-to-volume ratios are large, and the distances from their surfaces to their innermost recesses are short, so they are exceedingly vulnerable to chemical assault. The typical scheme for disposing of small prey, whatever the size of the predator, involves enticing them into some enclosed place and then pouring some digestive concoction on the intact creature. A predator may be unable to chew tiny food, but it need not do so anyway!

Gathering a lot of diverse items, most of which will reappear later, we can make a useful classification based on how things scale, that is, change size as the system as a whole changes size. It should be noted at the start that, for isometric objects, any length is directly proportional to any other, any area (surface, projected or shadow, etc.) is proportional to any other, and likewise for any measure of volume.

Some items vary more or less in proportion to the length of the system or organism—we say that they scale linearly with length. Height may seem a trivial case, but it does represent the distance an object will fall if it trips (to use an example developed earlier). The drag experienced by a small organism traveling slowly in air or water is proportional to its length. The perimeter of an organism or of one of its feet quite obviously increases with its length (still assuming isometry). As a result, the peculiar upward force of surface tension (Chapter 5) on a creature standing (not floating) on water rises with the creature's length. The distance from the edge to the center of a solid body also varies in direct proportion to its length—this distance is proportional to the distance material must move in order to be exchanged with the surroundings or between parts of the body.

Other things vary more nearly with the area, that is, the square of the length of the system. External surface is again the obvious example—its importance in the exchange of material or heat (recall Kleiber's furry mouse) needs no additional emphasis. Organisms of moderate size and speed of motion encounter a drag that varies nearly in proportion to their area. The maximum force with which bones, muscles, and tendons can resist stretching varies in proportion to their cross-sectional areas. And the capacity of a wing to produce lift or a tail to generate thrust varies similarly.

Still other items vary with volume. Density is nearly the same from organism to organism, so mass follows volume very closely. Indeed, mass is the common measure of volume since it is easier to use a balance than to immerse something in a full bucket of water and measure the overflow. The water a submerged organism displaces, and thus the upward buoyant force on it, is just proportional to its volume. And so is its momentum at a given speed.

Some things scale both more and less drastically than any of these geometric quantities. The frequency of sound produced by many animals varies inversely with their lengths. A bigger fruit fly or mosquito has a lower frequency of wingbeat, and thus a deeper-toned buzz—as is most common among animals, males are smaller (higher-toned buzz), and sex recognition takes advantage of the difference. Similarly, a female toad prefers to mate with the largest possible male and approaches the one with the bassest voice. But the males "know" what's going on and, since frequency also goes up and down with temperature, pick the coldest place, whether air or water, to sit while calling (Fairchild 1981).

On the other extreme, the speed at which a large organism hits the ground when it trips is proportional to the square of its length, or height (drag in air being minimal for short falls by large creatures). Since its mass is proportional to the cube of its length, its momentum when it hits the ground is proportional to the *fifth* power of its length (Went 1968). Literally and drastically, the bigger they are, the harder they fall. Infants who stumble learning to walk are far more likely to be frustrated than fractured.

But if a moderate-to-large organism falls far enough to approach terminal velocity, the speed at which the downward force of gravity (mg)

equals the upward force of drag, then size turns out to be less important. Gravitational force is proportional to mass and thus to the cube of length. Drag is proportional to area (square of length) times the square of velocity, so, equating the force of gravity with that of drag, at terminal velocity v,

$$l^3 \propto l^2 v^2.$$

Dividing each side by the square of length and taking the square roots we get

$$v \propto l^{0.5}.$$

Thus, terminal velocity is proportional to the square root of length—we speak of a "scaling factor" of 0.5. The square root of length is, by the way, proportional to the *sixth* root of volume or mass. Double the mass of an animal and you increase its terminal velocity not by a factor of 2 but by 2 to the ⅙ power, that is 1.1225, for an increase of 12.25%—certainly less than one would guess.

QUANTIFYING SIZE AND ALLOMETRY

Up to this point we've talked about size in fairly loose terms, using some slightly vague length as a measure. For respectably quantitative work it would be nice to have something a little more specific. What is the appropriate measure of size for present purposes? The choice of units is trivial; the real issue is the choice among quantities with different fundamental dimensions. We might measure length, width, diameter, or circumference; or we might use total surface or cross-sectional area; or we could use volume or even mass as an index of size. Which we choose matters less than appreciating the consequences of the choice.

In the physiological literature, body mass is the variable of choice, mainly because it's the easiest to measure. (A few problems crop up involving an organism's inert portions, such as thick shells or internal water, but they're rare enough not to be big issues.) I'd prefer to use something proportional to length rather than to the cube of length. Used as an independent variable a length makes most of the various scaling factors come out to be small numbers greater than unity rather than fractions. One can easily divide these factors by three to restore proportionality to body mass, if necessary.

But what measure of length shall we choose? We'll have occasion to use two, both of which differ from the definitive data of a tape measure. The first takes advantage of the nearly constant density of organisms and the practical ease of weighing things. It's the cube root of body mass, which we might call "the nominal length measure," a version of body mass made

45

into something proportional not to volume but to length. (Cube roots are easy to figure on a hand calculator using either an exponent key or a trial-and-error routine—you might try taking the cube root of ten by trial and error to see how quick it is when only two- or three-figure accuracy is needed.)

The second linear measure, which will appear in some later chapters, is something called "characteristic length," a purposely fuzzy concept. The fuzziness follows inescapably from our need to compare items of different shape—*exactly* how big a rectangular solid must be to be "larger" than a given sphere depends (as a moment's thought should convince you) on what common property we choose to measure. But where bodies differ in size by many orders of magnitude, shape differences usually dwindle into insignificance, and the arbitrary character of our choice of a length is no cause for concern. And for geometrically similar bodies (say a set of spheres) all "lengths" or linear measures remain in proportion to each other as size changes—double the radius and you double both diameter and circumference.

We now have our measure of size—our independent variable that we can put on the *x*-axis of a graph. There is yet another problem. It's easy to plot a graph for, say, the way the area of a cube varies with its length, putting length on the *x*-axis and surface area on the *y*-axis as in Figure 3.2a. The result is a set of points that defines a curve (in this case a par-

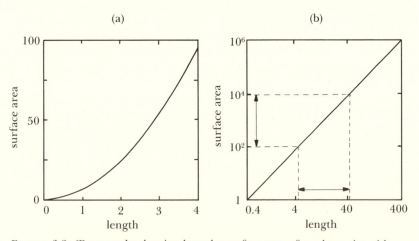

FIGURE 3.2. Two graphs showing how the surface area of a cube varies with changes in the length of an edge. (a) has ordinary axes; (b) uses geometrically scaled axes. From (b) one can get the useful information that a tenfold change in length results in a hundredfold change in surface area.

abola). But the graph is awkward. First, we'd like to read from the graph a single number that somehow specifies the way area varies with length. If we had plotted the length of a face diagonal of the cube against the length of an edge, the graph would have been a straight line, and the slope of the line would give the relationship between the two variables—the constant of proportionality (1.414 in this case, incidentally). But length against area gives a nonlinear graph, so its slope has different values at different points on the curve. Second, the scales don't cover much range—if one wanted to consider cubes of lengths from 0.000001 to 100 meters (the size range of organisms) most of the points would be indistinguishably clustered near zero. It turns out that we can get around both problems at once by a cleverly nonlinear set of labels for the axes.

Figure 3.2b has the axes relabeled according to so-called geometric scales. These scales have no honest zero, and the progression of numbers is what's commonly called "geometric"—the distance between 1 and 2 is the same as from 2 to 4 or from 4 to 8 or from 10 to 20. The distance between 1 and 10 is the same as from 10 to 100 or 100 to 1000 (or from 0.001 to 0.01, if an axis goes down that low). Let's consider what we've done in making this transformation.

The original graph, with linear axes, was convenient for plotting a function of the form

$$y = ax + b. \tag{3.5}$$

The function yields a straight line; one can get the value of b from where the line intersects the y-axis and the value of a from the slope of the line. The slope, to be tediously explicit, is the number by which an interval on the x-axis has to be multiplied to get the corresponding interval on the y-axis (Appendix 1). Or

$$(y\text{-interval}) = (a)(x\text{-interval}).$$

The graph with geometrically scaled axes is convenient for plotting functions of the form of equations 3.1 to 3.4, that is

$$y = bx^a. \tag{3.6}$$

With these axes one gets straight lines for such functions, and the value of a is the slope of the line. Except that the slope has to be read in a way appropriate for these axes. Intervals on the axes are read, not as arithmetic differences, but as factors of increase. For instance, an interval between 4.1 and 41 is a factor of 10; between 100 and 10,000 the interval factor is 100. For these geometrically scaled axes the slope is the *exponent* to which a given interval *factor* on the x-axis has to be raised to get the corresponding interval *factor* on the y-axis. Thus, the rule for obtaining a is

$$(y\text{-interval factor}) = (x\text{-interval factor})^a.$$

The exponent a may be obtained by trial and error with a calculator if it has a key for exponents (labelled "y^x"). If it has a key for logarithms (any sort of logs will do), one can avoid trial and error and instead use the formula

$$a = \log (y\text{-interval factor})/\log (x\text{-interval factor}).$$

We won't be concerned with b in equation 3.6—it's largely an accident of the particular units employed. If you ever need it, it's the value of y at an x of unity. (When $x = 1$, then $x^a = 1$ for any a; thus, $y = b$.) Incidentally, for the case above, of length versus surface of a cube, the slope is (as you might have guessed from the earlier discussion) 2—area is proportional to length squared.

The whole thing is well summed up by D'Arcy Thompson; the second chapter of his *On Growth and Form* (1942) is the great classic of the literature on scaling, and Thompson's splendid language tempts one to quote again and again.[1]

> It is a remarkable thing, worth pausing to reflect on, that we can pass so easily and in a dozen lines from molecular magnitudes to the dimensions of a sequoia or a whale. Addition and subtraction, the old arithmetic of the Egyptians, are not powerful enough for such an operation.

Our main use of a graph with geometrically scaled axes will be for finding the exponent of proportionality (the a, or scaling factor) for an unknown relationship. To be biological about it, consider the following example. A person wants to buy fish, paying in proportion to their weight, but has only a tape measure for routine use. To get the scaling factor relating length to weight, the person gets a set of fish that vary in size. The length of each fish is measured with the tape measure, and its volume is determined by dipping it into a full pail and catching the water that overflows. Volume is plotted against length on a geometric scale (one can, by the way, buy geometric graph paper—it's called "logarithmic"), and a best straight line is drawn through the points. The line turns out to have a slope of 3.0—what does the person conclude? That the fish are, for practical purposes, isometric—shape is the same whatever the size— and that weight is proportional to the cube of length. If a 10-inch fish costs $1.00, the person should pay no less than $8.00 for a 20-inch fish. (Incidentally, I tested the scheme, using it to pay my son for scaled fish

---

[1] Sir Peter Medawar refers to the writing in *On Growth and Form* as the best English prose in the scientific literature; there's a nice abridgement by J. T. Bonner (1961).

to feed the two of us during the summer of 1981. It is surprisingly dependable—a little longer means a *lot* heavier.)

Scaling factors such as that exponent of 3.0 are becoming matters of great interest to biologists, and a spate of recent books deals quite extensively with them—McMahon and Bonner (1983), Schmidt-Nielsen (1984), Peters (1983), and Calder (1984). The first two seem to be of the greatest general interest and accessibility.

SOME SPECIFIC SCALING FACTORS

The literature on scaling is particularly extensive for mammals, giving quite a complete picture of the size dependence of the construction and functioning of the animals most like ourselves. Various aspects of the topic will reappear later; however, discussing a group of scaling factors ought to help put the occasional additional factors into a proper context. The figures for a that follow have been extracted mainly from Schmidt-Nielsen (1984), but I've changed the x, or independent variable, from body mass to its cube root—the nominal length measure described earlier. The resulting numbers are summarized in Table 3.2. Bear in mind

TABLE 3.2. RELATIVE MAGNITUDES IN MAMMALIAN DESIGN: SCALING FACTORS FOR THE ALLOMETRIC EQUATION, $y = bx^a$. THE INDEPENDENT VARIABLE, x, IS THE CUBE ROOT OF BODY MASS; THE UNITS ARE SI (KILOGRAMS, SECONDS, METERS, WATTS). DATA EXCERPTED FROM PETERS (1983) AND SCHMIDT-NIELSEN (1984).

y	a	b
Surface area	1.95	0.11
Skeletal mass (terrestrial)	3.25	0.0608
Skeletal mass (cetaceans)	3.07	0.137
Muscle mass	3.00	0.45
Metabolic rate	2.25	4.10
Effective lung volume	3.09	0.0000567
Frequency of breathing	-0.78	0.892
Heart mass	2.94	0.0058
Frequency of heartbeat	-0.75	4.02
Kidney mass	2.55	0.00732
Liver mass	2.61	0.033
Brain mass (nonprimates)	2.10	0.01
Brain mass (humans)	1.98	0.085

that $a = 0$ implies size independence; $a = 1$ represents scaling in proportion to length; $a = 2$ means scaling with area; and $a = 3$ indicates scaling with volume. And bear in mind also that the scaling factors have variable but usually substantial uncertainty—the original data are often fairly untidy. So on to the as.

Surface

Are mammals isometric with respect to external surface area? That is, does our surface-to-volume ratio vary as it would for, say, a set of cubes, with a scaling factor of 2.0? Surface area proves to be an especially awkward quantity to measure, but some data do exist for animals ranging from mice to cattle; the factor seems to be between 1.9 and 2.0, or very near but slightly below the isometric value. The lower value implies that surface increases a little more slowly with size than the isometric expectation—that larger animals are a little more rotund and smaller ones a little more elongate or spindly overall. But not much, and data obtained within single species are very close to 2.0. (The size range within a species is, naturally, more limited, reducing confidence in the accuracy of the factor; but the data represent single techniques for making area measurements, which rather improves matters in compensation.)

Skeleton

The realization that large mammals have disproportionately thick bones goes back at least to Galileo. The problem faced by the large creature is that a structure stressed by twisting, bending, crushing, or anything other than simple pulling (tension) is relatively weaker as it gets larger. In compensation, the scaling factor for skeletal mass is above 3.0—it's 3.25. Over the range of mammalian size, the difference is appreciable. An 8-gram shrew is about 4% skeleton, a 5-kilogram cat about 7%, a 60-kilogram person about 8.5%, a 600-kilogram horse 10%, a 7000-kilogram elephant nearly 13%. You can tell from a drawing or photograph of a skeleton about how large the animal was merely from the proportions of the bones (Figure 3.3).

But the compensation for increased weight in large animals is only partial. We might reasonably assume that the force or weight a bone can bear is proportional to its cross-sectional area and that the animals, overall, are an isometric set. Then the weight that must be borne increases with the cube of length—double the length and you increase the weight eightfold, but the force a bone can withstand goes up only fourfold. Meanwhile, the length of each kind of bone must increase in proportion to the length of the animal. So the bones of big animals must be disproportionately thick—their cross sections must increase in proportion to the mass of the

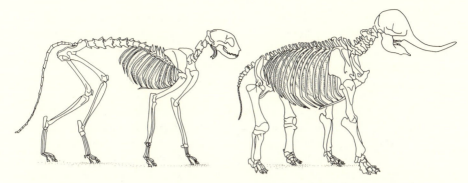

FIGURE 3.3. The skeletons of a cat (left) and an elephant (right), drawn approximately the same size. There's no trouble telling one from the other! Notice, in particular, the differences in both shape and position of the bones of the legs.

animals—and the bone masses must increase in proportion to their cross-sections times their lengths. Thus bones, overall, must increase in mass or volume in proportion to the *fourth* power of length—an *a* of 4.0, not 3.25. Built to the exacting standards of cats, a horse would therefore have to be no less than 34% skeleton and an elephant an impossible 78%. In reality, the big mammals are fragile fellows with limited margins of safety, as noted in the quotation with which the chapter began.

All these figures are for terrestrial creatures. On account of the long interest in whaling, data exist for marine cetaceans, for whom impact and gravitational loads have little immediacy. Not only do they have less bone than terrestrial animals relative to their body masses, but the amount of bone turns out to be more nearly proportional to length cubed, or mass—*a* for mature whales is 3.07. The value for fish, incidentally, is a statistically indistinguishable 3.09.

Muscle

A muscle functions by contracting, which ordinarily pulls toward each other the bones to which its ends are attached. The force with which the muscle pulls is proportional to its cross-sectional area; the work it does in pulling varies with that area times the distance it shortens; and that distance is proportional to the muscle's length. So the work a muscle can do ought to be proportional to its volume or mass. Similarly, the work a body's muscles are called upon to do ought to be proportional to the body's mass or weight—in short, to the load on the body. So things are nicely balanced. And, as far as data are available, muscle mass is proportional to length cubed or to body mass—mammals of all sizes are 40 to 45% muscle.

51

This constancy has an everyday implication—the edible fraction of a mammal ought to be independent of its size. Size-related considerations of growth rate, relative fat content, and so forth may influence our choice of cuisine, but edible fraction is not a factor.

Metabolic rate

This is the power input to an animal, the rate at which it consumes fuel to carry out its activities. It's usually measured (and commonly quoted) as a rate of oxygen consumption, but, with a few assumptions, oxygen consumption rate can be converted to watts of power—1 liter of oxygen represents 20.1 kilojoules of work, so 1 milliliter per second represents about 20 watts of power. Here the data for mammals, at least for animals at rest (the data cited here), is voluminous and of high quality.

First, though, the expectation. Virtually all of the power input should appear as heat (as will be explained in Chapter 15), and heat loss depends on surface area. Unless heat production is proportional to surface rather than volume, big animals will get excessively hot—as with Kleiber's mouse and cow mentioned earlier. If heat production, and hence metabolic rate, is proportional to surface, a should be 2.00. In fact, the value of a obtained from the classic mouse-to-elephant graph is 2.25, slightly but significantly higher. The difference has been the subject of much debate, conjecture, and calculation, the most noteworthy attempt to explain it being that of McMahon (1973). He made much of the slight nonisometry in surface area that we've already noted and predicted the precise value of 2.25 as a consequence.

Whatever the precise value or explanation of the scaling factor, the fact that it is well below 3.0—that metabolic rate increases with size more slowly than does mass—is profoundly important in determining differences in the ways of life of the small and the large. The oddity is baldly exposed if the previous value of a is converted to one for metabolic rate *relative to body mass* by subtracting 3.0—the scaling factor relating mass to length for isometric objects of uniform density. We get not zero but $a = -0.75$—*relative* food consumption is less for the larger animal. It can tolerate longer periods without food, it incurs a smaller relative price for maintaining a body temperature above that of its surroundings, and so forth. A bear sleeps through the winter with a body temperature near normal, and it's sensibilities are easily offended, as one physiologist did not bear in mind when attempting to measure core (rectal) temperature. By contrast, many smallish mammals hibernate with a temperature that may drop nearly to freezing—they could never store enough fuel to stay warm all winter without feeding. Some hummingbirds, tiny bats, and a few other of the smallest mammals undergo short periods of "torpor," permitting body temperature to drop for a few hours when they're not engaged in specific tasks.

Metabolic rate at rest is not the only relevant view of power expenditure. With more effort and complication, it's possible to measure an animal's maximum power expenditure—the trick is to induce the creature to work as hard as it can running on a treadmill or flying in a wind tunnel while you monitor oxygen consumption. Sustainable power levels for mammals prove to parallel resting levels, with about a tenfold consistent difference—that is, a typical mammal can increase its resting metabolic rate about ten times. But notable exceptions exist—the factor (called "metabolic scope" by physiologists or, as a unit, the "met" by physicians) is not 10 but 30 for dogs, and it is above 20 for horses and well-conditioned humans.

A few internal items

The big mammal needs more skeleton than does the small one, but on account of its relatively lower metabolic rate it should place lower demands on its *respiratory* system. Where might these lower demands be manifested? Effective lung volume ("vital capacity") is about proportional to body mass ($a = 3.09$) but the frequency of breathing is less in larger animals ($a = -0.78$); the algebraic sum of the two, in essence the oxygen supply, is 2.31—approximately the 2.25 of metabolic rate. Similarly, the *circulatory* system needn't do as much in a big animal. Heart mass scales with body mass ($a = 2.94$) but the rate of beating goes down in large animals ($a = -0.75$); the sum is 2.19. Organs that function in altering the composition of the blood seem to scale between the a's of 2.25 for metabolic rate and 3.00 for mass. For *kidneys*, $a = 2.55$, and for *liver*, $a = 2.61$. The smaller animal has, relatively, larger liver and kidneys, something that must be familar to butchers.

In addition, the larger animal has a relatively smaller *brain* ($a = 2.10$), but one wonders why the scaling factor is even this high (nerve cells being roughly the same size in all). The only sense organ that lends itself to straightforward measurement is the *eyeball*; I collected some data from the literature and calculate an a of 1.8. Again one wonders why eyes of the same size (equivalent to $a = 0$) will not suffice for all—perhaps bigger animals commonly look at more distant things.

If we divide one quantity by another where the two scale in about the same way, we get something that is size-independent. (Remember that scale factors are exponents, so to get the factor for a ratio you subtract the two values of a.) Heartbeats per breath is a fairly understandable ra-

53

tio—it's about 4.5 whatever the animal or level of exercise. Other ratios are more curious. For example, the life span of mammals relative to size has $a = 0.60$—the bigger live longer, on average. Also, since frequency of breathing has a factor of -0.78, the duration of a breath has one of $+0.78$ (time per cycle is the reciprocal of, or one divided by, frequency). The ratio breaths per lifetime then scales with an a of $0.78 - 0.60$ or 0.18—not far from size-independence! Heartbeats per lifetime ($a = 0.15$) is even more nearly independent of size. Perhaps the near constancy of these quantities is coincidental or incidental to some other factors—I've never heard of a direct, functional rationalization.

Some more obviously important things do not vary much with the size of an animal; a preeminent example is cell size. To a first approximation, a cell is a cell—its size varies far more (and not to an extreme) with cell type than with host organism. Any attempt at a functional explanation will have to wait for a discussion of diffusion (Chapter 8). In fact, the constancy of cell size is quite a lot stranger than you might guess. Recall that the relative metabolic rates of, say, mice and elephants are vastly different. If they have cells of the same size, then a typical mouse cell will have a metabolic rate (*not* just relative rate) more than twenty times as high as that of an elephant. It will need (and has) quite a lot more internal machinery—just as you can guess the size of the animal from a picture of its skeleton, you can guess the size of an animal from a photomicrograph of one cell of its liver! But why, with such a profound difference in metabolic rates, are the cells still the same size? A cell seems to be a pretty versatile bit of metabolic machinery. Thus the cells of a human fetus have metabolic rates appropriate for the maternal size; very shortly after birth they shift to the higher rates normal for cells within an organism of infant size.

SHAPE—A FEW GEOMETRIC MATTERS

Almost everything said so far about size and scaling hinges on a single geometric constraint—as size changes, the ratio of any aspect of surface to any measure of volume will vary unless shape simultaneously changes. In short, shape and a surface-to-volume ratio cannot be simultaneously conserved. The beauty of scaling arguments is that they transcend the host of details that we normally think of as constituting the shape of any complex object. Still, among those details lurk generalizations, and these (yet more promises!) will make sporadic appearances further along. To set the stage, though, we can look at a few matters pertaining to the geometry of life, inquiring about why certain shapes occur and why certain others either do not or are notably rare.

The pentamerous echinoderms

Starfish have five arms; sand dollars have five radial food grooves on their undersides—this arrangement of five elements radiating from a center point ("pentaradial symmetry") is widespread among the echinoderms but unknown elsewhere in nature. Why so and why not? Accident of ancestry seems unlikely—the fossil record is good, and the ancestral symmetry seems to have been trimerous, not pentamerous. More likely is an explanation based on the construction of early (and many contemporary) echinoderms and the geometric limitations of the scheme.

Early echinoderms were covered with a skeleton made up of discrete plates of calcium carbonate. Now one can pave a floor with triangles, squares, or hexagons, but using pentagons alone inevitably leaves gaps. One can't make an array of squares close on itself to form a hollow solid unless at eight special locations the apices of three rather than four squares touch, a distinct complication. And one can't make any array solely of hexagons close on itself at all. Conversely, one can get a closed, space-enclosing structure from triangles (a tetrahedron is the simplest, but others such as the icosahedron—twenty sides—are possible) and pentagons (the simplest being the dodecahedron—twelve sides). Among the pentagons (Figure 3.4) hexagons can be intercalated practically without limit, but twelve basic pentagons must remain. In the most symmetrical arrangement, the pentagons are in six pairs, with members of a pair positioned at opposite extremities of the solid. If we run an axis between the members of one pair, the ten other pentagons then arrange them-

(a)   (b)

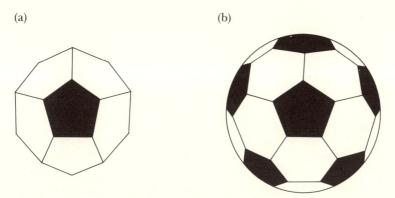

FIGURE 3.4. (a) A 12-sided solid made up of regular pentagons—the edges of the peripheral pentagons of this view radiate in an echinodermlike manner. (b) The 12 pentagons now are separated by hexagons—the radiating edges persist in the resulting 32-sided solid.

selves in two nearly equatorial rings. If enough hexagons are interca- lated, these can form the key elements of five arms. And a look at any book treating the paleontology of echinoderms reveals a host of hexago- nal plates. Perhaps pentaradial symmetry is, in fact, a "natural" or easy way to organize a radially symmetrical creature built of a shell of little solid elements!

And the conical mollusks

One of the most profound differences between an organism and any piece of technology is that the organism can, and indeed must, grow without ceasing to function—the possibilities for taking a time-out to en- large itself are exceedingly limited. And nowhere does this problem pose worse difficulties than in supportive systems. We've contrived bones ca- pable of growth, an impressive business, while spiders, crustaceans, and insects (among others) periodically molt their exoskeletons. Mollusks, a widespread and clearly successful group, have accepted a most peculiar constraint undoubtedly related to the need for functional continuity dur- ing growth.

Consider shapes that satisfy the following set of conditions. To provide both support and protection to the organism, the shape must be a hollow one, but an opening must exist somewhere. Growth can occur only by addition to the inner surface or the free edge. And the shape should change only minimally with increasing size. A cubic shell with an open face won't work—addition to walls will create more shell and encroach on the enclosed volume, and addition to the edge will make the rectan- gular shape increasingly elongate. Similarly, a cylinder doesn't meet the conditions—addition to the edge will change it from short and fat to long and (relatively) thin. What will work are cones, whether circular or ellip- tical. Add to the edge and thicken the walls and one gets a bigger cone, isometric with the original.

With only slight violations of the condition of isometry, all sorts of wild derivatives of cones are possible—and these are the shapes in which shelled mollusks occur. As shown in Figure 3.5, a broad cone, with an opening almost at a right angle to the cone's axis, is the shape of a limpet. An even broader cone, with a very oblique opening, is the shape of each valve (half-shell) of a clam. Elongate the cone and wind it up into a flat spiral, and one has a chambered nautilus or one of a great diversity of other cephalopods, now extinct. Wind the spiral with some amount of helical offset, and there's a snail. The notion of molluscan shape as re- flecting such isometric shell growth isn't new—Raup (1966) traces it back via D'Arcy Thompson to a paper in 1838. Raup, though, has initiated what is now a fine and biologically interesting game. A computer can be

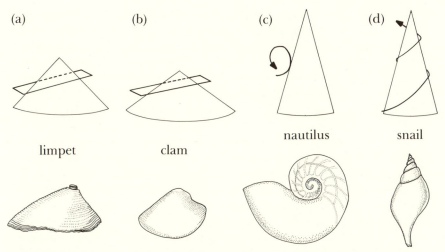

(a) (b) (c) (d)

nautilus   snail

limpet   clam

FIGURE 3.5. Deriving mollusks from cones. A limpet (a) is approximated by the surface above an oblique slice. A clam half-shell (b) is roughly a set of similar slices in which the tilt of successively lower slices increases a little. A nautilus (c) is a cone rolled up by stretching one side and squidging the opposite. A snail (d) is similar, but with the apex shifted to one side of the center of the coil.

programmed to describe and draw examples of a universe of hypothetical mollusks, some of which exist or have existed (shells fossilize very well), and others that don't appear to have evolved (the definitive reason for not appearing). Why are some but not others real? Are the limitations due to evolutionary accident or relative functional disability? If our biology is *really* competent it ought to be able to rationalize not only what has occurred but what might have but did not.

Right angles

As mentioned in Chapter 1, we seem to regard these as "right" for just about everything, but nature has some strange antipathy toward them. Still, right angles do occur in nature, and it's instructive to ask just where.

There are, of course, rarities and oddities. Walsby (1980) has described square bacteria from hypersaline pools in the Sinai Desert. Up to eight or sixteen bacteria form sheets like postage stamps when they divide and fail to part; with the aid of gas vacuoles they float flat on the surface of the water. On account of the salinity of the pools, the bacteria have no excess internal pressure, which would normally bulge their surfaces into curved shells; the squareness of the sides confers neither obvious benefit nor disability. These are not big organisms—they're about 5 micrometers on each side by about 0.2 to 0.5 micrometers thick.

CHAPTER 3

We detect acceleration by means of two sets of three semicircular canals in our inner ears. These canals are oriented at right angles to each other so that with them we sense acceleration in three mutually perpendicular directions. These directions correspond quite precisely to the roll (motion about the axis of progression), pitch (fore and aft rotation about the other horizontal axis), and yaw (motion in a horizontal plane about a vertical axis) of airplane maneuvers. And their identity with the three mutually perpendicular axes of our Cartesian coordinate system (two of which are used for every graph in this book) suggests that the Cartesian system may be more than an accident of (perhaps) the practices of ancient Egyptian land surveyors.

The fertilized eggs of many animals cleave into first two and then four and then eight cells; each of the three planes along which the cleavages occur is oriented at right angles to the others.

The most common occurrence of right angles in nature is exemplified by the angle a typical tree trunk makes with the horizon—it's particularly striking where (as where I live) the predominant forest is pine. The upright posture serves the obvious function of keeping the weight of the tree centered (properly, keeping the center of gravity) over the base and thereby minimizing any turning moment that the tree might find upsetting. Without the benefit of roots, our upright stance is even more uncompromising in demanding a right angle with the horizon lest we take a turn for the worse. Gravity, again, is important when you're large.

To reverse the question, why are there so many right angles in our technology and civilizations? Structurally, they are rather bad—we go to a lot of trouble to get rigidity through cross-bracing, and tacitly admit with just that term, "cross-bracing," that it's an afterthought. I see at least three factors. First, we *are* large and hence gravity-dominated. It's costly to climb when you're big, so level floors are a blessing (even a mild ramp is much more tiring than a corridor). With level floors come perpendicular walls for ease of construction and good use of floor space; and thence follow rectangular bricks and blocks and pipe ells; and from these latter come right angle intersections between walls. Second, rectangular solids pack well. Any collection of identical blocks or boards will stack with neither oblique interfaces along which gravity might cause elements to slip nor any internal voids. No other simple regular solids are so accommodating. And third, there is great convenience to a surveying system based on right angles—it is conceptually and operationally simple if one avoids deep philosophical issues such as whether Arizona borders Colorado, and if the scale is small enough so the curvature of the earth's surface can be ignored. Also, it permits straight access corridors (roads, for instance) between individual plots. Other animals don't have square or rectangular territories, but they don't make roads and may be more interested in minimizing the border to be patrolled or the remoteness of any bit of border from the center—such considerations argue for hexagons or rough circles rather than squares.

CHAPTER 4

Dimensions, gradients, and summations

"I never used a logarithm in my life, and could not
undertake to extract the square root of four without
misgivings."

George Bernard Shaw

IN TALKING about scaling, we kept oriented by bearing in mind the
dimensions of lengths, areas, and volumes. Still, that use of the notion
of dimensions was no more than a minor convenience, and someone may
be wondering why they were given quite so much notice in Chapter 2. As
it happens, the use of dimensions is powerful. Perhaps most of us missed
something important by learning mathematics from mathematicians who
unconsciously abhorred any intrusion of physical reality into their subject. And dimensions, as we'll use them, are part of a nice scheme to help
apply mathematics to practical matters. The scheme can be a real god-
send even (possibly especially) to our crude and cursory view of things.

At some point you were undoubtedly informed that an equation is
"true" or "correct" if the value of whatever is mentioned on one side is
exactly equal to whatever is mentioned on the other. Each equation we'll
use is a statement that the expressions on opposite sides of the "equals"
sign have the same numerical value for at least one set of values of the
variables. In addition to numerical equality, though, for the two sides to
be identical, *they must be dimensionally the same*. That means that if the left
side has dimensions of, say, length over time, the right side must also. In
fact, each term of each side must have the same dimensions. (Terms are
the parts of mathematical expressions completely separated by either a +
or a −.) Otherwise one might be adding chocolate chips to pork chops
or lengths to forces, both frowned upon. In the jargon, then, all terms
must be "dimensionally homogeneous" in any equation that correctly describes an aspect of the real world.

DIMENSIONAL ANALYSIS

So what of it? Dimensional homogeneity sounds like just another admonition to worry about, another place to go wrong. In fact, it's a fine
thing, a blessing informally discovered by countless students trying to remember formulas for physics exams—whether there's a v or a v^2 in an
equation can be determined if one knows the rest of the equation and

60

the dimensions of the variables. The essence of its merit is a powerful (if a mite inelegant) bit of applied mathematics called "dimensional analysis." Langhaar (1951), in a book on dimensional analysis, described it as "a method by which we deduce information about a phenomenon from the single premise that the phenomenon can be described by a dimensionally correct equation among certain variables." He continues, "The generality of the method is both its strength and weakness. With a little effort, a partial solution to nearly any problem is obtained. On the other hand, a complete solution is not obtained nor is the inner mechanism of a phenomenon revealed by dimensional reasoning alone."

How, then, to do it? For complex problems, formal theorems provide rules and guidance; here we'll get along with a semi-intuitive approach. The first step is to make an astute guess (yes, a guess) as to what variables are relevant to a physical phenomenon. Then one represents each variable by its basic dimensions (M, L, and T). Finally, one arranges the variables in the form of a dimensionally correct equation or proportionality. If the original guess was reasonable, then the resulting equation might just be useful.

A particularly good illustrative example was given by P. W. Bridgman (1961). He asked for a formula for the period of a simple pendulum—the time it takes to swing back and forth. One notices that the period is independent of the amplitude (width) of the swing, at least for small swings. And one guesses that the period (t) might depend on three variables—the mass (m) of the bob on the end of the pendulum, the length (l) of the string from bob to attachment, and the acceleration (g) of gravity pulling the pendulum towards its lowest position. We can obtain the dimensions of each of these variables from Table 2.1

$$t - T$$
$$m - M$$
$$l - L$$
$$g - LT^{-2}$$

What combination of the last three variables will equal the first? Somehow we have to get

$$t = m^a l^b g^c,$$

where a, b, and c are unknown exponents (which may be positive, negative, or even zero). Or, writing out the dimensions of the variables,

$$T = (M)^a(L)^b(LT^{-2})^c.$$

Simple cases like this can be done "by inspection" (dead reckoning or trial and error):

(1) To get the right side to come out to T just like the left, c must be − 1/2. It's the only way possible since gravity is the only variable on the right that has any time dimension.

(2) If so, the last L (in the gravity bracket) has an exponent of − 1/2.

(3) Since there's no L on the left, the Ls on the right must cancel, which means that the first L must have an exponent of + 1/2.

(4) Which is to say that b must be + 1/2.

(5) And a can only be 0.

One then goes back and reconstructs the original equation with the exponents ($a = 0$; $b = +1/2$; $c = -1/2$) inserted. A dimensionless constant factor (call it k) has to be assumed as well—remember Langhaar's comment about the limitations of dimensional analysis. Thus

$$t = km^0 l^{+1/2} g^{-1/2} \text{ or } t = k\sqrt{l/g}. \tag{4.1}$$

(In fact, $k = 2\pi$, but the value of that constant had to be found in some way other than by dimensional analysis.) Look how much could be derived just from the assumption that the dimensions had to come out right! Not only did the proper square root relationship emerge, but, amazingly, the zero exponent for a told us that the mass of the bob cannot matter in determining the period. The original inclusion of mass was revealed as a red herring.

It's worth checking all equations you use to see if they are dimensionally correct; conversely, one can use familiar equations to help one figure out dimensions. (I always think back to $f = ma$ to get the dimensions of force.) In addition, one can use the definitions or dimensions of physical quantities to jog one's memory of the interrelationships of different units. Thus by knowing either that power is the rate of doing work or the dimensions of the two, we know that the watt must be a joule per second; by knowing that work is force times distance we know that the joule is a newton-meter.

So all terms of an equation must have the same dimensions; in use, their values must be expressed in the same units. But the correctness of the equation does not depend on the actual choice among systems of units as long as an internally consistent set is used—even the constants are unaffected by the choice among systems. There is, though, a partial exception to the business about units. In the allometric equation $y = bx^a$ used in the last chapter, the value of b (but not a) depends on the specific units used for x and y. Which is why Table 3.2 had to specify SI units. The alternative, making these equations truly general by specifying the odd and approximate dimensions of every b, is judged to be more trouble than it's worth.

There's another way of viewing Bridgman's example. We can ask of the pendulum whether there is some dimensionless (pure) number that characterizes this sort of device. The question is procedurally trivial—just rearrange equation 4.1 to put k alone on the left; k becomes the pure number we sought. In this case little if anything is gained by the obtuse flourish.

DIMENSIONLESS NUMBERS

Peculiarly enough, though, these sorts of pure numbers can have quite a direct bearing on the real world. One might reasonably expect that decent physical variables would require some sort of dimension and (thus) unit—else why the designation "physical"? Oddly, an especially useful class of variables has no units at all—the values are pure numbers. These get their dimensionlessness by being ratios of two variables with the same dimensions. Indeed, one has been mentioned earlier—"strain" is the ratio between the distance a specimen is deformed and its original undeformed length (Chapter 2), thus a length divided by a length. And a good way to derive dimensionless numbers is through dimensional analysis, even if it sounds like a contradiction.

Some dimensionless numbers are named after the person who first either mentioned or extensively used them and given a two-letter symbol from the first letters of the perpetrator's name—Reynolds number, Grashof number, Womersley number (no apostrophe—it's Reynolds number but Stokes' law). Others are just called indices or coefficents, with some more-or-less descriptive designation—ponderal index, drag coefficient. And there is an ongoing tradition of concocting home-brew dimensionless numbers as smitten by need or inclination. We can illustrate the uses of dimensionless numbers by considering a series that covers a range of complexity, applicability, and frivolity.

Mechanical advantage

The efficacy of a lever, a system of pulleys, a windlass, or any similar device (Chapters 2 and 13) is most often a matter of the relationship between the force the device exerts against the load (output force) and the force you or some motor has to exert to operate it (input force). The ratio of these two forces is termed its "mechanical advantage" (MA) or "force advantage":

$$MA = F_o/F_i. \tag{4.2}$$

Since mechanical advantage is the ratio of two forces it cannot possibly have dimensions or units; it's not merely a constant but depends on the construction of the device; and it quite clearly has immediate physical applicability. (In practice it is inversely proportional to the ratio of the lengths of the two lever arms, again a dimensionless number.) Perhaps an item mentioned two chapters ago bears repetition: since a dimensionless number is automatically unitless as well, its numerical value does not depend in the slightest on what units were used in its calculation.

Efficiency

Most instances where efficiency is mentioned boil down to the same thing—a ratio of the work gotten out of some contrivance to the work put into the thing. Thus, efficiency is a measure of the energetic leakiness of mechanical arrangements. That the character of the input and output may be quite different makes no difference—electrical input and mechanical output for a motor, an input of the product of power and time versus an output of force times distance for some locomotory system. Efficiencies are always less than 1.0 (in the much lamented absence of perpetual motion machines), although they're usually multiplied by 100 to express their values as percentages:

$$E = 100W_o/W_i. \tag{4.3}$$

In practice, applying the term "efficiency" can prove most awkward, especially for living systems. Any of a number of inputs might be appropriate for, say, a jog around a track—total amount of food needed, food needed beyond requirements for ordinary body maintenance, extra work done by the heart and other muscles due to the jogging, or just the work done by the locomotor muscles? Output is even harder to define. If, ideally, it takes no work to keep an object moving at constant speed on a level surface, does car or runner have zero output and thus zero efficiency unless going uphill? Does a fish swimming steadily do work? Is the power output of a fish's muscles a good "output," or is the rate of momentum transfer to the water a better one? The moral is that all figures cited for efficiency have to be taken cautiously, and particular attention must be given to what items are being used for input and output.

Flatness index

The ratio of surface area to volume loomed large in the last chapter—as mentioned, its magnitude is certainly the most important single consequence of the size of organisms, a size that varies over at least eight orders of magnitude. Everybody talks about surface-to-volume ratios, but very few specific values are ever mentioned. The trouble is that the ratio, S/V, varies with size as well as shape so it's not a very useful measure of shape per se. Can we do better? That is, can we concoct some version of a surface-to-volume ratio that is independent of size and reflects only variation in shape? What we need is a ratio that does not have any dimension of L, and preferably no dimension of M or T either while we're at it, instead of the 1/L dimension of the ratio S/V.

There is no unique solution. Perhaps the simplest (you might think of others) is to raise surface area to the 1.5 power (equivalent to cubing it and taking the square root of the result) and dividing this by the volume—both numerator and denominator would thus have dimensions of length to the third power, and the ratio would be properly dimensionless. Since area is in the numerator, the further from sphericity the object, the higher the value—we might call the ratio the "flatness index":

$$FI = S^{1.5}/V. \tag{4.4}$$

You ought to check that this index is, indeed, size-independent by trying calculations for a few shapes over a range of sizes. I did a few myself—spheres give a value of 10.63, the minimum possible value since spheres of any size are the least flat solids by the present criterion. Cubes give a value of 14.70—they're a little flatter. Using the data of Martin (1981), a reasonable value for a mammal is about 26 (you can check this figure using the scaling factors of Table 3.2). Variation in the flatness index is a measure of deviation from isometry, so it provides an alternative to the scaling factor for surface area listed in Table 3.2.

Ponderal index

Here's one that may be too close to home for comfort. How corpulent are you? Weight is easy to measure, and it gives the sad story of one's past indulgences. But it fails badly in providing a comparison among different individuals—a tall person ordinarily has a greater mass than a short one even if they are of similar "builds." That one can do better was recognized in 1835 by the Belgian statistician Adolphe Quetelet, to whose work D'Arcy Thompson (1942) gave substantial attention. For geometrically similar bodies, volume is proportional to the cube of any *consistent* length; the handiest length for humans is our height. For two humans of the same shape, volume divided by the cube of height (h) ought to be constant and is, of course, dimensionless. A higher value would indicate a more corpulent corpus. Volume, though, is not nearly as easy to measure as mass, so Quetelet used the ratio of mass (or weight) to the cube of length; this ratio in the contemporary medical literature is called the "Quetelet index" or "ponderal index." (Confusingly, the term "Quetelet index" is sometimes used instead for volume divided by the square of height.)

But the conventional ponderal index isn't dimensionless, so one has always to specify the units for its variables, and to make matters worse the medical profession hasn't adopted SI. One can restore dimensionlessness by recognizing that all humans have nearly the same density, close to 1000 kg/m³—a little lower for the very fat, a little higher for the very bony (recall Table 2.2). So we might use mass over density (ρ) for volume and write a revisionist ponderal index:

$$PI = m/\rho h^3. \tag{4.5}$$

The assumption of constant density solely to make the dimensions come out right may seem a sophistry, but actually measuring density with some scheme based on submergence is likely to dampen the cooperativity of potential sources of data. You might at this point figure out your ponderal index, using the density above and consistent SI units (1 kg = 2.2 lb; 1 m = 39.37 inches) to get a figurative figure. In my experience, females seem to have slightly lower PIs than males, either with respect to the lowest values one encounters or relative to one's visual and cultural perceptions. The lowest value (for a nonanorectic individual) of the fifty I've collected is 0.0100; there's a long tail on the high end of the data. 0.0105 to 0.0140 seems to be about the range for "ordinary" looking people.

Versions of this index have been used by people in different fields. It is, among other things, exactly a hundredth of what is known as the "Rohrer body-build index" (Rohrer 1921) among anthropologists and other people interested in the human form—the Rohrer index (not dimensionless) puts mass in grams over the cube of height in centimeters. Sheehan (1986) describes its use in a fascinating book about the identification of the remains of victims of an old airplane crash. The commonest medical use of the index seems to be in exploring correlations between obesity and longevity or cardiovascular disease—after all, one must have some sort of quantitative data to put on the x-axis for the fatness of the subjects, and the data must be obtained from quick and nondestructive measurements.

Froude number

William Froude (1810–1879) was a British engineer and naval architect who took the first productive steps toward a rational science of hull design. His name is preserved in his important dimensionless number, abbreviated Fr, which governs the extrapolation from work on models to full-size ships. The number is derived from the ratio of two forces, inertial and gravitational. Inertial force is what it takes to divert a body from its rest or straight course at constant speed; it's the immediate thing described by $F = ma$. Gravitational force is given by the similar expression $F = mg$. For working with steady motion in a continuous medium such as water rather than with an accelerating solid, there are handier quantities than mass and acceleration. Mass is proportional to the cube of length times density, and steady acceleration dimensionally is velocity divided by time, so we can write Froude's ratio as

$$\frac{\rho l^3 v t^{-1}}{\rho l^3 g}.$$

Time is as awkward as acceleration, so we get it out by noting that length over time is velocity; that this latter velocity might refer to some other aspect of the system than the velocity already in the formula is only a minor worry. And densities cancel, so we have (still dimensionless, of course) the Froude number

$$Fr = v^2/gl. \tag{4.6}$$

(Occasionally the square root of the expression is used as the Froude number. If you're dimensionless, so are your roots.)

For a model to correspond to reality, the Froude numbers of the two situations must be the same. This implies the same relative values of inertial and gravitational forces. Under such a condition the wave patterns caused by movement of hull relative to water will be the same for both. (The g reflects these so-called gravity waves on the surface of the water.) Since we can't ordinarily tamper with g, if the length of the model is reduced by a factor of four then the speed at which it is tested must be reduced by a factor of two.

Besides its use as a guide to working with models, the Froude number has implications for performance comparisons among real ships. The difficulty of propelling a hull across the surface of a body of water depends, as you might expect, on the speed at which it travels. But the dependence is not as simple a function as one would guess. Near maximum speed most of the work that is being done is expended raising water in the form of gravity waves. There comes a point at which the bow wave and stern wave produced by a ship interact, and a ship can go faster only by climbing up its own bow wave—no cheap trick. The Froude number, though, gives an index to that point for ships of divers sizes—the practical maximum speed corresponds roughly to $Fr = 0.16$. In fact, few ships exceed $Fr = 0.12$—the cost of transport relative to distance covered is minimized at something lower than maximum speed. But as you can see from equation 4.6, at any given Froude number the larger ship (with longer hull length l) goes faster. Other things being equal (there are several complicating factors), a ship four times longer ought to be able to go twice as fast.

A few organisms are surface swimmers analogous to ships. Prange and Schmidt-Nielsen (1970) investigated the metabolic cost to ducks of swimming and found that very much the same constraints apply. A duck with a hull length of 0.33 m can go (or, really, *will* go) at steady speeds no higher than 0.7 m·s^{-1} (1.6 mph); the speed isn't impressive, but it corresponds to a Froude number of just about the maximal 0.16. Not that a duck gets breathless—its respiratory system is capable of the far greater demands of flight—what seems to limit speed is simple quantity of leg muscle. For a muskrat with a hull length of 0.25 m, that same Froude number gives a speed of 0.63 m·s^{-1}; Fish (1982) found that few muskrats on a pond exceeded that speed and that the typical swimming speed was just a little lower—0.58 m·s^{-1}, or Fr = 0.14. The low maximum speed imposed on small animals by surface waves might have some bearing on the relative scarcity of organisms that swim at surfaces—even air-breathers don't make a habit of surface swimming.

Recently, a completely novel and much more biologically relevant use of the Froude number has appeared. Alexander and Jayes (1983) have pointed out that terrestrial locomotion on legs is strongly dependent on the effect of gravity, involving both the pendulum action of the legs and the free fall between footfalls. The main change from talking about hulls and gravity waves is that length has to be redefined; they find that the distance from the hip joint to the ground is the most appropriate length to be considered for walking and running. Among their predictions are the following:

(1) At a given Froude number, reached at different speeds for animals of different sizes, there ought to be the same phase relationships between the legs. Phase relationship is, in practice, only a general term for what in locomotion is called "gait"—walking, trotting, galloping, etc. In short, one gait will be best at a given Fr.
(2) At a given Froude number, different animals will have the same stride length relative to their own hip heights.
(3) At a given Froude number, the legs of different animals will spend the same fraction of the time in contact with the ground.

These are fairly powerful generalizations about a complex set of phenomena. And they work surprisingly well. The transition between walking and running or trotting occurs very near Fr = 0.6. The transition, for quadrupeds, between trotting and galloping, happens at Fr about 2 to 4. Using my own hip height, I find that I should make the first transition when going 11.7 minutes per mile, and in fact I do spontaneously shift from a fast walk to an easy jog at about 12 minutes per mile. Crows and kangaroos, with whom I normally feel little affinity, shift from walking to hopping at that same Froude number. A nice summary of this work has been given by Alexander (1984).

Drag coefficient

A quantity called "C_D" or "drag coefficient" looms large in advertisements for slick and speedy cars; it will get more attention in Chapter 7. The coefficient amounts to a dimensionless measure of the retarding force of gas or liquid flowing past an object. But what's wrong with the direct use of force for drag? The limitation is that drag then refers only to the conditions operative for a particular measurement—it is handier to have an index that characterizes the "dragginess" of an object in a more general context. Thus drag coefficient is the drag of an object relative to a standard force; in this case the standard is the force that would be exerted on the object if the fluid directly upstream from it was brought to a halt upon impact. The standard is distinctly queer, since the halted fluid would have to disappear in order to keep out of the way of the principle of continuity (Chapter 2), but since it's only a mathematical trick reality needs no mollification.

So what is this peculiar standard force? Again we can begin with $F = ma$ and again note that neither mass nor acceleration are easy items for steady flows. This time let's replace volume by the product of length (in the direction of flow) and the frontal area, S, of the body, so

$$F = \rho S l v / t.$$

But l/t is also v, since l was specified as being in the direction of flow, so

$$F = \rho S v^2.$$

The drag coefficient, then, is drag D divided by the standard force F, with a factor of two included for reasons beyond the present discussion:

$$C_D = 2D/\rho S v^2. \tag{4.7}$$

Two comments must be made about the casual use of drag coefficients. First, at least for cars, aerodynamic drag is not a terribly crucial matter except at highway speeds. More importantly, the necessary inclusion of S in the coefficient can be misleading. If (and there are other conventions as well) it is taken as frontal area, a decrease in drag accomplished by reducing that area will not decrease the drag coefficient. Conversely, a well-designed bulgy body may have a low coefficient but a relatively high absolute drag.

69

Walking on water

Walking on water is a feat some creatures such as water striders can do. It's no trick at all to concoct a dimensionless index to the practicality of standing or walking (not floating or swimming) on the surface of water, held up by the force of surface tension on some hydrophobic (nonwettable) foot. All that is needed is the ratio of the force holding the creature up to that pulling it down. The upward force is simply the surface tension γ (which has the dimensions of force over distance, M^1T^{-2}) times the length of the air-water-foot contact line (usually the combined perimeter of the feet). The downward force is gravitational, the usual $F = mg$ or approximately the density (of the organism) times length cubed times gravity. We get, then, as an index (call it Je, the Jesus number):

$$ \text{Je} = \frac{\gamma l}{\rho l^3 g} \ or \ \frac{\gamma}{\rho l^2 g}. \tag{4.8} $$

Since the surface tension of water, the density of organisms, and gravity are all essentially constant, the index provides a simple scaling rule. The main variable is the square of length in the denominator, which shows that the practicality of walking on water becomes dramatically less as animals grow larger. (The choice of an appropriate length isn't especially critical for such order-of-magnitude judgments.) On the other hand, there should be nothing startling about the phenomenon in very tiny creatures, and its uncommonness is perhaps due merely to the comparative rarity of smooth water in nature. Still, the index gives only a distant view of the matter, to which we'll return later. Incidentally, 1/Je, the inverse of the Jesus number, is a recognized quantity called the Bond number and is used in chemical engineering.

With that mild bit of sacrilege (apologies to the offended), we have probably plumbed enough of these indices to persuade any skeptic that discrete dimensions are no prerequisite for applicability to the physical world. More importantly, it ought to be apparent that dimensionless numbers can prove useful in our attempts to uncover biological generalizations.

RATES AND GRADIENTS

Earlier we defined velocity as the rate of change of position—the rate at which the location of something changed as time progressed. Acceleration was then defined as the rate of change of velocity as time progressed. Thus acceleration is related to velocity in just the same way that velocity is related to position. The idea of a *rate* is clearly central to an orderly view of the physical world. It applies to much more than speeds, accelerations, and other time-related quantities—rates may refer equally well to how something changes with distance. How the elevation of your position on a mountain changes with distance from your base camp is such a rate, a rate of vertical ascent relative to horizontal progress. Such a rate is more commonly referred to as a "gradient," but gradients in ordinary usage are simply rates referred to distance instead of to time. Concentration or intensity gradients are exceedingly useful—how the concentration of a pollutant changes with distance from its site of discharge, how the intensity of a sound drops with distance from a caller. And, as we'll see, rates and gradients may refer to still other variables besides time and distance.

Rates and gradients really jump out at you from graphs. Consider a vertical cross section of a mountain (Figure 4.1). The rate of ascent with distance (gradient, or steepness) at any point is the slope of a straight line tangent to the mountain at that point—a line that just grazes the curve of the mountain without crossing it. The idea of slope applies as well to temporal gradients. On a graph of some variable plotted against time you merely take a ruler and draw a tangent line at any point you want. Just as with mountains, it's how far one progresses upward relative to how far one progresses horizontally.

All of this seems quite simple; in reality there's a subtle but profound problem. Previously, for the average slope of a curve, as in Figure 4.2, we've used some vertical distance (call it Δy) divided by the corresponding horizontal distance (Δx). Where the "curve" was a straight line (as when we figured out scaling factors) there was no trouble. But can we apply the same approach to get the slope of a curved line *at a point*, a trivially simple operation on a graph (Figure 4.1)? At a point, Δy is zero, which is

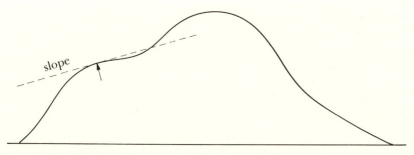

FIGURE 4.1. Side view of a mountain. The slope of a line tangent to the mountain is a measure of the steepness of ascent at the point of tangency.

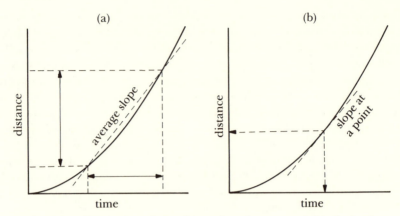

FIGURE 4.2. The average slope of a portion of a curve may be obtained, as in (a), by dividing an interval Δy on the y-axis (distance, here) by the corresponding interval Δx on the x-axis (time, here). If Δx is made infinitesimally small, one has the slope at a point on the curve, as in (b).

awkward enough. Worse, though, Δx also is zero, so our usual technique involves dividing by zero, a totally reprehensible act. Does this mean that the slope at a point is mathematically meaningless? One has to do a bit of mental sleight of hand and reduce the horizontal distance Δx until it is infinitesimally small (but not zero!). On a graph, as in Figure 4.2, the resulting "slope-at-a-point" becomes nothing other than the tangent line to the curve.

We owe this oddly difficult concept of slope at a point to Newton and Leibniz (independently) in the seventeenth century. The operation, though, remained a theoretical embarrassment (rather like sailing upwind) until Cauchy clarified things almost 200 years later. What we have here is nothing less than the basic idea behind the differential calculus—something we can't do without, even if we use graphical methods and other devices to avoid the formal complexity of the subject. As Leibniz, more generous than Newton, put it, "Taking mathematics from the beginning of the world to the time of Newton, what he has done is much the better half."

To reiterate, then. If you plot distance (y-axis) versus time (x-axis), the slope of the graph at any point is the "instantaneous" velocity; if you plot velocity versus time, the slopes are values of acceleration; if you plot height versus distance, the slope is the steepness—velocity, acceleration, steepness are all rates or gradients. As we'll see from a few examples, the idea of a *gradient* is as powerful as any notion in all of science.

Slipping down a gradient is as easy as rolling off a log. The direction of a gradient commonly determines the direction of processes that go by themselves—"proceed spontaneously" is the usual phrase. The stone rolls downhill, requiring Sisyphus only to restore its uphill position—strictly speaking it rolls down a gradient of what's called gravitational potential. If a fluid flows, you can safely presume the existence of a pressure gradient. If heat flows, there must be a temperature gradient. If a substance appears to diffuse, then there must be a concentration gradient. If electrical current flows, there clearly is a voltage or electrical potential gradient. If a chemical reaction proceeds, it reflects a gradient in chemical potential. And the magnitude of the gradient, its steepness, is an important factor in determining the speed of the spontaneous process.

Temperature gradients

What determines the rate at which heat passes out through the walls of my warmed house is not simply the difference in temperature between inside and out. The rate, in fact, depends on the magnitude of the thermal gradient. If a given sort of thermal barrier is made twice as thick, the temperature difference is unchanged but the rate of heat loss is halved. What has happened is that the *temperature change per unit distance* through the wall has been halved, the slope of the temperature-versus-distance curve is half as steep—the gradient, in short, has been halved.

The relationship between heat loss and temperature gradient is an enormously important constraint in the design of animals that maintain body temperatures substantially different from those of their surroundings. The role of fur is that of an air trap, preventing movement of air en masse from augmenting conductive heat transfer. Fur thus effectively lengthens the path heat must traverse in dropping from body temperature to the temperature of the local atmosphere. A layer of fat plays a similar role, quite critical for a creature moving through cold water where fur functions poorly. For a large animal in anything but a very cold climate, heat retention is not a serious problem, as mentioned earlier when we talked about surface areas and volumes. For a small animal, preventing heat loss may be crucial. Again there are oddly ordinary implications. Beef is well marbled with fat, skinned chicken is naturally lean. The difference is less a contrast between mammals and birds than between a large animal, which can afford to distribute its fat, and a small one, which makes any fat do double duty—as metabolic reserve and as thermal insulation beneath the skin.

Concentration gradients

These may be slightly harder to envision, but they work in exactly the same manner as do thermal gradients. If you put a crystal of dye in a beaker of really still water (not something easily accomplished), the dye will dissolve and over a few weeks spread through the beaker (Figure 4.3). In this diffusive process, there has been a net movement of dye from a region of high concentration, near the crystal, to areas of lower concentration. The operative rule, Fick's law (Chapter 8), has precisely the same mathematical form as Fourier's law for heat transfer by conduction. Movement of oxygen from lung alveoli to capillaries and from capillaries to active tissues of necessity involves diffusion down concentration gradients. These systems have a nicely self-regulating character—if blood enters the lungs already well charged with oxygen, then the gradient of oxygen concentration is reduced and less oxygen diffuses from the air into the blood.

Velocity gradients

When one solid moves across another solid, only people interested in lubrication pay much attention to any velocity gradient between the two. With moving gases and liquids, velocity gradients are an unavoidable part of practically every situation. The reason for their ubiquity is an oddly counterintuitive phenomenon called the "no-slip condition." It turns out that whenever a fluid flows across a solid surface the fluid immediately adjacent to the surface does not move—it has the same speed as the surface. The fluid slightly more remote from the surface has a speed slightly different from that of the surface, closer to the speed of

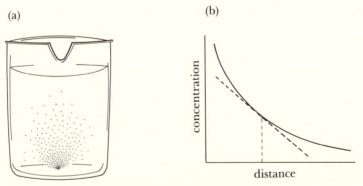

FIGURE 4.3. (a) Dye slowly diffuses away from a speck of dye on the bottom of a beaker. There's no edge to the cloud of dye but just a continuously changing concentration as shown in (b), a graph of how dye concentration varies with distance from the speck. At each point on the curve, the concentration gradient (slope) has a different value.

what we call the "free stream" some distance away. And fluid still further from the surface has a speed still closer to that of the free stream (Figure 4.4).[1]

As a consequence of the no-slip condition, there is a velocity gradient with respect to distance near the surface of every solid over which a gas or liquid flows. Have you ever asked yourself how dust and grime can accumulate on fast-moving fan blades? In reality, the speed of flow right near the surface where such tiny bits of debris are located is far slower than the general airstream. Or have you ever wondered why a dishcloth is so much more effective than running water in removing even water-soluble stuff like jam from a plate? An electric, nonscrubbing dishwasher has to use repetitive jets of strong detergent solution, and it still works best with prescraped plates. The faster the flow, the thinner the gradient region and the steeper the gradient, due to both the greater speed difference across the region and the shorter distance. If you sit in a sauna bath, air movements are very slow, and the gentle velocity gradient provides some insulation, slowing heat gain from the hot air. If you fan yourself you can get a burn—you've made the velocity gradient and hence the thermal gradient much steeper, and heat gain rises in proportion.

Temporal velocity gradients are more obvious than spatial ones. To return yet again to falling, we note that it's not the fall that hurts but the stop at the end—the velocity gradient associated with impact. Thus we make padded dashboards and collapsible bumpers on cars; and Vincent

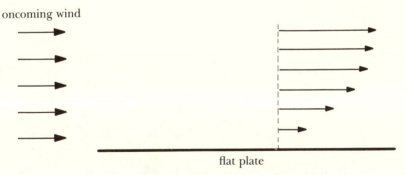

oncoming wind

flat plate

FIGURE 4.4. The velocity gradient adjacent to the surface of a flat plate exposed to a wind along its surface. As one moves away from the plate along the dashed line, the local wind increases, at first steadily and then more slowly as it approaches the overall ambient wind speed. The thickness of the gradient region has been considerably exaggerated in this figure.

[1] The reality of the no-slip condition was a matter of hot dispute in the mid-19th century. Goldstein (1938) has a special appendix that reviews the arguments.

CHAPTER 4

and Owers' (1986) hedgehogs, as mentioned in Chapter 1, have their crushable spines. Cats have no direct skeletal connection through a collarbone between the bones of their forelimbs (pectoral girdle) and those of the vertebral column—shock-mounted forelimbs, in effect, cushion the landing after a jump. None of these schemes reduces the extremes of velocities one bit; what they reduce are velocity gradients.

Stimulus intensity gradients

Even your perceptual world is a bunch of gradients, both temporal and spatial—the gradients are, in fact, grossly exaggerated beyond mere physical reality. Admittedly, in some instances an organism might find it useful to monitor the intensity or direction of some environmental factor over a long period of time—gravity detection in plants, measurement of day length (photoperiod) in all sorts of organisms, and so forth. But what is usually of immediate behavioral relevance is a change in the outside world, and, typically, the more rapid the change the more important it is that a response be initiated.

So we're equipped with various gradient detectors. Some are tactile and temporal—when you sit down you feel the chair but you're then unaware of it until you get up and feel the lack of chair. Others are tactile and spatial—you're more aware of the edge of something pressing against you than of the rest of its surface. Some are visual and spatial—you interpret each of the gray stripes in Figure 4.5 not as uniform in brightness but as varying, seeming brightest adjacent to the less bright next stripe. And others are visual and temporal—a sudden shadow causes you (or just about any other animal) to take notice, and a small

FIGURE 4.5. The gray stripe illusion. A series of stripes, each of uniform darkness, is misinterpreted by a visual system that accentuates edges or boundaries.

76

item in motion "catches the eye" amidst a jumble of stationary detail. It's an edge that matters, whether in space or time, and the prudent potential prey makes no sudden motion lest it be detected by edgy predators. Even with sound, a sudden change in intensity is especially effective as a stimulus. Music written before electronic amplification got most of its dramatic effect from changes in intensity, rather than from loudness itself, since the maximum amplitude of even a full orchestra is still not all that loud. In the auditory sense, though, there's an interesting asymmetry— for good behavioral reasons the sudden onset of a sound is more "noticeable" than any sharp termination. It is disconcerting to hear a recording of any percussive instrument, such as a piano, played backwards—the music seems both quieter (which it isn't) and rather drab. We've designed such instruments around the gradient-exaggerating properties of our auditory equipment!

SUMMATION OF CONTINUOUSLY CHANGING ITEMS

The first bit of math that we learn is counting; it comes simultaneously with memorizing the names and sequences of numbers. After that we learn to add, first combining sets of integers (two apples and three apples) and then numbers with fractional parts (three and a third plus two and a half plus five and a quarter pages of text). Even the latter, though, involves a discrete number of summing operations (three, here). We need to extend the notion of summation a bit farther. What if you travel at a varying speed, recorded on a speedometer, and want to know how far you've gone during a particular period of time? Or what if you have a crop growing, not by a fixed amount each day, but at a rate that is always proportional to the existing crop (a lawn or young forest, approximately), and you want to know a week's accumulation. You need to be able to do a summation not of a discrete set of items but of a so-called continuous function.

Again, let's view the matter in terms of manipulating an ordinary graph. Say we plot speed versus time (Figure 4.6a) and wish to ask how far we've gone over a given interval of time. Distance is, of course, the product of speed and time, but simple multiplication won't suffice— speed isn't constant. But just as length times width (two variables at right angles) gives area, on our graph area represents the product of speed and time, hence distance. What we're really looking for is the area under the curve for the elapsed time of interest—that area represents the distance traveled.

In practice, we run into a problem similar to that which bedeviled figuring slopes at points. For a trip of an hour we might figure the average

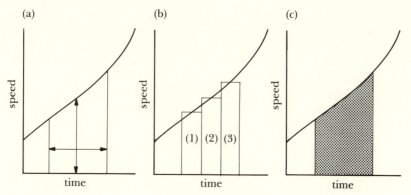

FIGURE 4.6. Distance traveled when one's speed isn't constant (a) can be figured by dividing the trip into intervals (b), taking the average speed for each,
and adding up the component distances (areas in the rectangles corresponding to the intervals). Ideally, one takes infinitesimally short intervals, obtaining, in effect, the area under the graph of speed versus time (c).

speed for each twenty-minute period and from these get the distance for
each period; we could then add the three distances (Figure 4.6b). Better,
we might take one-minute periods and add sixty distances. Best, we take
infinitesimally short periods and add a very large number of tiny distances (Figure 4.6c). Conceptually, the matter is simple (at least armed
with graphs); mathematically, it's tricky—again Newton and Leibniz got
people thinking correctly about the matter.

There are two easy ways to do these summations. The ancient and hon-
orable fallback is to weigh a sample of graph paper of known area, cut
out the offending area under the curve, and weigh the latter. The ratio
of the mass of paper under the curve to the mass of the known area
equals the ratio of the unknown to the known areas. With decent paper
and a steady hand, 1% accuracy, usually better than the original data, is
no trouble at all. The other easy way is with a computer. It has both speed
and patience, so it can add the areas for *very* small time increments and
thus get quite an accurate summation. Even better, one can instruct it to
try successively shorter increments and to stop when it finds that making
the increments still briefer doesn't change the result by more than some
tolerable uncertainty.

You'll recognize that the summation process is nothing more than the
opposite of the instantaneous value or slope-at-a-point problem. These
are, in fact, the essences of the two branches of calculus—differential (for
slopes) and integral (for summations). Change in position, for instance,
is the summation (the "integral") of tiny increments—instantaneous ve-

locities times infinitesimal intervals of time; change in velocity, similarly, is the integral of increments of instantaneous accelerations times intervals of time. And, as with gradients, the independent variable (on the *x*-axis) may be something other than time—often length, area, or volume. Two examples, both anticipating material further along, ought to add sufficient emphasis to the latter point.

Work of extension

Take a sample of some material—a string, a tendon, a length of arterial wall—and slowly stretch it until it breaks while keeping track of both its increasing length and the force (weight, usually) you apply. Data of this sort can be displayed as a force versus distance graph (Figure 4.7a). Alternatively, the data can be massaged into a more general form so as to apply to the kind of material rather than to a certain sample. Force can be converted to stress (force per unit cross-sectional area) and length to strain (fractional extension), and the results plotted as a graph of stress versus strain (Figure 4.7b). For most materials, neither graph takes the form of a straight line; and the form of the second graph, including the particulars of its curvature, are of especial interest—from it one can read (as we will later) strength (maximum height), extensibility (maximum horizontal distance), and stiffness (the slopes).

But those are not all. Work, remember, is force times distance. The area under the force-distance curve from one distance to another is the work it took to stretch the sample from the first to the second length.

(a)

(b)

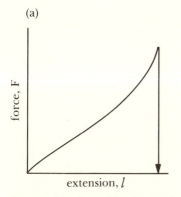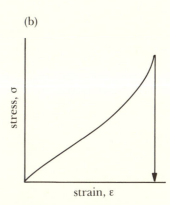

FIGURE 4.7. Something is stretched by an increasing force until it breaks. The basic data (a) may be generalized (b) by converting force to stress and extension to strain. The area under curve (a) is the work done on the sample; the area under curve (b) is the work per unit volume ("work of extension") for the material.

79

The area under the stress-strain curve—what is that? Stress has dimensions of force/area while strain is dimensionless, so the area has dimensions of force/area also. Force/area, though, is dimensionally equivalent to work/volume—we now have a form of work independent of the particular specimen just as are stress and strain. This property, work of extension (or for the full test, sometimes called work to fracture), can be important. For instance, if you want to store up work for later use by stretching a material, you pick one that has a high work of extension for loading. Hopping animals such as kangaroos use tendon for such purposes (Alexander 1983), and insects beating wings up and down use pads of a material called "resilin" to use the work of decelerating one stroke to start the other (Weis-Fogh 1960)—both materials meet our requirements well. By contrast, arterial wall (Figure 4.8) has quite a different functional role and a stress-strain curve of a different shape. So we have done two integrations, one with respect to a length, the other with respect to a (dimensionless) strain—one gave work as a result, the other work per unit volume.

Total flow through a pipe

Consider water flowing smoothly ("laminarly," not "turbulently") through a circular pipe, well downstream from where it entered. The no-slip condition must hold at the walls of the pipe, so the speed there is zero. Automatically, then, the speed of flow must be fastest in the middle (Figure 4.9). In fact, a graph of speed versus distance across the pipe comes out in the form of a parabola; a three-dimensional graph of speed across the area of the pipe would be a "paraboloid of revolution"—the parabola spun around the long axis of the pipe.

Question—if you know the speeds of flow, can you figure from them the total volume of water that passes (recall the principle of continuity)

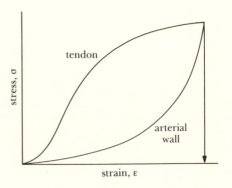

FIGURE 4.8. A mildly idealized tendon and an arterial wall take the same extremes of stress and strain but give very different stress-strain curves. The great difference in area under the curves is a matter of no small functional significance.

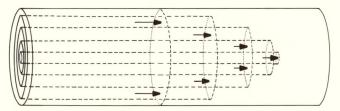

FIGURE 4.9. Flow through a pipe can be approximated as if through a set of concentric channels at a different speed in each channel, with fastest flow down the very center.

any cross section along the pipe's length? You can do it by integrating across the parabola in the appropriate manner. But here it isn't merely a matter of getting the area under the curve of speed against distance— since we're integrating with respect to area, not distance, we have to use concentric elements of cross-sectional area (these are annuli like the pieces of a telescoping radio antenna). We could make the right paraboloid of revolution out of clay and do a three-dimensional version of the paper-weighing game. Alternatively, we can take the average speed for each concentric annular area, multiply each speed by the area of the corresponding annulus, and add up the products. The dimensions come out right—speed times area is the same as volume per unit time.

WE HAVE now developed quite a bag of tools—physical quantities, vectors, scaling rules, dimensional analysis, differentiation, and integration. With these in hand it ought to be possible to do better than the potshots we've been taking at the biological world. We turn, then, to states of matter and a host of physical phenomena and, for each item, to its relevance for living things.

CHAPTER 5

Gases and liquids

"You may have inner tranquility, but you can't escape
surface tension."
V. Louise Roth

FOR THE next few chapters, the matter at hand will be just that, matter, and what matters about it. So far, little has been said about matter
beyond the assertion that it has mass and can be characterized, at least in
part, by a property called "density." We'll proceed from simple to complex—from gases to solids—even though that puts last the state that has
the most immediately intuitive reality to us big and earthbound animals.

STATES OF MATTER

Under normal circumstances matter exists in three distinct "ordinary"
states—solid, liquid, or gaseous (although sometimes other states such as
"dissolved" are recognized). We're familiar enough with these to need
little in the way of description; most of us will even recognize that under
everyday conditions there exist materials between solids and liquids but
not between liquids and gases. One can base a proper explanation of the
differences between states on molecular behavior; we'll follow our usual
practice of ignoring the latter as insufficiently intuitive. Alternatively, one
can distinguish among states of matter by the way each resists the application of forces or, more realistically, stresses. Which provides an excuse
to make some useful distinctions among several sorts of stresses, as done
in Figure 5.1.

The matter of stresses

The kind of stress depends on the kind of force and how it is applied.

Tensile stress. Tensile stress is force per unit cross-sectional area, as
usual, but limited to cases where the two ends of something are being
pulled away from each other by the stresses. Pull on a rope or on taffy
or even on a beam of wood or metal—the stress is applied whether or
not the material stretches appreciably. The area for converting force to
stress is, of course, at a right angle to the force—the cross section of rope,
taffy, or beam.

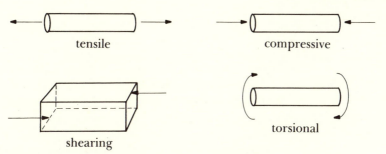

FIGURE 5.1. Four kinds of stress. More informally these are pull, push, slide, and twist.

Compressive stress. Compressive stress is like the above but pushes material together. Push on a wall, step on the top of a full can of beer, push on the handle of a bicycle pump with a finger over the orifice—you've applied a compressive stress to a solid, a liquid, a gas. Again, the areas are taken as being at right angles to the forces.

Shear stress. This stress is different. Apply a pair of equal forces to an object, forces in opposite directions but along lines of action that don't coincide. We applied such a pair of forces much earlier and observed rotation as the result—here assume instead that the object will deform rather than rotate. It makes it easier to picture if one of the forces (the resisting force) keeps a face of the object stationary. With a soft material the deformation will be noticeable—try a kitchen sponge, bread dough, a lump of jello. A cube will be distorted so some of its angles are no longer right. Whether the material is initially cubical, or whether we merely view it as an amalgam of coalesced cubes, a given shear stress will cause a certain angular deformation, a certain distance of movement in the direction of the force on the unanchored face of the object for each unit distance between the lines of action. Here the relevant area is different from that for tensile and compressive stresses. If the applied force is exactly above the resisting force, the area is that of any horizontal plane between them—pull on a bedsheet and you put a shear stress on the mattress.

Torsional stress. Grab a rod or rope and twist one end with respect to the other—the result is a torsional stress, included here just to complete the possibilities. In fact, the torsional stress on a rod can be expressed as a combination of tension, compression, and shear—twisting a rod stretches the outside, compresses the inside, and creates shear between concentric cylindrical layers.

The stresses of matter

Back to the states of matter and how they resist stresses.

Solid. A solid resists compression, tension, and shearing. No matter what you do to it, it fights back. As a rule, the more vigorously you aggress upon it, the more it deforms, but the relationship isn't necessarily linear—doubling the stress needn't exactly double the strain.

Liquid. A liquid resists compression and tension—squeezing and stretching. It's very hard to squeeze a liquid—squeeze it in a chamber and it will be forced out of any opening, a phenomenon underlying the design of quite a few pumps including our own hearts. It's also hard to stretch a liquid, although we rarely have occasion to try. Trees, as we'll discuss later, have long columns of water under very high tensile stress running up their trunks, so the matter is far from a mere curiosity. But liquids are magnificently disdainful of shear—the coffee fits equally agreeably into a cup of any shape.

Gas. A gas resists only compression. It may be easier to squeeze a gas than a liquid, but it still takes work. The gas doesn't mind being stretched in the slightest—indeed, it's always trying to stretch (expand) itself and will fill the volume of any container rather than lie limply low like a liquid. When you think you're stretching a gas—pulling on a piston in a cylinder, for instance—you're really only working against the external atmosphere. This less than self-evident point may seem more real if you consider that the *less* gas in the cylinder, the *harder* it is to pull the piston outward (Figure 5.2)—any gas inside is working with you, not against you. And the gas cares as little about shear as did the liquid. Tabor (1979) provides a more elaborate account of the states of matter from this particular viewpoint.

ABOUT GASES

In a surprisingly large number of situations, a gas is a gas is a gas. Properties that might vary widely among liquids or between one solid and another are either invariant in gases or change according to simple and regular rules. The gas of main interest biologically is, of course, that mixture we breathe and call air; but neither the fact that it isn't a pure substance nor the specifics of its composition are of fundamental importance to most bioportentous physical phenomena. As a result, it's a fairly easy matter to state some general rules for dealing with gases.

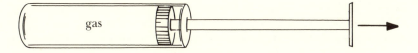

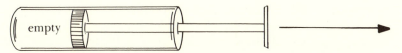

FIGURE 5.2. It's easier to pull the piston outward if some gas is present inside than if there's a vacuum in the cylinder. You're really pushing on the atmosphere outside, not pulling on the gas, so the pressure of any gas inside assists you.

(1) For a given gas at a fixed temperature and volume, either a particular pure gas or a mixture, the pressure it exerts on the walls of its container is proportional to its density. In short, if the amount of gas in the container, and thus the density, is doubled, then the pressure within the container (with rigid walls, we'll assume) is also doubled. But the relationship between pressure and density does vary from gas to gas—when one speaks of a "light" gas such as hydrogen or helium, one means that the gas exerts rather a high pressure for a given density or has a low density for a given pressure.

(2) For a given mass of gas at a fixed temperature, the pressure and volume vary inversely; that is, the product of pressure and volume is a constant. Double the pressure and you halve the volume. The relationship was shown back in 1660 by Robert Boyle—it's sometimes called "Boyle's law." The perceptive reader will protest its implication that matter can be squeezed out of existence by a sufficiently high pressure—there ought to be some irreducible volume that is much harder to compress. There is, but the gaseous state is so rarified that we can ignore the actual material volume of the underlying matter in all problems of present interest—we'll deal only with so-called ideal gases in which the rule is definitionally perfect. Incidentally, Boyle's law implies that you can measure atmospheric pressure by taking a volume of ordinary air and noting the pressure needed to compress it by half—you will have thereby doubled the pressure or added an additional one atmosphere (which turns out to be 101,000 pascals or Pa).

(3) In the absence of specific chemical reactions (as when hydrogen and

oxygen combine to make water vapor) the individual gases in a mixture act independently, and properties such as the pressure or density of the mixture are just the sums of the properties of the components. Thus, the air at sea level has a pressure of 101,000 Pa. Of its components, oxygen makes up 20.95% (by volume, not mass). Oxygen thus contributes 21,160 Pa to the overall pressure—its "partial pressure." Were all the other gases removed from a rigid container of air, the pressure inside would then be that of oxygen alone—21,160 Pa. In fact, in an atmosphere of pure oxygen at about a fifth of normal atmospheric pressure, your respiratory system would behave normally. The situation is equivalent to breathing pure oxygen at ambient pressure at an altitude of 12,000 m (39,000 ft).

(4) Gravity can be ignored as an immediate influence on the density and pressure of a gas in the atmosphere around any terrestrial organism. It is thus not dangerous to hold your breath while ascending in an elevator (although it may hurt your eardrums a bit). From the weight per unit volume of air of about 12 N·m^{-3} (which is the same as the air's density times gravitational acceleration) and the sea level pressure of 101,000 Pa (N·m^{-2}), it's easy to calculate that a column of air (of constant density) 101,000/12 or 8400 m high would be needed to exert an atmosphere of pressure. Ascending 84 m would therefore expand the air in your lungs by only 1%. Thus, the volume of a flexible container of gas will not vary much as an organism ascends or descends biologically reasonable heights in air. A soaring bird might be an exception—it ought to (and does) vent any internal air space to the outside.

(5) But the situation is quite different for a container of gas surrounded by water rather than by air—the analogous calculation yields another result altogether. Water weighs about 9800 N·m^{-3}. A column only 101,000/9800 or 10.3 m (33.8 ft), instead of 8400 m, high exerts an atmosphere of pressure. If you dive down about 10 m in water the pressure on your body increases by an atmosphere—from one to two atmospheres. The effect on the liquids and solids of your body is trivial, but the air in your lungs reacts by decreasing its volume by half. Note, though, that the fractional volume change of the air in the lungs per unit distance up or down depends on how deeply one is submerged. Going one meter beneath the surface will decrease lung volume by about 7%. That same 7% decrease at 40 m down takes an additional descent of almost 4 m. As divers (and fish) know, buoyancy compensation is most demanding near the surface.

Incidentally, it's possible to escape without any special machinery from a sunken ship or submarine up to about 100 m down—you just have to get out fast and start to ascend with your lungs filled with air at the local hydrostatic pressure. The expansion of air in your lungs will then maintain the ascent; you exhale continuously until you reach the surface. (When further instructions are needed consult Jensen 1986).

For essentially the same reason one can't use a snorkel tube to breathe at any great depth—air is drawn in at atmospheric pressure, but it is necessary to displace water at an additional pressure determined by the depth. A meter down, the extra pressure is 10 kPa, or a pound and a half per square *inch* of chest wall, about our limit. Schmidt-Nielsen (1984) punctured the prevalent presumption that long-necked brontosaurs lived in water using their necks as snorkel tubes, pointing out the unreasonable pressure differences involved. But mosquito larvae happily hang from the water surface with their internal air pipes (tracheae) connected to the atmosphere—the problem is tiny if you are also.

Buoyancy

If a body is submerged in water and its weight is less than that of the water it displaces, it will be subjected to an upward force. It "displaces," of course, a volume of water equal to its own volume, so there's a net upward force on it (it's "positively buoyant") if its density is less than that of water. Conversely, the body is forced downward (it's "negatively buoyant") if its density is more than that of water. The force, upward or downward, is just that of gravity—equal to the density difference times the volume of the body times gravitational acceleration:

$$F = mg = (\rho - \rho_o)Vg. \tag{5.1}$$

(Density with the subscript "o" is that of the medium, here water; the formula assumes that a downward force is positive.)

Here's the problem of the escaping submariner—with air-filled lungs body volume, and thus density, depends on depth; as ascent proceeds, the upward force gets greater and greater because the air in the lungs expands and thereby reduces the submariner's density. But the cure, exhalation, is simple. Not so simple is the problem of the submarine diving too deep—the pressure on the hull increases by an atmosphere for each 10.3 m. For moderate depths, the stiffness of the hull means that additional inward pressure makes little difference to the volume of the craft, and any small increase in its density can be offset by pumping water out of its ballast tanks. But if it goes too deep, its hull begins to compress too much, the craft gets too dense, and *the downward force increases, propelling it still deeper.* To the extent that either submarine or submariner has a nonrigid hull and a gas (air) inside, it (or he or she) is unstable with respect to depth. If it ascends, it wants to ascend faster; if it descends, it wants to descend faster.

The material of which organisms are made is almost always denser

87

than either fresh water (1000 kg·m^{-3}) or seawater (1024 kg·m^{-3}). Fat may be a little less dense—about 94% of water—but bone and other mineralized tissues are much more dense—170 to 240% of water. So organisms would sink were they not otherwise provided with some buoyant ballast. If the ballast is as incompressible as water itself, there is no problem of instability. The most space-efficient ballast, though, is the least dense material, and that means a gas; and a gas, quite independent of its composition, is unavoidably compressible.

How then do pelagic (non-bottom-living) animals maintain buoyancy? Schmidt-Nielsen (1979) gives a short but comprehensive account of the various mechanisms and their relative advantages and disadvantages. Some creatures assiduously avoid any dense materials used by their sessile relatives (jellyfish and nudibranchs or shell-less snails); some use body fluids enriched in ionic solutions of low density (a few large marine algal cells); some retain large reserves of fats and oils (sharks and some other deep sea fish); and some use gas-filled organs.

Two sorts of animals manage to store gas in structures of sufficient structural rigidity that the compressibility of the gas mixture is inconsequential (Alexander 1983); they thus maintain neutral buoyancy almost independent of depth. Both are cephalopod mollusks, but I'd guess that this evolutionary relationship doesn't reflect any common ancestor with a rigid gas tank. Cuttlefish (which look like bulgy squid) have a foamlike or corrugated cuttlebone containing a gas mixture at essentially constant pressure (Figure 5.3). And the chambered nautilus (Figure 3.5) has air pockets in the unoccupied chambers of its shell.

Many of the bony fish, though, keep gas in a flexible "swimbladder" in which the internal pressure cannot be appreciably different from that outside. The details of their arrangements vary. Some retain a connection between swimbladder and esophagus (the former develops as an outpocketing of the latter) and can gulp air from the surface or disgorge bubbles if need be. If, for instance, you put a goldfish in a large jar and reduce the pressure by attaching a vacuum source, the fish will give a mighty belch of gas. If the pressure is then raised by detaching the vacuum source, the fish sinks rapidly to the bottom. It may not be able to swim immediately to the surface for more air but can replenish its swimbladder from dissolved gases in its blood. Other fishes, especially ones that live at considerable depths in the ocean, have lost any connection between swimbladder and esophagus and depend entirely on secretion from dissolved gas in their blood—being brought to the surface entails a typically fatal expansion of the swimbladder. But one way or another, the volume of a swimbladder is subject to active control even if the control

(a)

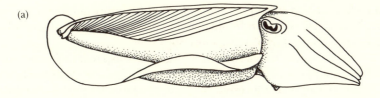

(b)

FIGURE 5.3. (a) A cuttlefish; the striped material near its upper surface is the nearly incompressible cuttlebone. (b) A slice of cuttlebone, with the air spaces shaded.

mechanisms don't operate instantaneously. Bottom-living fishes commonly lack a swimbladder, as do also a few (mackerel and sharks, for instance) that simply swim continuously—with swimbladders, one gets the odor of an awkward necessity.

Pressure and solubility

Carbonated beverages are familiar to us all, as is the observation that the bubbles don't appear until the bottle top is removed. The gas, carbon dioxide, is clearly soluble in water, and just as transparently obvious (for bottles if not cans) is the fact that the solubility varies with the ambient pressure. This is known as "Henry's law," named for William Henry (1775–1836) who pointed out that a given volume of water would take up a fixed volume of a gas whatever the pressure on the system. The fixed volume in solution, of course, contains a mass of gas proportional to the pressure (Boyle's law), so the higher the pressure, the more gas goes into solution. Again only the partial pressure of the gas matters— the same amount of pure oxygen from a tank will dissolve in a container of water at 21 kPa as will oxygen from atmospheric air whose total pressure is 101 kPa and in which about one molecule in five is oxygen.

Here chemistry does come in—different gases have different solubilities. On a volumetric basis, nitrogen is half as soluble in water as is oxygen, while carbon dioxide is about thirty times more soluble than oxygen.[1] The range is very wide—helium is a fifth as soluble as oxygen while ammonia is thousands of times more so.

There are major implications for organisms in Henry's law and the different solubilities of different gases. A deep dive means not only that an air store will be compressed but that its component gases will much more readily go into solution in either body fluids or ambient water. A diver breathing compressed air gradually absorbs quite a lot more nitrogen than usual in the bloodstream and tissues. A rapid ascent is like taking the top off a bottle of beer—gas bubbles (here nitrogen, not carbon dioxide) form. The effect ranges from unpleasant through crippling to fatal—circulatory systems don't like bubbles any better than they like blood clots. One partial remedy is to mix the oxygen for the dive with helium—much less soluble in water—rather than nitrogen. The common remedy (the solution, literally) is a very slow ascent, at a rate listed in tables that take into account the depth and duration of the dive. The alternative (the "bends") was first noted in workers preparing bridge footings from caissons of compressed air. A century ago, Eads, at St. Louis, recognized the necessity for slow ascents by workers leaving a caisson. His contemporary, Roebling (himself a victim), while building the Brooklyn Bridge in New York, took no such measures and lost quite a few workers.

Gas is already dissolved in natural waters—indeed, rapidly flowing streams may be slightly overloaded ("supersaturated") with atmospheric gases. But if a pond or ocean absorbs gases mainly from the atmosphere at the surface, dissolved concentrations will reflect the gases' different solubilities at atmospheric pressure. Due to biological activity and other causes, concentrations of some gases at substantial depths may exceed those derived from the atmosphere. But the concentration of oxygen at appreciable depths is rarely even as high as atmospheric—the abysmal darkness of the abyss precludes photosynthetic oxygen production.

The problem is quite real for a fish with a swimbladder. The partial pressures of the gases in the swimbladder of a deep sea fish must add up to the local hydrostatic pressure, which is equal to one atmosphere (surface) plus an additional atmosphere for each 10.3 meters of depth. But the swimbladder must be filled from dissolved gases in the blood and must not lose gas through redissolution into the blood. The dissolved gases in the water (and in fish blood, via the gills) will be equilibrated with

[1] The word "tension," in quite a nonmechanical sense, is often used for the concentration of a dissolved gas in a liquid—it specifies the partial pressure of that gas in an atmosphere with which the liquid containing the dissolved gas is in equilibrium.

air at pressures more like those of the atmosphere above than with the gas in the swimbladder. So there will be a terrific tendency for swimbladder gas to go into solution in the fish's blood and thence into the ocean. Two devices stand in the way. First, the swimbladder wall has a layer that turns out to be a very effective barrier to the passage of oxygen (Lapennas and Schmidt-Nielsen 1977). Second, blood leaving the so-called gas gland in the wall of the swimbladder passes through an exchanger (Figure 5.4) in which blood leaving the bladder loses excess dissolved gas specifically to blood moving toward the bladder (Scholander 1954).

In short, air under water is unstable stuff. Not only does it like to bubble upward, impelled by buoyancy, but it compresses into smaller volumes with depth and increasingly tends to dissolve in the surrounding water. Still, some organisms routinely carry external bubbles of air downward—air-breathing seems a hard habit to break, in an evolutionary sense. A number of beetles, for instance, breathe from submerged gas stores. But the scheme gets rapidly more difficult with depth—not only must the insect swim farther downward with its buoyant cargo, but a bubble has a shorter life at the higher pressure as the gases more rapidly dissolve in the water. One beetle, *Potamodytes tuberosus*, contrives to get the pressure in the bubble below atmospheric and thus render the bubble permanent, taking advantage of the pressure drop attending fluid flow (Stride 1955). But it is restricted to *very* shallow water (about 4 cm) and

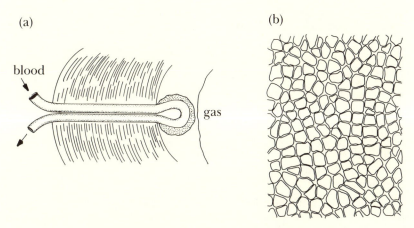

(a) (b)

blood

gas

FIGURE 5.4. (a) Diagrammatic view of the exchanger in the wall of a swimbladder. Blood vessels going inward through the wall are in close contact with those coming out, as seen in (b), a cross section drawn from a photograph. Half of these vessels carry blood inward and the other half outward.

rapid flows (over a meter per second). We'll return to it when we consider Bernoulli's principle in Chapter 7.

Perhaps the most impressive use of gaseous air under water is that of the European water spider *Argyronetes aquatica* (Foelix 1982). It is the only spider that lives constantly underwater—walking, swimming, feeding, mating, and raising young. It makes a bell-shaped web among the submerged vegetation of ponds and then travels repeatedly to the surface and down again, each time carrying down a coating of air on its abdomen (the allusion to silver in the name of the genus reflects this shiny layer). Air thus fills the bell; the fine mesh of the web and a coating on the strands prevents upward leakage, as will be explained in Chapter 11. From time to time the spider adds air to its diving bell to offset use and dissolution. Incidentally, it has one other, perhaps unrelated, oddity— uniquely among spiders (and unusual among animals in general) the males are larger than the females.

THE LIQUID STATE

A liquid may be a familiar matter, but the liquid state boils down to quite a queer business. It's something between solid and gas, although "between" implies nothing like a continuous quantitative scale. Rather, a liquid shares certain properties with each of the other states—for some problems a liquid behaves like a gas of exceptional density and coherence; for others it can be thought of as a broken-up solid. (Conversely, a fine powder is often better treated by the rules for liquids than by those for solids.) A useful way of viewing the distinctions is to recognize that, while all matter is characterized by mass, a gas has only mass. A liquid has, in addition, a specifiable volume, not just one imposed by the size of the container. A solid has, besides mass and volume, some specific shape, not just one imposed by the shape of the container.

Internal cohesion

Internal cohesion is the basic attribute distinguishing liquids from gases—a gas withstands no tension; a liquid does. Under conditions of weightlessness and windlessness, a glob of liquid will assume a spherical shape, spontaneously minimizing its surface area as it huddles together. But how strong is that internal cohesion? Measuring it requires that we pull on a liquid, a task superficially analogous to pushing on a rope. Ingenuity, though, has not been lacking, and the thing has been done by several techniques; the one shown in Figure 5.5 is perhaps the most straightforward. Water (that's the liquid we'll worry about) is loaded into a Z-shaped glass tube of fine bore open at both ends. The tube is then

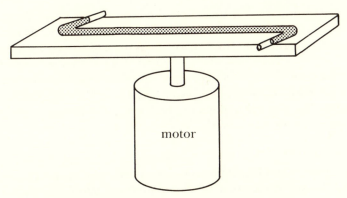

FIGURE 5.5. A diagram of the device of Briggs (1950) for measuring the cohesion of liquids—a motor spins a plate to which is glued a glass Z-tube containing liquid. From the speed at which the column of liquid ruptures in the middle the cohesion can be calculated. The bent ends of the tube ensure that the liquid is self-centering about the axis of rotation.

spun about its midpoint like a propeller at increasing speeds. By Newton's first law, the water prefers to go straight rather than to turn; that is, its mass requires that a force be applied along the axis of the tube to keep the water rotating, like the force in a string tied to a ball being whirled in a circle. But the only force is the internal cohesion in the water column, pulling both halves of the column toward the midpoint. By Newton's third law (action and reaction) half the column is pulling on the other half, connected together at the axis of rotation. When the tensile forces can no longer maintain the water's continuity, the column ruptures at the axis. From the rotation rate and the dimensions of the column, the tensile strength comes out to about 28 MPa—that is, to a stress equivalent to the pressure exerted by *280 atmospheres* (Hayward 1971; Sprackling 1985). Put another way, a rope one centimeter in diameter made of water, if you can imagine something so preposterous, would support a mass of about 225 kg (almost 500 lb).

This phenomenon of tension in liquids bothers people used to thinking about gases. Recall that you can't really pull on a gas since it has negligible internal cohesion. Negative pressure, for that's what tension is, is illegal in a gas. Still, it seems as if the stunt can be done—a pump at the top of a pipe can draw out air and raise water in the pipe. But by pumping out air it can't raise the water to a height beyond that set by the local atmospheric pressure—10.3 meters at sea level. The limitation arises because the pump, by removing air, is merely allowing the atmosphere to *push* the water upward (Figure 5.6), and no *pull* is actually involved. In theory, a pump dealing with a column of water only, with no air inside, can avoid this height limit. Instead, its limit is the 280 atmospheres mentioned above, which corresponds to a column of water approaching 3 km (2 miles) high. In practice, though, the water pump cannot do much better than the same 10 meters—the water fails in its adhesion to the walls of the pipe, dissolved gas in the water boils out at the low pressure, and the column of water is ruptured.

Are trees permissible?

The difficulties of adequately containing a liquid under tension are so great that the phenomenon of liquid tension is ordinarily dismissed as a curiosity without technological ramifications. It has received most of its attention from plant physiologists, for whom it is at once crucial and commonplace. Running upward in trees from roots to leaves are a set of pipes (xylem vessels and tracheids, technically) through which water ascends. (The ascent is far above the height to which water would rise due to surface tension, as in a capillary tube.) While there is some upward push from the roots ("root pressure"), it is now quite clear that the main agency impelling ascent is a pull from the top, as first suggested to a most skeptical world by Dixon and Joly, a botanist and a physicist, in 1894. If vessels are suddenly cut, the hissing sound of air entering can be detected. Water evaporates ("transpires") at the leaves, and the vacant space can only be refilled, as things are arranged, by raising water from the

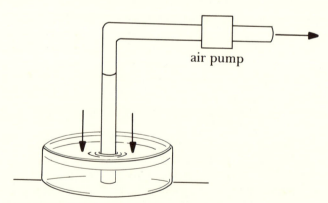

FIGURE 5.6. An air pump can raise water only to a height set by the local atmospheric pressure—in removing air from the pipe it merely permits the atmosphere to push the water column upward.

roots. An early bit of support for the existence of internal tension was the observation of Macdougal (1925) that when the transpiration rate increased, a tree trunk shrank. Trees are just a little skinnier at midday!

And trees exceeding 10 m in height have been around for the past hundred or so million years—trees get up to about 100 m, implying an internal tensile stress of 1 MPa, which corresponds to a pressure of −10 atmospheres. With a clever device called a "Scholander bomb" (Scholander et al. 1965; Figure 5.7) it's possible to check directly these expectations—in fact, the pressures are often still more extreme, with −20 atmospheres not at all uncommon high up in a tree. A negative 10 atmospheres of pressure might account for water staying up, but it turns out that still more is needed to move water through the narrow vessels and to detach it from the soil. Interestingly, the cells making up the vessels are not alive at what amounts to functional maturity. Perhaps the release of dissolved respiratory carbon dioxide from live cells into a vessel with a large negative pressure would cause a microcatastrophic bubble! While there is no doubt that sap is being pulled up and pressures are strongly negative, it must be admitted that not all of the details of the scheme are at this point clear—questions remain as to any special properties of internal vessel walls, how the continuous columns are initially established, how damage such as that caused by freezing can be repaired, and so forth.

Surface tension

For water deep within a large body of the stuff, internal cohesion gives rise to forces in all directions which, taken together, balance out so that they may have few immediate consequences. For a bit of water at the

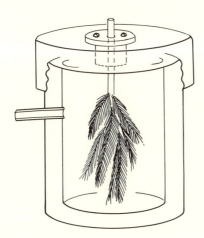

FIGURE 5.7. A Scholander bomb, with a stem protruding. Cutting the stem from the plant relieves the internal pressure, so the air-sap interface retreats inward as the vessels expand. Sealing the stem in the bomb and applying positive pressure through the side port mimics the effect of the original negative internal pressure. The applied pressure that just returns the interfaces to their original location at the cut surface is thus equal (if opposite in sign) to the original negative pressure in the sap.

surface of a liquid, the situation is usually quite different. Water below it pulls down, but the air above doesn't provide any balancing upward force. Since the bit of water quite evidently doesn't accelerate downward, there must be some other counterbalancing force—it is the resistance of the water to being compressed. To some extent, then, the presence of a surface increases the hydrostatic pressure within a body of water. In addition, the asymmetry of forces at the surface gives the surface itself a very special character—it may be nothing more than a no-thing, a non-material boundary, but it has (or confers on the liquid) the specific and most peculiar property called "surface tension."

Another way to look at the matter is to recognize that the internal coherence of a liquid makes a surface something unattractive which the system will arrange itself to minimize. Just as a droplet will spontaneously round up into a sphere, it takes work to force it into any other shape—it takes work to form surface. To put the phenomenon into quantitative form, we have either to measure the force needed to maintain a surface against the surface's urge to submerge, or to measure the work necessary to create some unit area of surface. It's easy to construct a semiquantitative apparatus to view the force or work; it's a little harder, though, to make precise measurements.

A film of soapy water amounts to two air-water interfaces with a very thin layer of water between. All one has to do is stretch such a film and you're stretching two air-water interfaces. And that requires no more than one wire sliding down two others (Figure 5.8). The results make a striking contrast to what would happen if you pulled on a sheet of rubber

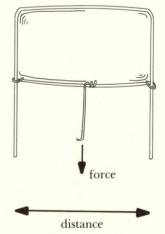

FIGURE 5.8. Expanding the pair of air-water interfaces that enclosed a film of soapy water by pulling down on a slide wire. It's really as simple as it looks.

or other elastic. The "material" is an elastic in one sense—it does pull back against the applied force. But here "stress" is apparently independent of "strain"—the force it takes to hold the slider against the surfaces' desire for oblivion, or to move the slider and create more surface, doesn't at all depend on the particular position of the slider. It takes work to make surface, but the work is independent of the size of the preexisting surface. What is happening is that the apparent stretching of the surfaces does elongate them, but by *creating new surface area*! Water from between the surfaces simply moves out to the interfaces to increase their areas, and the film gets thinner.

The force may not depend on the distance the interfaces have been stretched, but it does depend on the width of the two interfaces—double the length of the slider and you double the requisite force. The appropriate dimensions in which to express surface tension as a material property are therefore force divided by distance. Similarly, the work depends on that width and on how far the constant force has moved the bar—in short on the area of surface created. It does not depend on thickness—the interfaces have no thickness, and the water between them plays no role. So the appropriate dimensions for work-to-make-surface are work divided by area. But both quantities reduce to basic dimensions of MT^{-2}, so they are equivalent in dimensions (as well as units and numerical values). We usually think of force per unit distance and most commonly use a Greek lowercase gamma (γ) for a symbol.

For present purposes we need only one or two values for this physical quantity—the interface between water and air is the main one relevant to organisms. For pure water the value is 0.073 N·m^{-1}; for seawater it is 0.078 N·m^{-1}. For liquids against air these are unusually high values, another of the many peculiarities of water. Temperature has relatively mild effects on the values, but they are drastically lowered by soaps and detergents, the main reason we use the latter.

You've undoubtedly noticed that a small quantity of water seems to know something about the surface on which it resides. On a block of paraffin, a teflon-coated pan, or a freshly waxed car it forms little beads of the form of flattened spheres. By contrast, on scrupulously clean glass or on any surface with a trace of soap it spreads out. It isn't enough to worry just about an air-water interface—somehow, any solid partner in the interface has a decided influence on its behavior. The phenomenon goes under the heading of *wettability*—we say that water "wets" some solid surfaces but not others; the wettable surfaces are those on which water spreads rather than beads. Wetting occurs when the adhesion of liquid to solid is substantially greater than the internal cohesion of the liquid;

nonwetting indicates that internal cohesion is greater. There are degrees of wettability—the phenomenon is a continuous and quantitative one—but we'll mainly consider the two extremes.

Capillarity. The classic demonstration of the effect of wettability consists of immersing a narrow-bore glass tube in water. The air-water interface in the tube rises to a level above that of the surrounding water (Figure 5.9)—it is being pulled up by some force, and the fact that the periphery of the interface is highest points a clear finger at the location of the force. Attraction of water to glass draws water up at the edge; the resistance of the interface to expansion of its area (as we've just discussed) raises the rest of the water to nearly the same height. Wetting is nearly complete, as shown by a low contact angle between glass and water—at the point of contact, the interface is almost vertical; the low contact angle also means that water is being pulled almost purely upward and minimally outward. Pressure in the raised water is subatmospheric since it is being pulled up, not pushed from below. So you can't put a side tube on the apparatus and use the elevation due to surface tension to pump a continuous stream of water up into a reservoir higher than the original body of water and then let the water fall back through a generating turbine. Too bad—another perpetual motion machine grinds to a halt.

How high (*h*) will the water rise in the tube? It will rise until the upward force due to surface tension is just equal in magnitude (and opposite in direction) to the downward, gravitational force. The upward force is surface tension times a length, that of the line of contact between glass, water, and air—the circumference of the tube. The downward force is

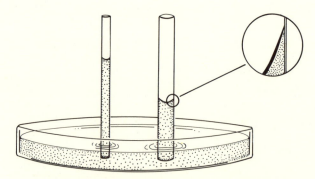

FIGURE 5.9. Water spontaneously rises in a clean, glass tube—the thinner the bore of the tube, the higher the column goes. The inset shows the "angle of contact" in air between glass and water.

the volume of water in the tube above the external water level times its density times gravitational acceleration. Setting the two forces equal for a cylindrical tube of radius r, we have

$$(\gamma)\,(2\pi r) = (\pi r^2 h)\,(\rho g).$$

Cancelling and solving for height gives

$$h = 2\gamma/\rho gr. \tag{5.2}$$

Clearly, water will rise higher in smaller tubes. The upward force is proportional to the first power of length, and the downward force is proportional to the second power; thus the rise varies inversely with radius or diameter of the tube. These consequences of surface tension are small-scale phenomena, suggesting that they ought to be of more immediate relevance to small organisms than to ourselves. But we do run into them. A bath towel works because it has lots of wettable surface, lots more than a wet person. Rub the two together, and capillarity among the fibers of the towel assures that the bulk of the water will end up transferred to the towel.

What if the solid isn't "wetted," if the cohesion of the liquid exceeds the adhesion between it and the solid? The usual demonstration consists of inserting a narrow-bore glass tube into liquid mercury; the interface in the tube is depressed beneath the level of the outside liquid. Mercury may not be the stuff of which life is made, but deliberate nonwettability may be quite important to organisms—it's as ordinary as water off a duck's (nonwettable) back. If you are small, as living things usually are, then being covered by a layer of water of substantial thickness isn't exactly a convenience. Contact with the surface of a body of water can be a perilous entanglement, best (but not inevitably) avoided by arranging to be as repulsive as possible to this clingy liquid. Water may be crucial to all life, but the air-water interface can hold a fatal attraction.

Walking on water. The surfaces of bodies of water aren't the commonest of habitats, but a decent number of creatures (bugs and beetles, mostly) are specialized for getting around on it. The possibility was raised in the last chapter, and an index was developed (equation 4.8) relating size and the practicality of being supported by the surface of a liquid. A more specific look is now possible—the problem is very similar to that of capillarity. The absolute prerequisite is a surface resistant to wetting, but a waxy coating is normal among insects, functioning both as a water repellent and a sealant against evaporative water loss. So an insect pushes down with all six feet against the surface, usually with the feet splayed out and thus not interacting—each foot makes a separate dent in the surface. How much insect can be supported? What matters is the combined perimeter of the feet, the total length of the air-water-insect contact line, although exactly how to measure that length isn't quite as clear as it was for the capillary tube.

A marine water strider, *Halobates*, has a mass of about 10 mg; using the surface tension of sea water, it would require a minimum of about 1.3 mm of contact line, about 0.2 mm per leg. *Halobates*, though, can jump upwards from the surface by several centimeters, so it must be able to exert a downward force of about ten times its weight (Andersen and Polhemus 1976) and should require 13 mm of contact at takeoff to avoid wet feet; it has, in fact, specialized long hairs on its legs. *Halobates* and the freshwater striders can propel themselves around on the surface by pushing on the local indentations beneath the legs. Andersen and Polhemus reported top speeds of 0.8 to 1.3 $m \cdot s^{-1}$, most impressive for such small animals. Once surface tension can be handled, such a surface must be quite a nice environment—it has neither the hilliness of land nor the high drag forces of water.

There's a less conventional way to get around on the surface, one that works *only* on an interface. A small beetle, *Stenus*, normally walks slowly on the surface. When speed is desired, though, it secretes a substance from its last abdominal segment that locally reduces the surface tension. The result is an asymmetrical force on the beetle, which propels it forward at up to 0.7 $m \cdot s^{-1}$ (Chapman 1982). The trick, though, is probably expensive inasmuch as the surfactant chemical cannot be reused. You can quickly demonstrate *Stenus*'s stunt in either of two ways. If a loop of string is loosely placed on the surface of a bowl of water and a piece of soap touched briefly to the surface within the loop, the loop will quickly be pulled outward and become circular. Or a small surface boat can be made of a piece of shiny paper and a tiny piece of soap stuck on its stern—the boat will be propelled forward.

What would be the maximum weight of a human who could walk on water? My size-9 sandals have a perimeter of 0.62 meters each; that length times the surface tension of water gives 0.045 newtons of force, or 4.6 grams (less than half an ounce) of weight—9.2 grams to stand (two feet in contact) or half that to walk. The theological implications are beyond the scope of the present book.

Mosquito larvae don't walk on water but instead hang downward from the surface, breathing, as already mentioned, with a snorkel system. Adding a substance to the water that reduces surface tension is highly detrimental to their well-being; it may be done deliberately or may be just a not-so-undesirable consequence of detergent pollution. Alternatively, some light and nonaqueous liquid can be poured on the water (kerosene, for instance), which will "wet" the waxes of the larval cuticle and plug the snorkel. If you live by surface tension you can die without it.

One additional use of surface tension ought to be mentioned. Quite a few aquatic insects breathe by means of a "plastron," a permanent coating of air made stable by its location within a covering of nonwettable hairs (Hinton 1976). The hairs are short but densely packed—from tens of thousands to millions per square millimeter. Pressure in the air stays well below the local hydrostatic pressure because water can't penetrate the interstices of the fur coat without creating lots of additional surface, which it doesn't take kindly to. Thus, gas is not forced into solution and, indeed, the respiratory depletion of oxygen is restored by spontaneous diffusive influx from the water.

Surface waves

The surface of a liquid is like the brow of a thinker—a perturbation wrinkles it into waves. Surface waves on a liquid, though, persist only as long as they move, so one can't separate the subjects of waves and wave motion. These are complex matters we'll only touch on here, ignoring really large waves, the special problems of waves in shallow (relative to the distance between wave crests) water, the rules for the motion of groups of waves, and so forth. The interested reader might look at Bascom (1980) or Trefil (1984) for simple and engaging accounts, or at Denny (1988) for both greater detail and a more biological slant.

The first and perhaps most surprising observation about waves is that, while they travel steadily, the water of which they're made travels only back and forth and up and down—a "wave" is an independent entity, not some specific mass of water in motion. Thus it's convenient to define a special variable for wave speed to avoid any confusion with water speed; the variable is called the "celerity," designated c. Beyond that, waves are characterized by their height, to which we won't pay much attention here, and the distance between adjacent crests or troughs, their "wavelengths," symbolized by the Greek lowercase lambda (λ).

A surface wave is an unstable item. The air-water interface really prefers to be flat and horizontal—flat on account of the surface tension of water and horizontal on account of gravity. But flat and horizontal are functionally the same thing, so a wave is subject to two quite distinct destructive forces acting in concert. And it is these forces due to surface tension and gravitational acceleration that, in practice, determine both the wavelength and the celerity of waves. Just which one dominates is a matter of scale. Surface tension (causing "capillary waves") predominates at small scales, and gravitation (generating "gravity waves") at scales more germane to human activities. The combination is clearer in the equation

101

below and a graph (Figure 5.10) than in just words. The celerity of an actual wave is equal to the sum of the predicted celerities of the capillary wave (first term) and the gravity wave (second term). (The equation is a bit idealized, presuming a mathematically tidy wave shape, water much deeper than the wave length, and so forth—they were established by Lord Kelvin a century ago, when Britannia ruled the waves.)

$$c = \left(\frac{2\pi\gamma}{\lambda\rho}\right)^{0.5} + \left(\frac{g\lambda}{2\pi}\right)^{0.5}. \tag{5.3}$$

Incidentally, but not surprisingly, the ratio of the two enclosed terms in equation 5.3, less a numerical constant, is the Jesus number of equation 4.8. Equation 5.3 and its graph, the top line of Figure 5.10, show the complexity resulting from two fundamentally different bits of physics operating simultaneously. For gravity waves, the wavelength increases with speed. For capillary waves, wavelength varies inversely with wave speed. And there is a minimum speed below which real waves cannot travel; it

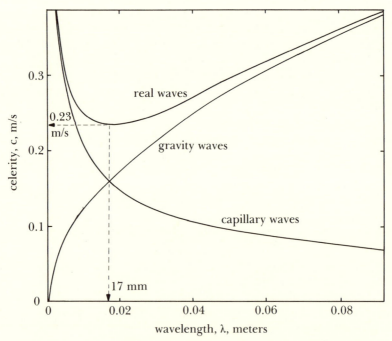

FIGURE 5.10. The relationship between wavelength and wave celerity for real waves results from the combined effects of gravity waves and capillary waves.

is about 0.23 m·s^{-1} (0.5 mph) and occurs at a wavelength of about 17 mm (0.67 inch).

There's a lot of biology riding on surface waves, most of it involving large gravity waves and the consequences of their destructive dissipation on shores. A few studies, though, have explored the world of small waves and how organisms capitalize on them. One is an investigation of wave-making on ponds and slow streams by the whirligig beetle, *Dineutes* (Tucker 1969). *Dineutes* swims on the surface, supported about half by its buoyancy (gravity) and half by surface tension. Swimming slowly, it makes no waves—that minimum celerity is no abstraction in its world. Swimming faster, it propagates waves outward, and Tucker calculated swimming speeds from photographs showing the angle made by the arms of the "V" of the propagating waves (Figure 5.11). (An analogous "V" is formed by shock waves in supersonic flight; the maximum design speed of a supersonic aircraft can be calculated from the sweepback angle of the wings. Which is why the military often censor photos showing top or bottom views of new designs.)

One can go back to the material of the last chapter (equation 4.6) and calculate the maximum speed of these beetles. For a hull length of 10.6 mm, the predicted speed comes out to only 0.13 m·s^{-1}, well below their relatively impressive observed maximum of about 0.4 m·s^{-1}. In fact, the hull speed limitation of a maximum Froude number doesn't apply, since

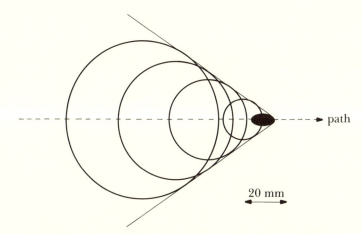

FIGURE 5.11. A whirligig beetle swims fast enough to create surface waves. If the beetle were stationary (just bouncing up and down), the waves would be concentric. Since the beetle is swimming, the center of each circle is displaced ahead of the previous one, forming a V-shaped envelope.

the beetles are short enough to be in the range where wavelength is mainly determined by surface tension rather than gravitational forces. Indeed, Tucker notes that there are no common whirligig beetles with lengths exceeding 17 mm, the point above which gravity waves begin to predominate. Again size enters in determining the practicality of a way of living, and again the underlying physical reality both raises opportunities and imposes constraints.

There's apparently a world of unfamiliar phenomena available to the fauna of liquid surfaces. Whirligig beetles seem to be able to "echolocate" through reflection of their ripples—it must be a queer sort of sense that requires body motion to send the signal. And Wilcox (1979) has shown that males of at least one species of water strider can tell one sex from the other by sensing surface waves—males add on a high-frequency flourish, a kind of tremolo, that conveys the message to the male receiver that they are not suitable mates for each other.

CHAPTER 6

Viscosity and flow

"Blood is thicker than water."

Sir Walter Scott on viscosity

THE EVERYDAY meaning of "fluid" is that of "liquid"—the stuff we drink or use to fuel our cars. Properly, though, a fluid is any substance that flows, making the term a generic for both gaseous and liquid states. The lumping may seem odd just after a chapter that dwelled heavily on the differences between gases and liquids, but the two do have the important common quality of responding to unbalanced forces by flowing. Indeed, for problems involving flow the two are remarkably similar, more so than one has any a priori reason to suspect.

On account of their lower densities, gases are easier to push around—but that's a quantitative rather than a qualitative difference. One might worry about the vast differences in compressibility between gases and liquids—double the pressure on a gas and you almost exactly halve its volume; do the same to, say, water, and the reduction in volume is less than one part in 20,000. But it turns out that in the sorts of flows organisms encounter gases are functionally incompressible as well. Consider a wind of 20 m·s⁻¹ (45 mph) brought locally to a halt by hitting the beak of a flying bird—the increase in pressure will be 240 Pa (Chapter 7) at the upstream tip. This trivial increase (0.24%) over the ambient atmospheric pressure will cause only a comparably trivial compression. Compressibility becomes significant only when a gas encounters something at an appreciable fraction of the speed of sound (about 320 m·s⁻¹ or 720 mph). Or one might worry about the different effects on liquids and gases of gravity and surface tension. These agents are certainly significant—they make waves, for instance—but their effects are almost entirely limited to fluid-fluid interfaces.[1] So such effects can be ignored for natural flows that don't involve interfaces between fluids, which is to assert that they don't much matter for most situations of biological interest—again gases and liquids behave quite similarly.

Very young children really appreciate fluid flow—just watch a tot intensely absorbed by the transfer of a liquid from one container to another. But we (most of us, anyway) grow into a world dominated by the

[1] That is, unless the fluid is stratified by a density gradient as might result from temperature variation in an atmospheric inversion or to salinity variation in an estuary.

more staid and apparently practical domain of solid mechanics, losing sight of both the beauty of fluids in motion and the variety of their counterintuitive behavior. I must admit, though, to a persisting infantile prejudice in favor of fluids, a partiality that has for many years supplied me with notions to investigate and that abetted the predecessor to the present book (Vogel 1981b). The beauty of flow, no mere intellectual aesthetic, is nowhere better presented than in a book of photographs, *An Album of Fluid Motion* by Van Dyke (1982).

We've already looked at the flow of fluids in several contexts. Chapter 2 introduced the very important principle of continuity and made the point that one could derive forces just by measuring speeds of flow. Chapter 3 casually mentioned that drag varied either with the first or the second power of length, depending on the sizes and speeds involved. Chapter 4 discussed the Froude number and introduced the drag coefficient and the crucial notion of velocity gradients. And Chapter 5, of course, took a brief look at surface waves.

Solids versus fluids—viscosity

Gases and liquids share a property, *viscosity*, that is absent in solids and lack another, *stiffness* (associated in popular usage with the idea of elasticity), that is characteristic of solids. It's worth emphasizing that the distinction we make isn't just one of degree but a matter of fundamental applicability of these analogous but clearly different properties. Solids resist being deformed in shear—fluids simply don't. Fluids, though, aren't totally indifferent to shear—what is different is *how* they respond to shear stresses.

Consider (assume weightlessness, for convenience) rectangular blocks of solid and fluid, each distorted in shear (Figure 6.1). For the solid, the greater the force, the more the distortion. One can write a rough formula for what the "shear stress" does—more stress means more strain (distortion), and the ratio of stress to strain (what a biologist might call the ratio of stimulus to response) is the "shear modulus," a measure of stiffness under shearing loads:

$$\text{shear stress} = \text{shear modulus} \times \text{shear strain.} \qquad (6.1)$$

For the fluid, by contrast, it doesn't take a greater force to produce more distortion—the stuff is for all practical purposes capable of infinite distortion at any force since it takes on any shape equally happily. But what does depend on the force is the *rate of distortion* of the fluid cube—the resistance of the fluid depends only on how *fast* it's distorted. This resistance to *rate* of shear is what we mean by viscosity; in short,

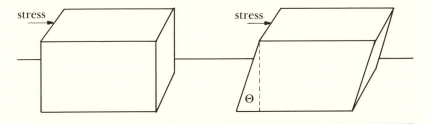

FIGURE 6.1. Pushing across the top of a block whose bottom is fixed distorts the block—two faces of a rectangular solid become nonrectangular parallelograms as they are tilted to an angle Θ.

shear stress = viscosity × shear rate. (6.2)

Since the flow of a fluid along a solid surface inevitably involves shearing or distortion (I'll defend that bald statement below), viscosity comes down to a measure of the resistance of fluids to flowing across surfaces or through conduits. Thus, the more viscosity, the less "fluid" the fluid (which may seem backward, but that's how it's defined).

This notion of shear rate is worth a few more words. It amounts to a measure of how rapidly "layers" of fluid are sliding with respect to each other—as if you push across the top sheet of a pile of paper and each sheet moves a little faster than the one beneath it. It's thus a rate of change of velocity with distance—a velocity gradient, but one in which velocity varies, not in the direction of travel (as with acceleration), but at a right angle to flow. It's the gradient you'd experience when wading out into a creek through the quiet waters near shore toward the torrent in midstream. As mentioned in Chapter 4, the lack of "slip" of a fluid at a solid surface implies the presence of such a shearing region and velocity gradient wherever a fluid flows past a solid.

But we need a more quantitative view of viscosity. Imagine our previous cube now transformed into two thin, flat plates with fluid between (Figure 6.2). The lower plate is fixed, a force moves the upper one horizontally, and the no-slip condition applies to both inner faces. As a result, the fluid between them shears, producing a smooth variation in flow rate, that is, a constant velocity gradient. With this diagram we can put equation 6.2 in more formal terms, using the Greek mu (μ) for viscosity (sometimes called the "coefficient of viscosity" or "dynamic viscosity"):

$$F/S = \mu v/l. \qquad (6.3)$$

From this equation one can figure out the basic dimensions of viscosity— they come out to $ML^{-1}T^{-1}$, a combination that has no obvious intuitive

107

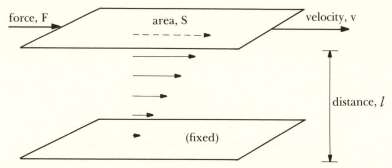

FIGURE 6.2. Flat plates with fluid between. With the no-slip condition applicable to both plates, a constant velocity gradient develops between them; the viscosity of the fluid determines how much force it takes to give the top plate a given velocity for a certain area of plate and distance between them.

significance. The SI unit is the kilogram per meter-second or, shorter, the "pascal-second" (Pa·s).

Looking up viscosities isn't especially difficult but is rarely necessary, since two materials, air and water, dominate the biological scene. Table 6.1 gives their viscosities at a temperature of 20°C, along with the densities and the ratio of viscosity to density, the latter called the "kinematic viscosity" (problems involving viscosity very often involve density as well). Notice that the substances have very different viscosities and densities but the ratio of the two properties differs only fifteenfold. It's surprising (and will prove a considerable convenience) to find such a similarity between two such different materials.

BOUNDARY LAYERS

Since the no-slip condition is universal, so are velocity gradients adjacent to surfaces—that's why viscosity is so important. Pipes and arteries develop deposits on their walls, it takes a strong wind to remove a spore resting on a leaf, and water without suspended grit has no erosive action

TABLE 6.1. VISCOSITIES AND DENSITIES

	Viscosity, μ (Pa·s)	Density, ρ (kg·m^{-3})	Ratio, μ/ρ (m^2·s^{-1})
Air	18.1×10^{-6}	1.20	15.0×10^{-6}
Water	1.00×10^{-3}	1.00×10^{3}	1.00×10^{-6}
Seawater	1.07×10^{-3}	1.02×10^{3}	1.05×10^{-6}

on rock. If a fluid is very resistant to (rate of) shearing, then the velocity gradients in it will tend to be gentle, and there will be a relatively great distance between the stationary surface and the region of moving fluid where the motion is at full speed (the so-called free-stream velocity). If, conversely, the effects of viscosity are slight, then the gradients will be steeper and the distances shorter. Figure 6.3 graphically depicts how speed changes within the gradient region for flow parallel to a flat plate at moderate speeds.

Even more useful than a measure of the steepness of the gradient would be a figure for the thickness of the whole gradient region, from zero speed at the surface to the final free-stream speed. There's a difficulty, though. While the inner limit (zero speed) presents no problems, the outer one is, strictly speaking, indefinite—the local speed of flow approaches the free-stream speed asymptotically. So if we are to quote a figure for the thickness of such a gradient region we must arbitrarily define the outer limit. The classical definition was established early in this century by Ludwig Prandtl (Von Karman 1954 gives an engaging historical account) who set the outer limit where the speed achieved 99% of the final free-stream value, so we can speak of the gradient region defined by that convention as the "Prandtl boundary layer."

Perhaps the clearest way to show how the thickness of a boundary layer varies is by citing Prandtl's formula for its thickness in the simplest possible case—a flat plate oriented parallel to the direction of flow (Prandtl and Tietjens, 1934). Thickness here is designated by the Greek lower case delta (δ):

$$\delta = 5\sqrt{\mu x/\nu\rho} \ . \tag{6.4}$$

oncoming wind

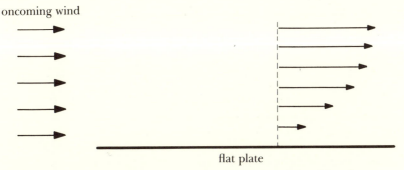

flat plate

FIGURE 6.3. The velocity gradient on a flat plate oriented parallel to flow; the length of each arrow above the plate is proportional to the velocity at its base. The thickness of the gradient region has been grossly exaggerated for the purpose of illustration.

The thickness of the layer increases with the square root of x, which is the distance behind the upstream edge of the plate—thickness increases parabolically (Figure 6.4), that is, rapidly at first and more gradually farther downstream. It is thicker, as we'd expect, if the viscosity is greater; and it's thinner for greater values of density and free-stream speed. It should be noted that the formula, taken strictly, refers to a rather specific geometry rarely encountered in nature—surfaces aren't commonly rigid, smooth, just parallel to flow, and with sharp upstream edges.[2]

The arbitrary element in the definition should be emphasized. The figure of 99% has biological, not physical, roots—the exact value is an accident of our bipedal and pentadactylic anatomy. Prandtl wanted a generous definition—he was interested in calculating drag and wanted to omit no region where viscosity might be doing anything significant. If, instead, one is interested in whether the spore-bearing stalk of some fungus extends out into substantially full wind, one might pick a value of 90% rather than 99% of free-stream speed for the outer limit. The resulting boundary layer will be only 3/5ths as thick—a constant of 3 will replace the 5 in equation 6.4.

People sometimes ignore the picture given by Figure 6.3 and focus instead on a diagram such as Figure 6.4. They thus retain a fuzzy notion that the boundary layer is a discrete region of nonmoving fluid rather than the discrete notion that the boundary layer is a fuzzy region in which there is a strong velocity gradient. Gradients, again, are all impor-

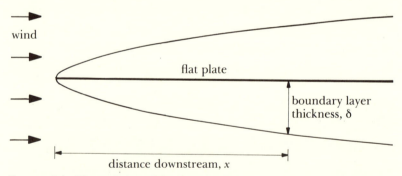

FIGURE 6.4. The variation of the thickness of the boundary layer on a flat plate with distance from the upstream edge. Note that the velocity gradient gets less steep as one moves downstream. While the shape of the curve is real, the line as drawn is somewhat arbitrary; it certainly represents no physical discontinuity in the flow.

[2] The formula is limited, as well, to Reynolds numbers beween about 500 and 500,000; more about these numbers in a few pages.

tant; boundary layers are just regions of velocity gradients and thus places where viscosity exerts its pernicious effect on flows and forces.

These boundary layers provide both trouble and assistance to organisms. Getting down near a surface provides protection from the forces of fast flows of air or water (although nothing is so flat as to have no drag at all—at the surface itself the velocity may be zero, but the all-important velocity gradient is not). Conversely, an organism feeding from the water passing by must somehow raise its feeding equipment or intake pipe up high enough to encounter adequate flow speeds. In short, there's lots of biology here, enough for an entire book.

A moth's antenna

The gentle velocity gradients at low speeds are no small matter even where the geometry of the situation precludes using tidy formulas such as equation 6.4. For example, consider the problem of male saturniids (giant silkmoths). Females draw males for mating by releasing a volatile attractant chemical, but these moths typically live at very low population densities. Males have equipment to smell out females at distances of miles—large, feather-shaped antennae (Figure 6.5) on which some 70% of the sensilla are sensitive to nothing other than the females' perfume— truly an olfactory sensation. A nineteenth-century French entomologist, Fabre, did a few experiments on the phenomenon but simply could not believe that any sense of smell could work so well (Teale 1949)!

In order for an odorant to be picked up from the airstream, air must be made to pass through an antenna. The tacit presumption among people interested in olfaction has commonly been that air directly upstream

FIGURE 6.5. A male *Luna* moth and one of its antennae.

from an antenna would pass through it. But gases as well as liquids have viscosity, and passing through a dense array of sensilla involves shearing and velocity gradients, so the air might well prefer to go around an antenna as if it were a solid plate. The moth thus faces an odd problem in design—more sensilla would give more sensitivity to odorant but at the same time would reduce the passage of odorant-laced air. For another investigation I had built a very tiny device to measure airspeeds, so I looked at the speeds immediately behind antennae, where air had just passed through (Vogel 1983). Comparing these speeds to the free-stream gave figures of from 8 to 18% for the fraction of the air approaching an antenna that went through and not around—for the "aerodynamic transmissivity." The exact figure depended on the particular free stream speed. Lower speeds gave lower transmissivities, as one would expect from the presence of velocity in the denominator of equation 6.4—lower speeds mean thicker boundary layers even where that equation doesn't precisely apply. The figures are far below the generally assumed 100% and well below even the antenna's optical transmissivity of 43%. With less air passing immediately by them, the sensilla must be even more sensitive than previously imagined—such are the vicissitudes of viscosity.

The moth's problem is only an instance of something quite widespread—a great number and diversity of organisms, creatures such as sponges, barnacles, some sea anemones, and various aquatic worms, live by separating edible material from large volumes of water. The game is called "suspension feeding" and almost always involves some sort of filtration mechanism and thus the attendant difficulty of getting out tiny particles without excessively discouraging the flow of the water in which the particles are suspended. But the rules and possible moves in the game are complicated—simple sieving is only one way of doing the necessary separation (Rubenstein and Koehl 1977).

Stirring the substratum

One occasionally notices snow swirling around on the downwind side of a tree or smoke vortices behind a chimney. The same phenomenon occurs on a much smaller (but not microscopic) scale behind small cylindrical organisms protruding from a solid substratum into a flow. But here there's a second component of the motion. Near the top of the protrusion the local speed of flow is greater than it is near the bottom (substratum)—the cylinder is, of course, sticking up through a boundary layer. Fluid, as repeatedly mentioned, doesn't like to shear rapidly, and the fluid behind the top of the cylinder is subjected to more shear from the faster-moving fluid passing by it than is the fluid near the bottom. So fluid behind the top is drawn, essentially by its viscosity, into the passing

112

flow; and that fluid is replaced by fluid moving up behind the cylinder from the bottom. In short, there is a net upward flow behind a protruding vertical cylinder—the effect is occasionally referred to as "viscous entrainment." It is sometimes misattributed to Bernoulli's principle (Chapter 7); the principle, though, doesn't apply to velocity differences due solely to position within a boundary layer.

Protruding vertical cylinders are extremely common in nature—all sorts of plants grow that way. They occur, as well, as attached animals sticking up into moving water. Until recently it was assumed that arrays of cylinders inevitably had a stabilizing effect on any erodible substrate, that they caused what is termed "skimming flow" in which the cylinders shelter the water between them from the current which then "skims" over the top. But it's now clear that skimming requires a very dense array of cylinders, and that lower densities have the opposite effect of eroding sediment by creating the the vertical aftercurrents just described (Eckman et al. 1981). The protruding tubes of several kinds of marine worms effectively resuspend detritus—edible dirt, loosely speaking (Carey 1983)—one worm switches from feeding on detritus to suspension feeding with its pair of tentacles when the water velocity increases (Taghon et al. 1980).

Some black fly larvae that live in shallow, rapid streams have an interesting variation on the device. They are attached at their posteriors and have a pair of food-trapping fans on their anterior ends. A larva twists lengthwise so one fan is uppermost and feeds from material suspended near the outer edge of the boundary layer (Figure 6.6). The other fan is then lower and a bit downstream; it feeds on material resuspended in the vortices rising behind the body. Upward flow of water in the vortices is further encouraged by a downstream tilt of the body as a whole. Groups of larvae, positioned as they are in nature, don't compete for food but instead enhance each other's feeding efficiency (Chance and Craig 1986). There's more than sex to being social; sometimes it's just a way to stir up some dirt!

REGIMES OF FLOW AND THE REYNOLDS NUMBER

Even with fixed geometry of the solid elements across or through which fluid passes, flows vary widely in character. Sometimes fluid moves smoothly, with each bit following a course nearly parallel to adjacent ones. Under other circumstances fluid moves less regularly, with the overall direction of flow mainly a statistical matter and the individual bits tracing all sorts of erratic courses. We speak of these situations as "regimes" of flow—"laminar" (smooth) and "turbulent" (irregular) in these

113

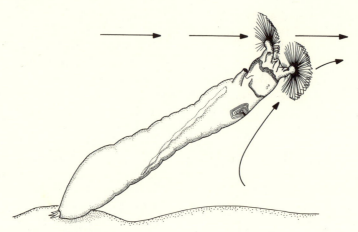

FIGURE 6.6. The larva of a black fly on a rock in a stream. One of its fans feeds from mainstream plankton, the other from resuspended detritus.

particular cases. You surely have had experience with both—syrup pours in a laminar fashion, water emerges turbulently from a nozzle, and deliberately induced turbulence disperses the cream in the coffee.

The consequences of the existence of such distinct regimes underlie most of the apparent complexity of the phenomena attending flow—the definition of viscosity and the Newtonian laws may not change, but the manner in which they apply to particular situations is certainly variable.

The first systematic investigation of the phenomena seems to have been that of Osborne Reynolds (1883). He knew that flow through pipes was sometimes laminar but could abruptly become turbulent, so some queer discontinuity clearly existed. Using a variety of pipes, fluids, and flow speeds, he was able to show that the transition from laminar to turbulent flow happened at about a certain value, 2000, of a dimensionless, composite variable to which we now give his name—the Reynolds number. This proves to be an extraordinarily useful quantity, perhaps the best single index to the character of flow, not just in pipes but in streams, lungs, ocean currents, blood vessels, or industrial vats, in fact just about everywhere that real fluids flow steadily at substantially subsonic speeds reasonably far from fluid-fluid interfaces. Which includes most biologically interesting flows—without mention of Reynolds numbers, no sensible treatment of flow in living systems is possible.

What the Reynolds number does is give an indication of the importance of viscosity—actually the unimportance, since low Reynolds numbers are associated with very viscous situations. One might expect that the value of viscosity would suffice, but other factors have equal influence in determining its relative role. We've noted that what viscosity does is to oppose the existence of steep velocity gradients, so viscous force (as in equation 6.3) measures what one might term the "groupiness" of a fluid, the tendency of bits of fluid to flow coherently, in unison. What does the opposite, promoting gradients? Recall that these gradients are across the direction of overall flow, not along the flow. The inertia of a bit of fluid will keep it moving steadily in the face of the retarding or accelerating effects of adjacent bits of fluid. Thus, inertial resistance to acceleration, the need of a force to change speed, opposes viscous forces and reflects the "individuality" of a bit of fluid.

We thus have two opposing sorts of forces, loosely called "inertial" and "viscous." The Reynolds number is simply the ratio (dimensionless, of course) between these two. It is entirely analogous to the Froude number, which you may recall is the ratio of inertial to gravitational forces. Inertial force is represented by the familiar $F = ma$, although disguised a bit into a form useful for steady flow of fluids. Mass is replaced by density times area (across the flow) times length (with the flow), and acceleration is re-placed by velocity over time. Since length/time is velocity,

$$F_i = \rho lSv/t \text{ or } \rho Sv^2. \tag{6.5}$$

For viscous force, we adjust equation 6.3, introduced earlier to define viscosity:

$$F_v = \mu Sv/l.$$

The result is surprisingly ordinary, without even an exponent to chal-lenge the calculator:

$$\text{Re} = F_i/F_v = \rho lv/\mu. \tag{6.6}$$

None of the factors in the Reynolds number presents much of an eval-uation problem for practical situations. Density and viscosity refer to the fluid and (if not found in Table 6.1) are usually matters of public record. Notice that the ratio of the two appears in the Reynolds number—the ratio matters more than the individual values. For a given size and speed, then, flows of water have Reynolds numbers fifteen times greater than those of air—flows of water are, oddly, less viscous in character. Velocity is that of the mainstream fluid relative to the object, or substratum, or the wall of a pipe. Length is another "characteristic length" of an object chosen largely by convention—maximum length in the direction of flow for flows across solids, or the diameter of pipe or channel for internal flows. As with scaling rules (which the Reynolds number is), one is deal-ing with wide-ranging situations and a fairly crude index. In fact, one almost never cites a Reynolds number with more than two significant figures, and one can't compare the Reynolds numbers for different shapes to even that rough level of precision. It's a ballpark business.

And organisms encounter an especially wide range of values (Table 6.2). The extreme range results from the fact that small creatures typically move slowly or encounter slow flows deep within boundary layers; large creatures experience more rapid flows; and the formula includes the product of length and speed. The main consequence is that a good shape for one organism may be quite inappropriate for another if the two operate at substantially different Reynolds numbers, even if they are engaged in what seem similar activities.

At low Reynolds numbers (roughly 10 and lower), flows across solids are laminar and orderly; if vortices occur at all, they are relatively large and distinct. Velocity gradients are gentle and, in general, drag is proportional to the first power of velocity and doesn't depend very strongly on shape. As we'll explain shortly, streamlining is almost without value as a scheme for reducing drag at low Reynolds numbers.

As Reynolds numbers rise (from roughly 100), flows become increasingly disordered, with turbulent chaos setting in, first in one situation or location and then in others. Velocity gradients are increasingly steep, and drag may be proportional to the square of velocity (but the relationship is rarely tidy) and becomes strongly shape-dependent. For instance, total drag may be reduced as much as fiftyfold by shifting from a sphere to a well-streamlined form where both have the same frontal area perpendicular to flow.

Incidentally, the identity of most of the factors of equation 6.4 with those of equation 6.6 is neither accidental nor irrelevant. For the range of Reynolds numbers over which equation 6.4 applies, the thickness of a boundary layer relative to the length of a flat plate varies quite systematically with changes in the Reynolds number—thickness is inversely proportional to the square root of the Reynolds number. Thus, the higher the number, the thinner, relatively, the boundary layers. There's an agreeable consistency to these matters.

TABLE 6.2. REYNOLDS NUMBERS—EXAMPLES

Bacterium swimming	0.000001
Pollen grain falling, or sperm swimming	0.01
Fruit fly (fuselage) in flight	100
Small bird flying	100,000
Squid fast jetting	1,000,000
Large whale swimming	200,000,000

Modeling

Like the Froude number, the Reynolds number provides a useful tool for making models. For two geometrically similar situations, equality of Reynolds numbers assures equality of patterns of flow, whatever the individual values of length, speed, density, and viscosity. We biologists seem always to be dealing with extremes of size so a good scaling rule is a big boon, although it must be admitted that biologists are often skeptical of its legitimacy—those *mu*s and *rho*s look like Greeks bearing gifts.

I've had recourse to the use of the Reynolds number for scaling experiments on quite a few occasions. At one time I was interested in induced air flow through the burrow systems of prairie dogs at low ambient winds (Vogel et al. 1973). A burrow in nature is 3 m deep, 15 m long, and about 0.1 m in diameter—quite an awkward item to bring into (or reproduce in) a laboratory. But tenfold reduction of lengths permitted use of a simple model made of copper water pipe beneath a plywood "ground" that would fit in the local wind tunnel. In compensation for the reduction in size (l) all I had to do was increase speed (v) by the same factor, so 1 m·s^{-1} became 10 m·s^{-1}. As a bonus, the internal speeds were raised to a more easily measured range. More recently, I felt I needed to know how pressures varied along the length of a rapidly swimming squid (Vogel 1987). My best pressure-measuring device works only in air, and I wanted to get up to rather high speeds. The local engineers kindly loaned me a wind tunnel that went up to 75 m·s^{-1} (170 mph), and I made my model half again as large as life. As a result I could simulate a squid jetting at nearly 8 m·s^{-1} in water, as the reader can easily verify.

For such modeling it's only necessary that the conditions for applying the Reynolds number be met—no fluid-fluid interfaces, decently subsonic flow—and that any shift from, say, air to water not cause abnormalities (due to the change in forces) in the shape of the organism or model. The changes can be extreme—use of corn syrup instead of water permits more than a thousandfold increase in size and thus macroscopic modeling of microorganismic phenomena. By contrast, modeling surface ships with the Froude number is far messier—ships protrude downward from the surface, so the Reynolds number matters as well. Comparing equations 4.6 and 6.6, it's easy to see that an alteration in size or speed cannot be made without changing either one dimensionless index or the other, so serious compromises are inevitable.

Streamlines

The principle of continuity, presented in Chapter 2 in terms of flow through pipes, states that the rate at which volumes of fluid enter a rigid

117

system of pipes must exactly equal the rate at which volumes of fluid leave the system or pass any cross section of it. For an incompressible fluid, that requires that the product of velocity and cross-sectional area must be the same for all cross sections of the system.

The principle turns out to be quite as useful for flows across objects and surface—open fields of flow—where there are no rigid pipes to provide boundaries. Fluid (for our purposes) is still incompressible, and mass is still conserved. But to apply the principle, it is necessary to provide some analog of the walls of pipes; in practice the scheme is a simple one. Consider a two-dimensional (to fit on paper) field of flow (Figure 6.7) where the speed of flow varies from place to place but does not, at a given place, vary from time to time. (This condition defines "steady flow," which, somewhat confusingly, may involve acceleration and is thus not analogous to "steady motion.")

Assume that you have some device to tell the direction of flow at any point (a thread on a stick, a stream of dye, a time-exposed photograph of a particle passing through the field, etc.). The directions of flow can be mapped by drawing short lines that follow the local flow directions—lines bits of fluid will pass along as they flow. With enough points mapped, the short lines can be connected into longer ones running completely across the field in the directions of flow. Mapped this way, fluid is automatically prohibited from crossing the lines, so the lines amount to nonmaterial partitions and play exactly the same role as the walls of rigid pipes! Between any pair of lines, the principle of continuity must hold. If a pair of lines diverges, the fluid must be slowing down; if a pair converges, the fluid is accelerating. The lines are called "streamlines" since

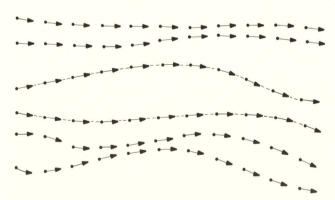

FIGURE 6.7. Streamlines (dotted lines in the center) connect upstream points with others successively farther downstream with respect to the local directions of flow.

they follow the stream of fluid at all points. In three dimensions the set of lines just takes the form of a set of tubes, and the principle applies as well. "Streamlining" in the popular sense means shaping something so fluid flows smoothly around it and causes a minimum of drag; it's only indirectly connected with these streamlines.

Armed with streamlines, we can take a more physical view of the way the Reynolds number delineates flow regimes. Figure 6.8 gives the streamlines for flow around a cross section of a long circular cylinder—the real messiness of the world with which fluid mechanics must deal is immediately obvious, but so also is the way the Reynolds number at least orders the mess.

THE WORLD OF VERY LOW REYNOLDS NUMBERS

The orderliness of flow at low Reynolds numbers is evident, if unremarkable, in Figure 6.8. Flow undergoes no further awkward transitions below Re = 10, not just around spheres but around practically any shape at all. Such simplicity doesn't imply familiarity, though. Perhaps nowhere else do our experiences and prejudices as big animals who move rapidly distance us further from immediate reality than in the world of low Reynolds numbers—a new domain of intuition must be cultivated to deal with this novel regime of flow. A few general examples ought to point up the strangeness.

Flow without inertia

We normally and uncritically regard flow as a disordering process—even linguistic conventions reflect the view, as someone "stirs up trouble." At Reynolds numbers below unity, flow is not necessarily disordering—it's very hard to mix two viscous fluids, such as corn syrup and molasses or the two components of epoxy glue, in terms of both the work it takes to stir *and* the amount of stirring needed to have much effect. With a little care one can demonstrate in such syrupy combinations the phenomenon of stirring without mixing—the effect of three turns clockwise with a spoon can be pretty well undone by three turns counterclockwise.

And turbulence is simply unimaginable—turbulence is associated with high rates of shear and steep velocity gradients. Sustaining the velocity gradients of turbulent vortices is beyond the poor power of the fluid's inertia as it pales before the militance of viscosity. While vortices have been made at Re = 0.01, they were created with difficulty and could be sustained only by continuously doing work.

This reversibility of flow extends as well to the forces involved. At higher Reynolds numbers one can get a net force (and thus propulsion)

119

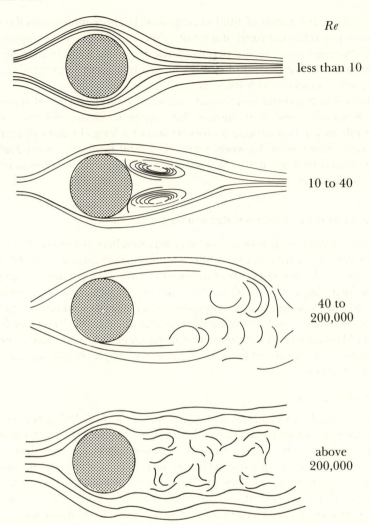

FIGURE 6.8. The character of flow around a circular cylinder (shown in cross section) depends very strongly on the Reynolds number, from orderly flows at low values through several transition regimes—attached vortices and periodically shed vortices—to thoroughly disorderly flows at high values.

by moving a fluid with a rapid stroke in one direction and a slow one in reverse. The total "impulse" of the force (the product of force and time) differs for the two directions, and so there can be "power" and "recovery" strokes. The difference traces to the fact that the force of drag for each stroke varies (very roughly) with the square of its speed, whereas the time for each stroke varies inversely with speed to the first power—the product of time and force is therefore different for the two strokes. By contrast, at low Reynolds numbers drag is proportional to the first power of speed, and time is still inversely proportional. While a faster stroke may still mean more drag, the time is proportionately shorter, so the magnitude of the impulse is always the same for two strokes that cover the same distance. Each stroke, whatever its speed, always just cancels the effect of the other.

Nor can one do terribly well by "feathering" a paddle for a recovery stroke. Drag no longer depends very much on shape, and it turns out that a paddle broadside to flow (for a power stroke) has only about 50% more drag than the same one edge on to flow (for a recovery stroke). About the best arrangement possible along these lines involves the use of a cylinder, oddly enough. The drag of a cylinder with its axis perpendicular to flow is nearly twice that with the axis parallel to flow, so some progress is possible by erecting a cylindrical paddle for a power stroke and then folding it down for a recovery stroke. And folding down gets the cylinder further from the mainstream. This arrangement is just what is used by the cilia of many tiny organisms (Figure 6.9). It's not wonderfully efficient, but it does work, and there are few alternatives.

Low Reynolds numbers may create an orderly world, but the dominance of viscous forces makes it a gooey one. Howard Berg views a tiny organism in motion as like a person "swimming in a sea of asphalt on a summer afternoon." Inertia has no practical meaning—Berg calculated (Purcell 1977) that if a bacterium could instantaneously stop spinning its

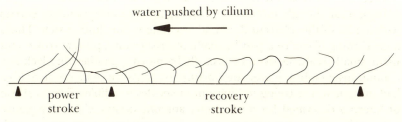

water pushed by cilium

power
stroke

recovery
stroke

FIGURE 6.9. The stroke of a cilium, progressing from left to right, consists of a short (fast) power stroke and a longer (slower) recovery stroke. For a cilium beating as is this one, the liquid above will be pushed from right to left.

flagellum it would almost as instantaneously stop moving through water—it would coast a distance equal to only the diameter of a hydrogen atom! To move is to distort the fluid, not just immediately around you but to a remarkable distance away. Among other consequences, the difficulty of moving is affected by the presence of walls far from the immediate vicinity of the mover. As an extreme example, the drag on a cylinder dropping in a viscous fluid at Re $= 10^{-4}$ is *doubled* by the presence of walls *five hundred cylinder diameters* away (White 1946). Biologists have not been au courant with these small-scale, slow flows even though we ordinarily work with small items—our standard icon is a microscope. We describe the motion of microorganisms abnormally crowded between glass slide and cover slip; we have tables of falling speeds for spores obtained by watching them descend within unreasonably tiny capillary tubes.

Attention has recently returned, beginning with Strickler and Twombly (1975), to the long-neglected consequences of the wide field of disturbance caused by the passage of an object or organism. For a small creature such as a microcrustacean to swing its appendages and swim is, in effect, to announce "here I am." A nearby predator can easily sense its presence. Conversely, it is at least theoretically possible that simply by monitoring the forces needed to propel itself, such an organism might detect walls, food, or even predators. Even an object settling under gravity announces its passage. An analysis by Wu (1977) suggests a partial evasion of the problem. An organism (a paramecium, for instance) swimming by means of a coating of cilia must make far less "noise" than even an inert object falling at the same speed. To work, the cilia must extend from the organism's surface out into fluid not moving with the organism, so the velocity gradient region on the surface of a ciliated creature cannot be thicker than the length of its cilia. Thus the region of distortion of flow is greatly reduced compared with that associated with swimming with great paddles or even with sinking passively. There's perhaps some indirect benefit from the inefficient system of ciliary locomotion!

In general, though, movement of an organism or an appendage carries with it a lot of the adjacent fluid and anything in the fluid as well. That's a good thing if you're a paddle made of bristles on a power stroke, and many small creatures swim or even fly by moving paddles that lack any continuous membrane over their bristly struts. As we've mentioned, it's a bad thing if you're trying to filter out molecules or edibles. The problem of filtering described for a silkmoth's antenna occurs with a vengeance when a tiny crustacean tries to capture food with its appendages, as studied in detail by Koehl and Strickler (1981). This functional gooiness of water also has other peculiar consequences. A tiny attached protozoan on a contractile stalk can contract the stalk when a small but voracious predator is about to bite off its head, but contraction might just draw the predator forward at the same time—the predator's inertia is inconsequential. If only the Reynolds number were higher! Faster contraction has just this effect, and, as it happens, the most rapid contraction of any bit of animal is that of the so-called spasmoneme of the stalked, colonial protozoan *Zoothamnium* in Figure 6.10 (Weis-Fogh and Amos 1972). On the other hand, a predator also has to get to its prey, not push the prey forward while still out of reach. Again a burst of speed does the deed— one predatory microcrustacean, a millimeter long, can briefly get up to Re = 500; to do so, though, it must move at the remarkable speed of 200 body lengths per second (Strickler 1977).

Stokes' law

If, as at low Reynolds numbers, drag depends on viscous and not on inertial forces, it's possible to do a simple dimensional analysis to get the suggestion of a formula for drag. Assuming that the relevant variables are the size and relative speed of an object and the viscosity of the fluid, it turns out that drag ought to be proportional to the product of the three, each raised only to the first power. (Ignoring inertial forces has the

FIGURE 6.10. Contraction of the spasmoneme of *Zoothamnium*—the straight stalk through which the spasmoneme passes hunkers down into a tight helix.

effect of removing density from consideration—if density is included, the analysis is no longer so simple, as you can easily verify.) As usual, there's a numerical constant about which the analysis is silent; it turns out to depend on shape and orientation. For a sphere, George Stokes (1819–1903) long ago showed that the full formula (Stokes' law) can be given simply as

$$D = 6\pi\mu av, \tag{6.7}$$

where a is the radius of the sphere.

The formula is tidy and useful, at least up to Re = 1—it gives the drag on a fog droplet or a spore sinking in air or on a buoyant algal cell rising in the ocean. Since drag is not strongly dependent on shape, the formula gives a decent approximation of drag for shapes other than spheres (although formulas for other ordinary shapes are available). Several cautions are necessary in connection with its use—first, as should be clear by this point, it doesn't work when a wall is nearby; second, it is unlikely to be useful for locomoting organisms, for which the production of thrust complicates the local flow; finally, it assumes no slip at the surface of the object. For a droplet of gas rising in a liquid, the liquid induces motion in the gas, and their joint motion has the effect of a slipping surface— the factor of 6 in the formula should be replaced by a 4, according to Happel and Brenner (1965).

The most common use of Stokes' law is in the derivation of a formula for the terminal velocities of spheres. At terminal velocity, you'll recall, the upward forces (here drag and sometimes buoyancy) equal the downward force of gravity. So one just sets the Stokes formula for D equal to mg and solves for v. Where buoyancy is significant, or if it's handier to use densities than masses, there's a minor additional complication—for mass one substitutes the product of the volume of the sphere and the difference in densities between sphere and fluid medium, to get

$$v = \frac{mg}{6\pi\mu a} = \frac{2a^2 g\,(\rho_o - \rho_f)}{9\mu}. \tag{6.8}$$

It's important to emphasize that equation 6.8 gives no universal scheme for calculating terminal velocities, even for spheres. All of the limitations on the use of Stokes' law apply, in particular, the limitation to Reynolds numbers not substantially above 1.0. On the other hand, one ought to recognize that the rule refers to sinking or rising with respect to the local fluid and is applicable whatever the larger-scale downdrafts or upwellings.

The formula for terminal velocity provides a scheme for measuring viscosity—it isn't always applicable but is far more practical than the imaginary apparatus of Figure 6.2. One just releases a ball that is sufficiently small or has a density near enough to that of the fluid to ensure a sufficiently low Reynolds number. The density and size of the ball and the rate of ascent or descent go directly into equation 6.8. I once got a rather surprising result with this technique. I had filled a small recirculating tank with corn syrup and wanted to know the viscosity. So I timed the fall of a plastic sphere. Looking up the viscosity of a saturated glucose solution, I found that my calculated viscosity was fully two orders of magnitude higher—a hundredfold error is hard to explain away as inaccuracy of balance, ruler, or stopwatch. But an answer emerged after a bit more investigation, working from the density of the corn syrup and the calorie content on the label. It turned out that commercial corn syrup is mainly *supersaturated* glucose that just doesn't crystallize, so the viscosity I had looked up was inappropriate. In the process I found out why the stuff is used in cooking. Table sugar (sucrose) crystallizes readily—if you want sticky frosting or fudge, use lots of corn syrup; if you want solid stuff, use mainly dissolved table sugar; read a recipe and anticipate its result.

LAMINAR FLOW THROUGH PIPES

Let us turn from flow around an object such as a sphere to the opposite, an object around a flow, in particular, flow through a pipe. Drag isn't the most useful quantity any longer, but the drop in pressure per unit length as a fluid flows down a pipe is completely analogous. Again we'll consider a situation in which the only force resisting flow is that due to viscosity. It ought to be mentioned that so far we've talked a lot about velocity gradients but only once had a decently straightforward one, the constant gradient between the two plates used in defining viscosity (Figure 6.2 and equation 6.3). The gradient in the boundary layer was nearly constant ("linear") near the surface, but velocity asymptotically approached that of the free stream, giving something quite messy mathematically. And we silently stepped over the complexity of the gradients adjacent to spheres in the so-called creeping flows of low Reynolds numbers.

For laminar flow through circular pipes circumstances are nicer. At least, once fluid gets well downstream from the entrance to a pipe, from a bend, or from any abrupt change in bore, the gradient of velocity across a pipe turns out to be quite thoroughly ordinary. Each bit of fluid proceeds at an unchanging distance from the walls; resistance is entirely a consequence of the no-slip condition and viscosity; and, since it's then just as hard to force fluid through one part of the pipe as another, the pressure loss per unit length remains unchanged down the pipe. A simple derivation to be found in almost any textbook of fluid mechanics shows that the "velocity profile" takes the form of a paraboloid (Figure 6.11) with, naturally, the highest speed in the middle and zero speed at the walls. The equation describing the profile is

$$v_r = \frac{\Delta p(a^2 - r^2)}{4l\mu},$$
(6.9)

where a is the radius of the pipe, r is the distance of some position on the radius from zero at the center, and v_r is the speed at that position.

From the velocity profile it is a minor step to get a well-behaved equation giving the relationship among total volume flowing per unit time, Q, pressure drop per unit length, $\Delta p/l$, the radius of the pipe, and the viscosity of the fluid. In fact, integration of speed (dimension LT^{-1}) over the cross-sectional area of a pipe (L^2) to get volume flow rate (L^3T^{-1}) was mentioned near the end of Chapter 4. In terms of equation 6.9 (or a graph of it), we're integrating from $r = 0$ (the middle) to $r = a$ (the wall). The result is straightforward:

$$Q = \frac{\pi \Delta p a^4}{8l\mu}.$$
(6.10)

The only difficulty is in pronouncing its name, "Poiseuille's equation," or the "Hagen-Poiseuille equation." (The name of the old unit of viscosity, the poise, was a merciful truncation.)

This equation asserts that volume flow rate and pressure drop per unit length are mutually proportional—it takes more pressure to push fluid along faster—and that volume flow rate is inversely proportional to viscosity—stickier fluids flow more slowly. All this is nicely intuitive. But it says something more startling as well. Volume flow rate varies with the *fourth* power of the radius of the pipe, which is pretty extreme stuff. Part of the latter is obvious—if the radius is doubled, cross-sectional area is

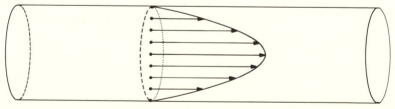

FIGURE 6.11. The paraboloid of revolution formed by the local velocities in laminar flow through a circular pipe.

quadrupled, so even without viscosity, volume flow rate should be proportional to the square of the radius. In addition, though, a bigger pipe has less wall relative to its area (recall the reasoning in Chapter 3), so it has (relatively) much less resistance to flow on that account as well. Thus, if a big pipe is divided into a pair of small ones without change in total cross-sectional area or pressure drop, the overall volume flow rate is halved. Or if, to save material, you halve the diameter of a pipe in a plumbing system, you need sixteen times as much pressure to keep the flow rate the same. These matters of scale will prove decisive in explaining many features of the plumbing within organisms (Chapter 8).

Some other useful relationships come out of these equations. From equation 6.9 comes the maximum speed of flow—just look at the center by setting $r = 0$. From equation 6.10 comes the average speed of flow—just divide volume flow rate by the cross-sectional area of the pipe, πr^2. Then something else spills into our laps. Maximum flow speed is just twice as fast as average flow speed, independent of the values of any of the specific variables. How nice! Sometimes it's handy to measure the maximum, using the time of first arrival of some tracer substance such as a dye. At other times it's easier to measure the average, catching and timing the discharge, say, with beaker and stopwatch. No matter—one can always convert one into the other.

Still, you may ask, just how useful is all this talk about laminar flow in pipes? After all, laminar flow around spheres as described by Stokes' law is good only up to Re = 1, which corresponds to rather small-scale operations, to an object a millimeter in diameter going a millimeter per second in water or a little over a centimeter per second in air. But recall Reynolds' original observation. Laminar flow in pipes extends up to Re = 2000, where the length in the expression for Reynolds number is taken as the diameter of the pipe. That corresponds to a pipe of 1-cm bore carrying water at 0.2 m·s^{-1}. It may not be much of a piece of commercial plumbing, but it includes virtually all internal fluid transport systems in organisms. We approach the limit in our aortae, but their careful design normally ensures the absence of turbulence with its resulting higher pressure drop. Turbulence makes noise, and the character of this pathological noise, audible with a stethoscope, is a useful diagnostic tool.

Manipulating the parabolic profile

For many purposes a pipe of circular cross section and laminar flow is just fine. Its construction uses less wall material relative to its inside volume than does any other shape, and it withstands internal pressure with less bulging or cracking. (We'll examine the mechanics of such cylinders in Chapter 11.) No other shape gives as low a pressure drop per unit

127

length for a given flow. The pressure drops in laminar flow are usually less than those associated with any amount of turbulence. And flow is minimally sensitive to the texture of the inner pipe walls—small bumps, scratches, and minor bends make little difference.

For functions that depend on having a lot of surface relative to volume there may be no worse shape than a pipe with a circular cross section. In particular, flow through pipes and channels within organisms is commonly associated with exchange processes. Heat, dissolved respiratory gases, ions, and organic molecules pass back and forth between internal fluid and external material, going through the walls of the pipes—this exchange is the main reason for pumping liquids and gases around within organisms. Exchange, obviously, is facilitated by the most extensive contact between fluid and walls, and a combination of geometry and fluid mechanics that minimizes this contact is clearly a disability. The single most effective way to increase contact is through a proliferation of small pipes; the point was made in connection with surface areas in Chapter 3. But an organism can play various games with the profile of flow itself.

If flow is between parallel plates instead of in circular channels, there is more wall and less channel; on the average, a bit of fluid flows closer to the wall relative to the overall thickness of the channel (Figure 6.12). Gills of many aquatic organisms, both vertebrate and invertebrate, are arranged in the form of parallel flat plates, as are the so-called "book lungs" in spiders. The nasal passages of mammals are extensive and flattened in cross section—contact between air and wall serves a variety of functions even beyond the exchange involved in olfaction. In small mammals and birds, heat from an exhalation may be used to warm the walls; the same heat then warms the subsequent cold inhalation instead of being lost to the atmosphere. And in the same process water vapor from the lungs may condense out on the cold walls where it can be used to

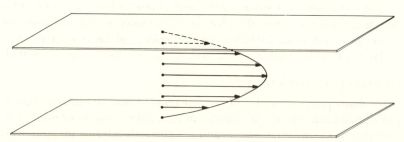

FIGURE 6.12. The parabolic distribution of velocities from flow in the narrow gap between two very wide flat plates parallel to each other.

moisten the next inhalation as another conservation device (Schmidt-Nielsen 1972). Interestingly, adult humans exhale air only slightly below body temperature, but infants exhale air closer to ambient temperature (at least my son did so)—their different surface-to-volume ratios within the air passages probably make the difference.

Other schemes also promote intimacy between flowing fluid and channel wall. Blood is transported from heart to capillaries in large vessels with a nearly parabolic velocity profile. (Some distortions arise from the pulsatile flow, from branching, and from the peculiar irregularities of blood viscosity.) In capillaries the flow is far from parabolic—red blood cells just squeeze through (Figure 6.13), so the plasma between successive cells has to flow faster near the walls than it would otherwise. Thus the blood cells, whatever else they may accomplish, must passively enhance exchange between plasma and capillary walls.

Recall that the parabolic profile comes about because the resistance to flow is entirely due to the presence of the material walls of a pipe. What if the walls provide propulsion instead of resistance? A ciliated surface does just that, greatly increasing the steepness of the local velocity gradient. We're a bit big to make much use of the device (or so one guesses)—cilia in our respiratory passages move surface mucus, not the air itself. But lots of invertebrate gills have ciliated surfaces as the basic pump, and the digestive tubes of many small creatures are lined with cilia, so the arrangement is really quite widespread. Such ciliary propulsion ought to facilitate exchange between fluid and walls of the conduits.

FIGURE 6.13. Red blood cells flowing through a capillary. The fit is sufficiently tight so the retarding effect of the walls on adjacent flow ends up distorting these normally disc-shaped cells into something closer to deflated basketballs.

Pressure and flow

"Fluid flow is not currently in the mainstream
of biology."

Maliciously attributed to the present author

I N THE last chapter we saw that the character of the motion of a fluid
was largely determined by the opposing tendencies of bits of fluid to
go their separate ways and to march forward together. The relative importance of the two was what the Reynolds number (equation 6.6) told
us. At low Reynolds numbers it turned out to be useful to ignore inertia
or any residual individuality of bits of fluid; the consequences of this simplification were some crisp, precise, and useful rules—Stokes' law (equation 6.7) and the Hagen-Poiseuille equation (6.10). At moderate, or at
least high, Reynolds numbers we might try doing the opposite—ignoring
viscosity to derive formulas. It can be done, and a large literature deals
with the behavior of such inviscid (and so-called ideal) fluids. But often,
ignoring viscosity gives quite preposterous results—for example, a mathematical demonstration that no shape should have any drag. As Lord
Rayleigh noted, in such a fluid a ship's propeller would no longer work;
but on the other hand, in the absence of drag it wouldn't be needed anyway.

The crux of the problem is that while there are Reynolds numbers low
enough so inertial forces are negligible, the numbers are never so high
that viscosity can be completely neglected. What decreases with increasing Reynolds numbers is the relative size of the region in which viscosity
has its effect; the size, in effect, of Prandtl's boundary layer. For viscosity
to be really negligible, the no-slip condition would have to stop operating,
which just doesn't happen. Put another way, at higher Reynolds numbers
the gradient region may be thinner but the steepness or intensity of the
gradient becomes concomitantly greater.

That prefatory remark ought to emphasize the peculiarity of the topic
that follows. For a few pages the defining property of fluids, viscosity,
will be assumed negligible, and we'll look at that grotesque misnomer, an
ideal fluid. Still, simplification is central to the success of science—we follow the philosopher William of Ockham, trying the simplest things first
and adding complications only as absolutely necessary. Ignoring viscosity
is analogous to treating the interactions among solids as frictionless—
while the world doesn't work that way, it is a most useful abstraction that

permits us to see some general principles underlying a lot of secondary complications.

THE GREAT PRINCIPLE OF BERNOULLI

Conservation laws have already been declared Good Things, and the principle of continuity was pronounced a Useful Version of the idea of conservation of mass. There's another principle that amounts to a kind of conservation law, attributed to one of the Bernoullis (Daniel, 1700–1782). It can be derived from the notion of conservation of energy or from consideration of forces. An honest derivation is a bit awkward without sophisticated mathematical tools, but the principle isn't especially hard to explain. Bear in mind, though, that it presupposes an inviscid fluid, which means that one has to be cautious in assuming its applicability. Zero viscosity implies perfect slipping between a flowing fluid and the walls of pipe or channel, thus no boundary layers with their velocity gradients at surfaces, and no transfer of momentum from moving fluid to walls of pipe or channel.

For simplicity we'll make still another assumption—that gravity also can be ignored. In effect, we're presuming that fluid doesn't move upwards or downwards with respect to the earth to any appreciable extent. Pressure does vary with altitude, as explained in Chapter 5—the change is 9806 Pa·m^{-1} in water and about 12 Pa·m^{-1} in air[1]—but it's easiest to treat these variations as items requiring at most minor post hoc adjustments.

Consider a fluid flowing across a flat plate (Figure 7.1). We connect a tiny hole flush with the surface of the plate (called a "static orifice") to a device underneath that measures the pressure of the fluid. How does that pressure vary with the speed of flow? One might expect the pressure to rise as speed increases, but in fact the opposite happens! Higher speed demonstrably entails lower pressure, at least with this arrangement—that's the essence of Bernoulli's principle. Hold the edge of a small piece of paper against your lower lip and blow across its upper surface—the paper should lift up as a result of the reduced pressure of the moving air above it.

Venturi tubes and prairie dogs

Now consider the other device of Figure 7.1, something called a "Venturi tube." It consists of a constriction in a pipe with an aperture in its

[1] The difference, 9794 pascals per meter, is, in fact, the factor used to convert height to pressure in an ordinary U-tube manometer filled with water and working in air.

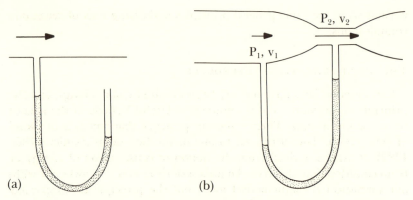

FIGURE 7.1. (a) Pressure, shown on the manometer, drops as a result of flow across a static orifice. (b) Similarly, faster flow caused by a constriction in the pipe lowers the local pressure in a "Venturi tube."

wall and a similar aperture in an unconstricted portion just upstream. A pressure measuring device is connected between the two apertures. How does the pressure differ between the locations? Pressure turns out to be least where the flow is fastest—in the constriction, of course, by the principle of continuity. The greater the total flow through the pipe, the greater will be the speeds in both portions, the greater will be the difference between these speeds, and the greater will be the resulting pressure difference.

If the relationship between speed and pressure difference together with the degree of constriction are known, then a measured pressure difference can be used to figure the speed of flow in the pipe outside the constriction. In short, this deliberate but slight and local constriction is the basis for a fine meter for flow through a pipe. You merely have to pay a little attention to the shape of the constriction, you have to keep the two pressure-measuring holes reasonably close to each other to minimize the effects of viscosity, and you can't trust the thing at really low Reynolds numbers. But what exactly is the relationship between speed and pressure? The relationship amounts to a statement of Bernoulli's principle. It is that (1) the magnitude of the pressure difference is proportional to the difference between the *squares* of the velocities, and (2) the algebraic signs are reversed so that, as expected, lower pressure is associated with higher velocity:

$$p_1 - p_2 = \rho(v_2^2 - v_1^2)/2. \tag{7.1}$$

The presence of the squares of velocity in the formula reduces the practicality of the device for low speeds—decrease the velocities tenfold and the pressure difference you must measure drops a hundredfold. The formula, of course, includes two velocities; but that's no problem. If the cross-sectional areas of both normal (subscript 1) and constricted (subscript 2) pipes are known, the principle of continuity allows the speed in the constriction to be eliminated:

$$S_1 v_1 = S_2 v_2. \tag{7.2}$$

The basic principle underlying the Venturi tube works as well on a surface beneath an open field of flow—it's just not immediately handy as a measuring tool. Let's remove the indicating fluid from the pressure-measuring manometer so pressure differences will drive fluid through the manometer's pipe. And let's simply erase the line in Figure 7.1 that represents the upper surface of the pipe. We now have an arrangement that occurs in nature in several places, most notably and straightforwardly in the burrows of prairie dogs (Figure 7.2).

These rodents, which live on the relatively windy plains of North America, are consummate burrowers even by the high standards of their order—a typical burrow is two-ended, about 15 meters long, with a max-

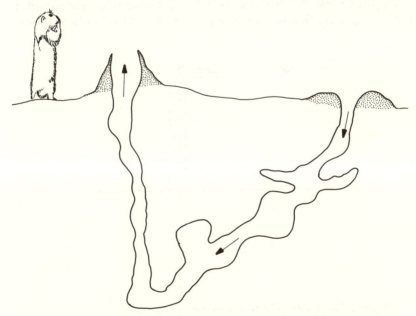

FIGURE 7.2. A prairie dog and its burrow. The width of the burrow has been considerably exaggerated—it's really less than 1% of the distance between the openings.

imum depth of about 3 meters. It's not difficult to calculate (using Fick's law, Chapter 8) that if oxygen is supplied by diffusion alone, a prairie dog in the bottom of a burrow will asphyxiate, so some system of forced ventilation is needed. But work with some models in a wind tunnel showed that all an animal has to do is maintain a slight difference between the geometries of the two openings of its burrow for even a slight breeze to do the rest. Tests in the field with a smoke generator showed that air moves through burrows whenever there is any noticeable wind above, and various observers have noted that the animals assiduously maintain the difference between the geometries of their two openings. While the scheme is direct and effective, there are a few complications in the physics—in practice, Bernoulli's principle acts in concert with viscous entrainment, and flow through the burrow to some extent relieves the predicted pressure difference (Vogel et al. 1973; Vogel 1978).

Pitot tubes and an aquatic insect larva

What if one wants to measure speed not as volume flow in a pipe but at a point or in a small region in a large field of flow? A device related to the Venturi tube, a "Pitot tube," puts Bernoulli to this task by pointing one opening into the flow (Figure 7.3). At this upstream-facing orifice the local speed of flow is zero—flow has been halted by the obstacle, caus-

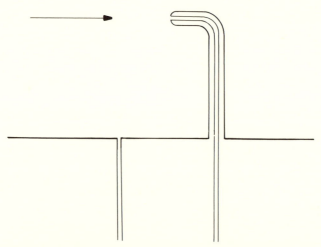

FIGURE 7.3. A Pitot tube facing, as it must, upstream, with the required static orifice adjacent. The thickness of the tube has been exaggerated severalfold.

ing a local increase in pressure. The extra pressure (called the "dynamic pressure") must therefore reflect the speed of the fluid just upstream from the orifice or, very nearly, the speed which would have existed at the orifice if the device had not intruded. To reveal the extra pressure, some other location must be used as a reference point of undisturbed pressure for comparison—a "static orifice" that does not obstruct the flow is located as near the upstream-facing orifice as convenient. The arrangement may sound more devious than a Venturi tube but generates a simpler equation, since one velocity has been reduced to zero, and continuity need no longer be invoked. Thus,

$$p_1 - p_2 = \Delta p = \rho v^2 / 2. \tag{7.3}$$

Admittedly it does seem odd that the device ends up measuring the speed of flow at a point where its presence reduces that speed to zero! But at least it's intuitively comfortable to detect an increasing flow speed by having it increase the pressure on something (even a hole) that faces into the flow. As an air-speed indicator (anemometer), a Pitot tube is standard gear on all small airplanes—a tube protrudes, facing forward, and a little hole on the body has the cryptic legend adjacent, "static orifice, do not block."

The pressure difference between the intrusive and static orifices can as easily drive an internal flow as deflect some manometer. And again we see such devices in nature. One is the silk-bound bent tube spun by the aquatic larva of a caddisfly, *Macronema* (Figure 7.4). It lives in streams and by this arrangement capitalizes on the current to drive flow (and food) through its catch-net, spun across the middle of the overall tube. As Wallace and Merritt (1980) put it, *Macronema* "perfected the Pitot tube long before Henri himself."

An equivalent scheme serves the same function in at least one tunicate, or ascidian, sessile marine animals about as distantly related to caddisflies as are we. These "sea squirts" have the relevant plumbing inside themselves rather than in some external domicile, but the fundamental problem of suspension feeding is unchanged—how to move water through a separation device in such a way that the yield from food decently exceeds the cost of moving the water. The particular ascidian, *Styela*, takes advantage of local water motion to pass water through itself, but to do so it must contend with the rapidly changing direction of coastal wave surge. It has, though, quite a flexible stalk, which it manages to reorient in a manner analogous to a weathervane, so that the input opening always confronts the flow (Young and Braithwaite 1980).

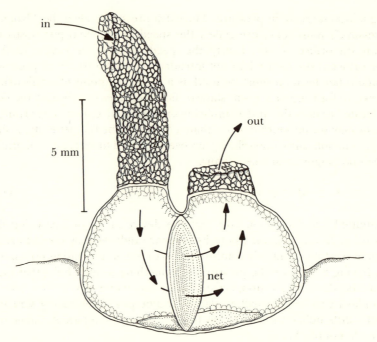

in

out

5 mm

net

FIGURE 7.4. The Pitot tube and static orifice of the larva of the cad-disfly *Macronema*. The animal lives in a small pipe that bypasses the silken catch-net and that has been omitted here.

PRESSURES ON OBJECTS IN FLOWS

The pressure measured with a Pitot tube is the so-called dynamic pressure, in contrast to the static pressure. Using a standard reference hole for static pressure and an upstream-facing orifice for the other end, a Pitot tube indicates velocity. Pressure measurements, though, are much more interesting than just as data en route to velocity. Consider what happens if we set up a remote static aperture such as that in Figure 7.1 and, as the business end, use first one and then another and then still another orifice along the length of some object in the flow. What we get is a graph or map of the way pressure varies along the object. This is of interest in several ways. First, this pressure distribution is in the most immediate sense a set of forces trying to distort the object; if the object is nonrigid as organisms often are, the pressure distribution tells us a lot about how shape will change as a result of exposure to flow. Second, for real (as opposed to ideal) fluids, the magnitudes and directions of these pressures add up to some overall force on the object—a resultant force in the direction of flow is part of the drag on the object; if it has a component perpendicular to flow it contributes a lifting force as well.

How might such a plot of pressure versus location most conveniently be presented? What we mainly care about is the influence of the shape of some body on the pressures. It would be nice to be able to compare pressures around objects of a variety of sizes—that's easily done by using a dimensionless x-axis, dividing the distance from the upstream end of the object to a given point by the overall distance from upstream to downstream ends. Such an axis, like Procrustes' bed, fits all. There's a good universal y-axis as well. For any object in any flow, pressure at the extreme upstream hole will be the full dynamic pressure of the flow, $\rho v^2/2$, as in a Pitot tube. Thus, if each datum of pressure is divided by this dynamic pressure, the plots for all objects and flow speeds will be dimensionless and begin at the left at 1.0. To this new variable, measured pressure divided by dynamic pressure, the name "pressure coefficient" is given; it is formally stated as

$$C_p = 2\,\Delta p\,/\,\rho v^2. \tag{7.4}$$

Velocity is, of course, the free-stream speed of air or water. Objects operating in different media or at different speeds or of different sizes all yield data that fit comfortably on the same axes with the same scales. Restoring a properly dimensioned pressure, for instance, requires nothing more than the reuse of equation 7.4. The overall plot now displays one dimensionless variable against another, but with that one quickly becomes contemptuously familiar.

Figure 7.5 gives such a graph for pressures around a sphere at a Reynolds number of 28,000. It shows the expected value of unity at the front and negative values further back where (by continuity) flow has had to speed up beyond its free-stream value to get around the obstacle. Bernoulli's principle is clearly operating—pressure drops as flow speeds up and vice versa. Two points are especially noteworthy from our viewpoint. First, although one result of flow is an inward pressure near the very front, the overall effect is an outward pressure across the surface of the object. Second, it is clear that the net force on the object is strongly rearward. A region of high pressure near the front pushes it backward, but no equivalent region of high pressure near the back exists to push it forward. In short, a sphere has a relatively high drag.

As a contrast, Figure 7.5 also gives the analogous graph for an object with a slightly elongated but still rounded front along with a much elongated and pointed rear. We loosely refer to this teardrop shape as "streamlined"—this one was a toy water rocket with its fins filed off. Not only are the pressures on the sides less negative (it is, after all, relatively

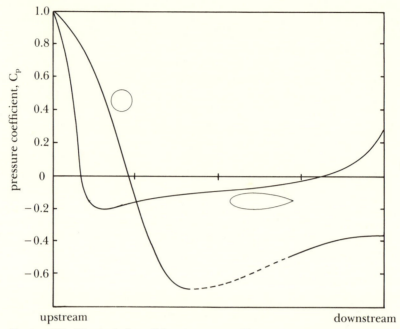

upstream   downstream

FIGURE 7.5. The variation of the pressure coefficient with location for a
sphere and for a streamlined object. The dashed portion of the line for
the sphere corresponds to the region at and behind the separation point
where the pressure varies erratically.

more slender), but there is now a decently positive pressure pushing on
the rear to offset in part the rearward push on the front. Clearly, this
latter object has less drag than the sphere. We'll return to that shove from
behind shortly.

Blimps, fast fish, and jetting squid

For an engineer, calculation of overall drag and noting the effects of
the details of shape are the main uses of these plots of pressure coefficient versus location. I mentioned earlier that for flexible objects the
pressure distribution itself took on immediate importance. Blimps, in
fact, have reinforcing battens near their front ends to offset the potentially distorting forces—it wouldn't do to have the front flattened by air-
flow or for shape to vary appreciably as a function of flying speed. But
among the products of our technology blimps are unusual in their lack
of rigidity. What effects might these variations of pressure with location
have on the design and operation of organisms?

Several years ago, DuBois, Cavagna, and Fox (1974) managed with some difficulty to make a set of pressure measurements on a bluefish at about 2 m·s⁻¹—it's a lot tougher to run pressure-conveying conduits through a fresh fish than through a model! The results (Figure 7.6a) were clearly worth the trouble even though they proved gratifyingly similar, overall, to expectations from data such as that in Figure 7.5. One cannot, strictly speaking, prove that such pressures were decisive in de-

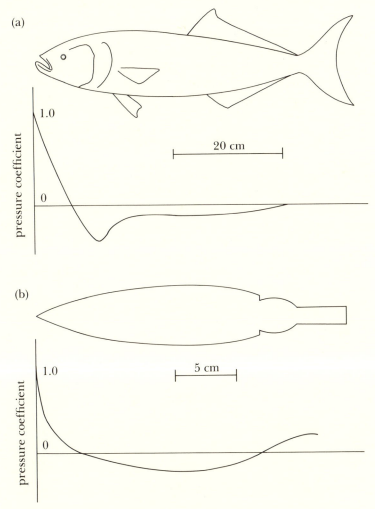

FIGURE 7.6. The variation of the pressure coefficient from upstream to downstream extremities of a bluefish (a) and of a model squid (b).

termining the design of fish, but they certainly are at least consistent with the usual overt rationality of evolutionary products.

Fish respire by taking water in their mouths, passing it across the gills, and ejecting it behind a lateral shield, the operculum. The mouth is obviously at the point where pressure is most positive (inward), while the operculum is located where the pressure is most negative (outward); no greater pressure difference to augment a fish's pumping could be easily contrived. Indeed, fish such as mackerel do no pumping, but swim continuously and depend on this so-called ram ventilation to keep from suffocating (Randall and Daxboeck 1984). The eyes of the bluefish (and, almost certainly, of many other fish) are located at the point where the pressure coefficient passes through zero; this is the unique location at which pressure is independent of swimming speed. Focus of an eye depends in large part on the curvature of the cornea; it might be awkward to have the latter vary with swimming speed. The heart is located beneath the area of minimum pressure; hearts are made of muscle, which can actively contract but which needs some special device to ensure reexpansion. The low pressure might aid reexpansion and thus increase the heart's stroke volume during rapid swimming, which, after all, is the only heavily aerobic exercise of fish. As a possible corollary, fish commonly increase stroke volume during activity more than the frequency of beating, the opposite of our own compensation (Jones and Randall 1978). Finally, the skull of a fish reinforces it against inward pressure throughout the anterior area where such pressures can occur; further aft its largely inextensible skin ought to be adequate bracing against forces that are dependably outward. It does appear that there's nothing fishy about the design of fish.

When startled or when pursued by predators, squid squirt water from a tube just beneath the head and move rapidly rearward (Figure 7.6b). Water is contained in a cavity between the viscera and the muscular body wall (the "mantle"); this mantle cavity is refilled through openings at each side of the head. To refill, though, the mantle must be expanded. In part, expansion involves very short muscle fibers running radially in the mantle—when not in motion a squid has little to use other than these muscles—and some residual elasticity of the mantle. During motion, though, flow-induced pressures can be brought to bear. Water enters back where pressure begins to turn inward again; the greatest area of the mantle covers the region of greatest outward pressure. As a result, a difference in pressure coefficient of about 0.25 is available for refilling. At a swimming speed of 3 m·s^{-1} this difference can provide about half the total refilling pressure while at 9 m·s^{-1} (about 20 mph—a rough estimate of top speed) flow-induced pressure ought to account for somewhere around 90% of the total (Vogel 1987; determining these numbers was why, as mentioned in the last chapter, I fussed with a model squid in a wind tunnel).

In Chapter 5 the problem of depletion of underwater bubble-lungs was mentioned, along with a beetle (*Potamodytes*) that uses moving water as an evasion. Its scheme is based on the fact that the pressure on a submerged body in a flow, averaged across the body's surface, is outward—the net pressure coefficient is negative as is evident in Figure 7.5, especially if the body is well rounded. So flow reduces the pressure in its bubble (Stride 1955); if there's high enough flow and low enough hydrostatic pressure (shallow water), the internal pressure can get subatmospheric. If, in addition, the water is well saturated with oxygen, then the dissolved oxygen will spontaneously move into the bubble to restore that used up by the beetle. The conditions are demanding, though—speeds over a meter per second and depths of less than four centimeters—so the trick is limited to fairly rare habitats. Again that square of velocity in Bernoulli's principle is daunting.

PRESSURE AND DRAG

A bit earlier I mentioned that one of the consequences of the pressure distribution across the surface of an object was drag. And still earlier (Chapter 3) a rough relationship was claimed to hold between drag and the square of velocity at moderate and high Reynolds numbers. The origin of that v^2 is clearly related to Bernoulli's principle—where drag is more a matter of pressure than viscosity, drag should vary with the second rather than the first power of velocity. (We'll ignore, for simplicity, any implicit contradiction in using a notion derived for ideal fluids to say something about drag.) In fact, though, the business of drag is even more complicated.

Two kinds of drag

Figure 7.5 presented a severely asymmetrical graph of pressure against location around a sphere—speed was minimal and pressure maximal at the front; speed was maximal and pressure minimal near the portion widest with respect to the flow; but pressure did not greatly rise again in the rear, so clearly there was nothing in the back like the zero-speed point in the front. In physical terms, flow follows around to somewhere near the widest part—easily, because pressures are dropping, and flow goes naturally from high to low pressure. But the flow fails to follow the surface farther back—here pressure begins to rise again, so, to follow the surface, flow would have to go toward higher pressure. Which it doesn't

141

do—instead, it "separates" from the surface near the widest point (Figure 7.7), and we notice a wide "wake" of disordered flow behind the object. The result is the front-to-rear asymmetry seen in Figure 7.5 and thus the high drag of a sphere.

Just where separation occurs determines, to a large extent, the relative magnitude of the drag. And where separation occurs depends largely on the momentum of the fluid near the surface. At higher speeds (really, higher Reynolds numbers) boundary layers are thinner and momentum higher—separation may be later, the wake narrower, and the relative drag reduced. For a sphere or cylinder this occurs abruptly at a Reynolds number somewhere between 100,000 and 250,000 (Figure 6.8), the value at which turbulence invades the boundary layer. We can blame all of this drag on pressure, specifically on the fore-to-aft pressure difference, but this difference is due mainly to separation of flow, and the separation is an indirect consequence of viscosity. Without viscosity, the momentum of a moving fluid near the surface would not be taxed by a velocity gradient, and the flow would follow the surface without separation.

We really have to contend with two elements of the force we call drag. What we have just been talking about is *pressure drag*—it may come ultimately from viscosity but can be calculated from the pressure distribution and varies (assuming the separation point doesn't change) with the square of velocity. In addition, there is another sort of drag, something called *skin friction*, the direct effect of viscosity through the fluid's resistance to the unavoidable shearing motion near any surface. This drag varies with the first power of velocity. At low Reynolds numbers, skin friction was mainly what mattered, so while the world of the last chapter may have been unfamiliar, it was at least tidy. At moderate and high Reynolds numbers—that is, at values great enough for separation, which is to say above 10 or so—pressure drag may be far greater than skin friction. But the friction is always there—the no-slip condition cannot be evaded.

How, then, might drag be minimized at moderate and high Reynolds numbers? The best approach is to delay separation—to make flow follow the surface farther rearward. A gentler curve near and behind the widest

FIGURE 7.7. The streamlines for flow around a sphere or a circular cylinder, with the separation point marked by a pair of arrowheads.

point is very effective, and we can see in Figure 7.5 how the pressures behind rise further with such a streamlined shape. As with so much in life, it's how things end that matters most. Or, as Shakespeare put it, "All's well that ends well." Conversely, how might drag be maximized? One makes the wake as wide as possible. A flat plate broadside to flow or a hollow hemisphere concave upstream (a parachute, for instance) has a wake substantially wider than the object itself—flow heads outward and keeps on going on account of its momentum even beyond the edge of the object. A rounded front, although far from fully streamlined, gives less drag than a flat front, as truck and car designers have recently begun to appreciate.

Drag is drag—a force on an object carrying it back in the direction of flow, a rate of transfer of momentum from fluid to object. But one sort, skin friction, depends mainly on total exposed surface; as to the other, pressure drag, it would be nice but inaccurate to assert that it depends just on frontal area, but it actually is a complex function of shape and Reynolds number. At low Reynolds numbers streamlining does almost no good—any reduction in residual pressure drag is offset by a greater total exposed surface. Microscopic swimming creatures don't often have the familiar rounded front and tapered rear. Fish, squid, whales, the larger swimming insects and crustaceans—these by contrast are inevitably streamlined.

The drag coefficient

Because of all the complications, a simple and general formula for drag as a function of speed at moderate and high Reynolds numbers is available for almost no shape—there just aren't easy analogs of Stokes' law. So we're driven to the use of graphs and tables, and the important task is that of figuring out the most efficient presentation. It's obviously impractical to plot drag separately against all the factors that might affect its value—speed, size, density, and viscosity—for each shape of interest. The Reynolds number collects these quantities into an appropriate composite variable, so we might plot drag against Reynolds number. But it turns out that even for a given shape there is no unique value for the drag at a particular Reynolds number.

Still, something related to drag does vary uniquely with the Reynolds number. Not to be coy about the variable, it is essentially a version of the pressure coefficient introduced earlier (equation 7.4). Recall that the pressure coefficient, by the trick of dividing by dynamic pressure, gave a dimensionless pressure adjusted for the density and speed of the medium. Drag is, of course, just pressure times area; the drag on an object is the net pressure across its projected area times that area. ("Projected area" is the area of the shadow cast by the object in a light shining in the direction of flow.) So if we divide drag per unit area by dynamic pressure we get something dimensionless; this not-so-new dimensionless variable is the "drag coefficient," touched on in Chapter 4. Thus,

$$C_D = \frac{2D/S}{\rho v^2}, \text{ or}$$

$$D = C_D \rho S v^2 / 2. \tag{7.5}$$

This drag coefficient turns out (for a given shape) to vary with nothing else besides the Reynolds number. Thus a graph with the drag coefficient on the y-axis and Reynolds number on the x-axis "says it all" with regard to the drag on an object of a particular shape. It ought to be noted that equation 7.5 is not an "equation for drag" in any way analogous to, say, Stokes' law; rather it is just a formula for converting drag to drag coefficient and vice versa.

A plot of drag coefficient versus Reynolds number for a sphere (Figure 7.8) has bumps and discontinuities that satisfyingly correspond to the changes among the regimes of flow illustrated in Figure 6.8 and mentioned a page or so back. The graph itself, though, takes some getting used to—one dimensionless variable has been plotted against another. The excuse for such obscurity is the generality obtained. Given the drag

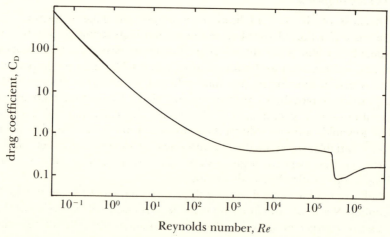

FIGURE 7.8. The drag coefficient (calculated using frontal area) versus Reynolds number for a sphere. The portion below Re = 1.0 follows Stokes' law—it can be calculated by substituting equation 6.7 into equation 7.5. The great drag crisis at about Re = 200,000 is clearly evident.

coefficient and the Reynolds number for a given shape you can quickly figure the drag for any combination of the variables in these expressions—any size, any speed, any fluid medium—just plug 'n chug with equations 7.5 and 6.6. And these drag coefficients (at least for regular shapes) are matters of public record.

One final element of complication must be admitted. The explanation of drag coefficient presumed that the area involved was the projected area. Any area, though, preserves dimensional propriety; and sometimes other reference areas are used, such as total surface ("wetted area") or even the nominal area created by taking the two-thirds power of volume. So any use of equation 7.5 really must specify the area of reference.

Drag and flexibility

For some objects, as Figure 7.8 shows for spheres, the relationship between drag and speed or between the drag coefficient and the Reynolds number is a bit volatile. For others, such as a round disc broadside to flow, the drag is almost precisely proportional to the square of speed. For such a disc the relationship proves to be regular enough ($C_D = 1.2$, using frontal area, for Re between 10^3 and 10^6) so its drag can be a reliable and regular indicator of speed. Relevant data for all sorts of shapes and situations have been gathered into a most useful compendium by Hoerner (1965).

For streamlined objects, drag coefficients are lower by roughly five- to fiftyfold, depending on the Reynolds number and the area used as reference. Moreover, the drag increases less sharply with speed—it's no longer proportional to the square but to the 1.5 power of speed. As mentioned, streamlined shapes are very much the rule among macroscopic swimmers and all but very small fliers. Indeed, the first deliberately streamlined shape, contrived by Sir George Cayley (1773–1857), was based on girth measurements of a trout (Von Karman 1954).

What is at least initially puzzling (like Sherlock Holmes's dog that didn't bark) is the comparative *un*commonness of streamlined shapes among sessile (attached) organisms, even those clearly subjected to potentially hazardous forces from winds and water currents. One obvious difficulty is that a streamlined shape achieves its low drag only in a very specific orientation with respect to wind and water currents, and an attached organism commonly has to contend with erratic flows. And the flows are not only directionally erratic but very variable in intensity—a low-drag shape may be crucial in a storm or surge but needlessly compromising of other functions otherwise. Obviously drag reduction cannot be an end in itself—it isn't exactly a common keystone of reproductive success.

145

I know of only one streamlined attached organism whose drag has been investigated—a marine alga, *Halosaccion*, of the rocky intertidal areas of sheltered shorelines in the Pacific northwest. The body ("thallus") of the alga is about the shape and size of a cigar; it's known (at least at the Friday Harbor Laboratories in Washington) as the "sea condom." The creature manages the problem of orientation in variable flows with a short but flexible kite string (the "stipe") so it repositions continuously in the manner of a weathervane. Its drag coefficient (using frontal area) is 0.084 at a Reynolds number of 40,000, only one-fourteenth that of a disk of the same frontal area (Vogel and Loudon 1985). Put to it, nature clearly can do a good job with conventional streamlining in a sessile organism.

But the usual approach for an attached organism involves a clever combination of solid and fluid mechanics. A tree responds to the wind with what one is wont to call a "deformation." Since the deformation results in a dramatic reduction in the drag of the tree, perhaps we ought to stop using a term with unavoidably pathological overtones—"reconfiguration" would be better. The key is flexibility—if an organism is occasionally subjected to strong forces from flow, then a reasonable response is to accept some temporary loss of other activities (photosynthesis, feeding, and reproduction come immediately to mind) as shape is briefly altered. This seems to be the basic tactic of trees—Figure 7.9 shows the leaves of a small holly tree changing configuration on a branch as the wind in a wind tunnel is increased to 20 m·s⁻¹ (45 mph). The process is completely reversible, even for leaves with such short, stiff petioles ("stems") as those of the holly.

A few years ago I collected what data I could find in the literature and did some additional measurements of my own to see what were the exponents relating drag to speed for flexible organisms. These exponents are a kind of scaling factor, derived in the same manner as the factors cited in Chapter 3. Ignoring data for very low speeds on the grounds that drag would present no problem there, I found that the exponents are surprisingly similar, mostly between about 0.9 and 0.8, for a very wide variety of plants and animals, both terrestrial and marine—pine and holly trees, a large sea pen and a tiny colonial hydroid (coelenterates), and several very diverse-looking marine algae (Vogel 1984). These exponents should be compared with the values of 2.0 for nonstreamlined and 1.5 for streamlined rigid bodies—for the flexible forms, drag increases very much more mildly with increases in speed. To put the matter another way, instead of remaining roughly constant, the drag coefficient drops dramatically with increases in flow speed.

There are further interesting flourishes to the business. Telewski and

0 m/s

10 m/s

20 m/s

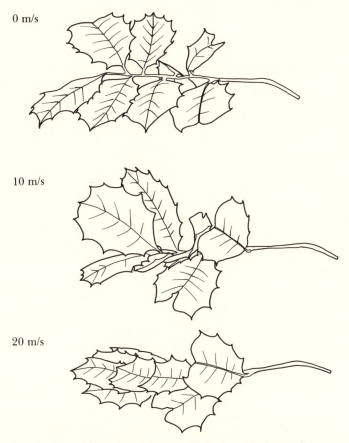

FIGURE 7.9. A branch and leaves of holly in a wind tunnel at
three different speeds; the wind here blows from the right. Notice how the leaves bend over and, at the highest speed, reduce
their exposed surface area by pressing against each other and
against the branch.

Jaffe (1986) showed that high-altitude fir trees, which are normally exposed to substantial winds, are stiffer than low-altitude trees of the same
species. The result is puzzling unless one bears in mind the possible loss
of other functions due to reconfiguration—perhaps the tree afflicted
with chronic wind, like those of us with chronic illnesses, must manage to
be adequately functional (photosynthetic, for them) in all but quite extreme episodes. Incidentally, flags suffer higher drag than weathervan-
ing flat plates (Hoerner 1965), and I've found that the drag of individual
leaves increases with speed with drastic exponents of between 3 and 4 in

147

much the same manner. It appears that the low exponents previously cited result from the interactions of the leaves on a branch—quite a lot of adaptive tuning of shape and mechanical properties is probably just waiting to be investigated.

ABOUT LIFT

Drag was a fine mess of complication, even in our limited and simplified treatment. Lift is more obscure than complex; again we're onto a subject whose existence has been long known but whose physical explanation was worked out only in the early part of this century. The first thing to settle is a definitional matter—we'll call "lift" any force at right angles to the direction of flow of the overall fluid. Thus, from the viewpoint of a craft in level flight, fluid approaches horizontally, and any lift it produces is normally directed upward. The peculiarity of the definition emerges when flow is not horizontal. Consider the local wind striking a glider that's slowly losing altitude—to the extent that the glider descends, the wind approaches along an upward slope. The lift force, then, is no longer quite upward but is directed slightly forward of vertical. As we'll see shortly, that slight forward tilt of the lift vector is what keeps the craft from slowing down and eventually tumbling earthward.

Lift and circulation

More important than any definition is the matter of physical origin. Let's imagine a truly bizarre sort of sailboat with one or more cylindrical, vertical funnels ("smokestacks") on the deck (Figure 7.10). The funnels are mounted on bearings and connected to motors below that cause them to rotate around their vertical axes. Naturally, the nearby air moves around with the funnels, on account of the no-slip condition, with the air nearest the funnels going fastest. (Technically, the bits of air translate in a circle, or "circulate" rather than truly rotate, around the funnels—recall the distinction in Chapter 2.) If the boat encounters a wind from one side moving across the deck in the same direction as the front of each funnel is turning, then the ship moves forward ("reaching," a sailor would say).

What's happening? We have superimposed a circulating motion of fluid (caused by the rotation of the funnels) and a unidirectional motion (the wind); as a result, a force at right angles to the wind has been produced, which is, by our definition, lift. If the wind came from the other side, the boat could progress forward by reversing the rotation of the funnels. It can go backward as well as forward; the only thing it can't do is sail *directly* into the wind—it's still a sailboat. (Only the Flying Dutchman

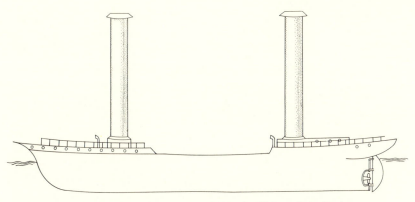

FIGURE 7.10. Not really so imaginary—this is the 600-ton brig *Buckau* equipped by Flettner in the early 1920s with a pair of rotating funnels, each 16 m (52 ft) high and 3 m (10 ft) in diameter.

could do that and only while the wind whistled Wagner.) Anton Flettner's craft may have been a technical success, but it was a commercial failure, a sailboat that needed a motor. Boats with Flettner rotors emerge from time to time, impelled by hope and journalistic hyperbole (wind, of a sort). In the same way, one can design a functional airplane using length-wise-rotating cylinders in place of wings—the cylinders must rotate so their tops move rearward and their bottoms move forward; but it's both more complicated and a poorer flier than a conventional plane.

But what's really happening? Where the wind and the flow induced by rotation of the cylinder are in the same direction (in front of the funnels), they add, so the overall fluid motion is more rapid than either alone. Where they are in the opposite direction, they oppose, and the net motion of the fluid is slower than either alone (Figure 7.11). By Bernoulli's principle, more rapid flow is associated with lower pressure. So there will be a pressure difference fore and aft on the ship, impelling it forward, and a similar one above and below the rotating pseudo-wings of the airplane, holding it up—in both cases, *lift*, as we've defined it.

Quantitatively, it turns out that the lift is nearly proportional to the product of the surface area of the cylinders, the density of the fluid medium (here, air), the rate of rotation, and the speed of the wind (boat) or forward movement (airplane). All these are straightforward direct proportionalities—no exponents other than unity.

While Flettner rotors are, as technology, unattractive, they have evolved in nature on several occasions. You're probably familiar with the spiral descent of maple seeds ("samaras" properly) whose sinking speed is reduced by the action of a lift-generating airfoil in the manner of an

149

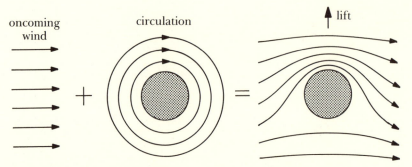

oncoming wind   circulation   lift

FIGURE 7.11. If a rotating body (here a sphere or cylinder) encounters a wind, the resulting asymmetry of flow around the body gives rise to an asymmetrical force—lift.

autogyro or unpowered helicopter. Overtly similar to the descent of maple seeds is that of tulip poplar or ash samaras, but examination shows here a curious symmetry about the long axis—there is no difference between upstream and downstream edges as on a maple seed. And that, as McCutchen (1977) pointed out, is because there simply is no functional distinction between the edges—such a samara rotates around its long axis at the same time that it spins about a vertical axis of descent. One doesn't notice the rotation because it's obscured by the spinning descent, but these are indeed Flettner rotors. If you're not convinced, flip a rectangular strip of index card (about 1 x 5 inches) and watch it descend. If it rotates, it will not descend vertically but will instead glide off with some horizontal component. If you use a triangular instead of a rectangular strip, it will descend in a helix while rotating lengthwise just as do these samaras.

Airfoils

Most lift producers, however, use something familiar to us as an "airfoil," the shape used for the cross section of an airplane wing, a propeller blade in air or water, or a billowing mainsail (Figure 7.12). None are in the habit of rotating rapidly around their long axes. Now what's happening? It turns out that the special functional feature of such airfoils is that they give the same sort of circulatory flow pattern as did the cylinders and samaras, *but without themselves rotating.* Either on account of their own asymmetry or by facing the flow at a slight inclination (or both), they cause the flow to be more rapid on one side (on the top of a well-behaved wing) than on the other.

Airfoils, though, display an indicative quantitative difference. The lift

150

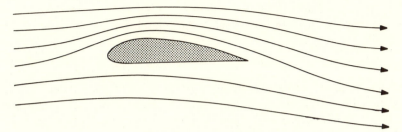

FIGURE 7.12. The streamlines around an asymmetrical airfoil (shown in cross section). More closely spaced streamlines indicate faster flow; a flow faster above than below the airfoil results in lift.

is again directly proportional to the density of the fluid and to the area of the wing (top view, or "plan form," area is the usual area of reference). But it's now proportional to the square of the speed of the flow encountered by the airfoil rather than the first power. Apparently, the airfoil really makes the air around it combine the circulating and translational motions of the air around Flettner's cylinders. But here the circulation is caused by the translational wind interacting with the peculiar shape of the airfoil—instead of being proportional to the product of separate translational and circulating flows, lift is now proportional to the square of translation. In obedience to Bernoulli's principle, pressure is below ambient above the wing and above ambient below the wing.

For a craft that is not actually accelerating upward or downward, the lift must just equal the weight. If lift is approximately proportional to the square of speed, then there's big trouble at low speeds. Two compensatory changes are common. First, the area of the wing can be changed. Birds certainly do this, using the broadest possible expanse when taking off, landing, or flying slowly; and large jet airplanes unroll a remarkable amount of extra surface for takeoffs and landings. The other adjustment is in the angle between wing and oncoming air—the "angle of attack." Up to a point, increasing the angle increases the lift proportionately. Still, two problems arise in increasing the angle of attack. First, a critical "point"—at some not immense angle (typically 20°) flow separates from the top of the wing (as with the sphere mentioned earlier), and the lift drops off catastrophically. So operating near the point of maximum lift can be hazardous. Second, increasing the angle of attack increases the drag as well, and the long-term best operating point, that giving the maximum lift-to-drag ratio, is usually at a fairly low angle.

The easiest way to view the differences in the behavior of airfoils is on a so-called polar plot, introduced by that towering figure Gustav Eiffel.

151

To offset the effects of size one defines a lift coefficient comparable to the drag coefficient; the area used is the "plan form," or top view, of the airfoil:

$$L = C_L \rho S v^2 / 2. \tag{7.6}$$

One then plots, for a given airfoil, a graph of lift coefficient versus drag coefficient, writing on the angles of attack (Figure 7.13). A straight line from the intersection of the axes is a line of constant lift-to-drag ratio; the steepest such line that touches the curve gives the maximum ratio for that airfoil configuration; the point at which it touches is the optimal angle of attack.

The effect of size is dramatically evident on the polar plot—for the small flier, the curve is shifted far to the right. At lower Reynolds numbers, drag turns out to be greater relative to lift—skin friction is the main sort of drag for streamlined airfoils, and skin friction is substantial at low Reynolds numbers, so the best lift-to-drag ratios are correspondingly lower. For an airplane wing the best ratio may be 25; for a locust wing it

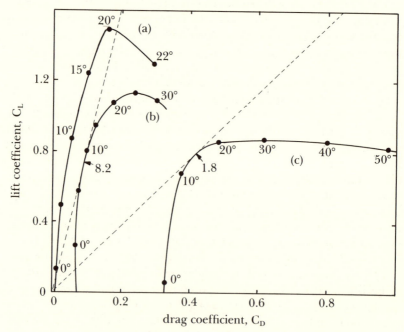

FIGURE 7.13. The polar plots for three airfoils; (a) is the wing of a light aircraft, (b) is the hindwing of a desert locust, and (c) is the wing of a fruit fly.

is about 8; for a fruit-fly wing it's less than 2. (Vogel 1981b). Moreover, at low Reynolds numbers drag is less dependent on orientation, so inclining an airfoil to extract lift incurs a relatively lower additional drag. There's an odd consequence—for a small flier the best lift-to-drag ratio is not only lower but occurs at a very high angle—essentially at the angle that just precedes separation or stall. And the fact that no microscopic creatures fly by the conventional scheme falls into place—the great viscous forces at low Reynolds numbers gives inescapably high drag and low lift-to-drag ratios.

What price lift?

Airplanes are profligate consumers of fuel, and the highest known metabolic rates are those of birds and insects in flight. Lift, though, is a force, and work (which is what fuel is really needed for) is required only when a force moves something some distance. So what's being moved upward or downward for a craft or creature in level flight that requires such a high rate of work, such a high power consumption? It is, in fact, the fluid itself that has to be forced downward to sustain the craft's altitude. Drag, you'll recall, can be viewed as the rate at which an object removes momentum from the flow across itself. Similarly, lift is equivalent to the rate at which momentum is given to the fluid at right angles to the ambient flow—for a craft in level flight, it's the rate of creation of a downward component of momentum:

$$L = mv/t \text{ (where } v \text{ is downward).} \tag{7.7}$$

Now momentum is the product of two variables, mass and velocity; thus a given rate of momentum creation could be achieved either by giving a small mass of fluid per unit time a big downward speed or a large mass of fluid per time a small downward speed. Which should be done?

Consider how much power has to be expended to create this momentum. Power is force times distance divided by time, or *force times velocity*. Force is mass times acceleration, or *mass times velocity over time*. Combining the two statements, we see that power is thus the product of mass per unit time and the *square* of velocity (times 1/2 for reasons irrelevant at this point):

$$P = mv^2/2t \text{ (} v \text{ is again downward).} \tag{7.8}$$

From the point of view of power expended, it's clearly better to take a large mass and give it a small downward speed—one doesn't want to pay the high price implicit in that square of velocity! In the most extreme case, one might take an unlimited mass of air and give it an infinitesimal downward push. Lift would then be created at a negligible cost. Some-

153

thing for nothing? Not really—it's a force we want, and it doesn't necessarily take work to generate a force. After all, the chandelier is held above the table by a force even when the electricity is turned off, thank goodness.

But how to produce cheap lift? It's easy in theory—we just ask for an airplane with infinitely long wings, leaving the difficulties to the structural people. We have, though, explained why, when power is really limited, wings are long and skinny—on gliders, for instance, whether living or manufactured. And why a helicopter has such ungainly long rotor blades instead of a small but rapid propeller facing upward. And why you can't easily take off vertically just by pointing the ordinary jet or propeller upward. And why hovering is difficult for a propulsive system designed for rapid forward motion—to attain any decent speed, a propeller or jet must impart a high rearward speed to the fluid, so it has to be a relatively high velocity, low mass-per-time momentum maker. Incidentally, the result of achieving lift with wings of less than infinite length appears as an additional component of drag, called "induced drag." This drag times the speed of an airplane is that extra "induced" power.

How to glide

According to Newton's first law of motion a body remains at rest *or in steady motion* in the absence of any external force. For an aircraft flying steadily, the law implies that all forces on it must balance out to zero, taking into account both their magnitudes and directions. What muscles or motor must manage, via wings and propeller, is to provide enough force to counterbalance the weight of the craft and its drag. In airplanes, the propeller or jet provides a horizontal force opposing drag and thus propelling the craft through the air; the forward motion then inveigles the wings into generating lift. In most flying animals, the functions of propeller and wings are combined, much in the manner of a helicopter. Most commonly, wings don't beat quite up and down in the manner of propeller blades that reciprocate rather than rotate. Instead they beat in a plane inclined a little off vertical, so they take air from above and in front and accelerate it downward and rearward.

It's perhaps easiest to approach this set of forces—lift, weight, drag, and sometimes thrust—through a quick look at "simple gliding." This is defined as a situation in which a craft descends just fast enough to maintain a steady speed and a straight glide path. In simple gliding, then (Figure 7.14), there has to be an aerodynamic resultant force just equal and opposite to the weight of the flyer—weight is the only force independent of the craft's activity. The resultant force can come only from the combination of lift and drag, with lift defined (as before) as the force at a

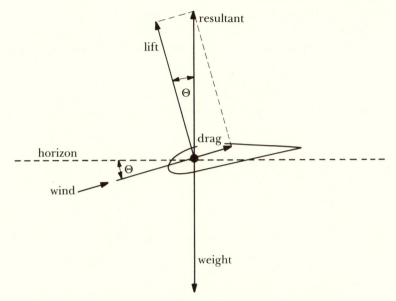

FIGURE 7.14. The forces and angles in simple gliding. As a result of its descending path, the craft (here represented by an airfoil section) encounters an ascending wind. To progress steadily the craft must descend rapidly enough so there is no net force on it. The descent is exaggerated here—a 40:1 ratio corresponds to 1.4°.

right angle to the glide path. For the craft in the figure, one consequence is clear. The angle between the resultant and the lift must be the same as that between the horizon and the glide path—what is called the "glide angle."

As a consequence, the glide angle is set by the lift-to-drag ratio—if the ratio is, say, 40:1, then the craft will descend one meter for each forty it goes horizontally. So for a good low angle or nearly horizontal glide the craft needs a high lift-to-drag ratio, which, incidentally, 40:1 is. This relationship places serious limitations on the gliding of very small creatures—as we saw earlier, small size is inevitably concomitant with a much-reduced best lift-to-drag ratio. If a fruit fly were to glide (it apparently does not) it would descend at an angle of almost 30° to the horizon. Very small biological objects carried on the wind—wind-dispersed seeds, baby spiders, etc.—don't glide but rather go for drag maximization with fluff, silk, and such devices to slow their descent, while wind currents provide horizontal motion.

The situation isn't quite as dismal for biological gliders as it may seem. Angle of descent is set by the lift-to-drag ratio of the craft, but speed of descent is determined by the combination of that angle and the speed of progression through the air. A monarch butterfly may not have a very good angle, but it moves slowly, so its actual speed of descent (as opposed to say, maximum horizontal movement in still air) is about the same as that of the best sailplane (Gibo and Pallett 1979). Also, the smaller and more maneuverable living gliders can take better advantage of local variations in the speed and direction of the wind—a large number of schemes are possible for what is generally known as "soaring," as distinguished from simple gliding.

The uses of lift

The simplest use of a lifting airfoil is the one made by a fixed-wing aircraft in level flight—the propeller provides thrust to offset the various sorts of drag, and the wing does nothing more than provide lift. The arrangement is rare in nature, although the forewings of beetles (the elytra) in some cases function as fixed wings. The commonest use of lift-producing airfoils is for the generation of thrust. And this application works in essentially the same manner whether we consider the wing of a hummingbird, the fluke of a whale, the forelegs of a sea turtle, the tail fin of a tuna, or the wing of a swimming penguin.

Getting thrust is much like an extreme form of gliding. If the descent angle of the airfoil is even greater than needed to get a vertical resultant, then the resultant force of lift and drag will be directed forward of a

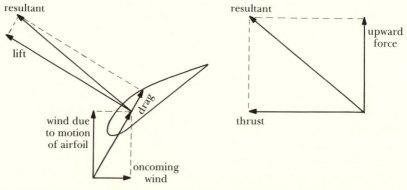

FIGURE 7.15. Thrust is possible if an airfoil is moved perpendicularly to the oncoming wind—the airfoil thus "sees" something other than a pure headwind, and its resulting force (resolved at right) has a forward-directed component. The upward force isn't generally useful—for a propeller it's balanced by a downward force from the opposite blade; for a beating wing, it may be balanced by a downward force on the subsequent upstroke.

156

vertical line through the craft; in short, it will have a component of thrust (Figure 7.15). The way this is done without the craft as a whole having to dive steeply is to have the airfoil in motion with respect to the craft—in short, to use some form of motorized propeller. The wind striking this flapping or rotating airfoil will thus be far from horizontal. But beyond this summary the analysis gets quite complex, needing an amount of space quite unreasonable for the present book. Many of the same considerations as for lift still apply—for instance, the quality of the airfoil, even if now a flapping wing or propeller blade, is still measured by its best lift-to-drag ratio. And, as in gliding, the scheme doesn't work at all well at very low Reynolds numbers, so the swimming of all microorganisms must be based on different mechanisms entirely.

Finally, it ought to be pointed out that the consequences of lift extend beyond locomotion and that lift isn't inevitably a blessing. For a plaice (similar to a flounder), for example, lift is apparently an embarrassment. Lying on the bottom, the fish presents a convex upper surface and develops lift if the water is in motion. At flow speeds inadequate to push it downstream it begins to lift off and has to dig actively into the substratum in order to remain in place and not flounder (Arnold and Weihs, 1978).

Diffusion versus convection

"As aromatic plants bestow
No spicy fragrance while they grow;
But crushed or trodden to the ground
Diffuse their balmy sweets around."

Oliver Goldsmith, confusing
diffusion and convection

QUITE a lot of noise was made in Chapter 3 about the diversity of sizes in which organisms occur—roughly 10^8-fold in length or, less comprehensibly, 10^{24}-fold in mass. In view of this wide range, anything biological that doesn't vary in size ought to strike us as noteworthy. One such item has already been remarked upon—the cells of animals are almost all within a factor of two of being ten micrometers across. This constancy has been at least implicitly recognized for a century and a half, and for D'Arcy Thompson (1942), at least, it was no mere parenthetical point. Plant cells are typically an order of magnitude longer than those of animals, but they too are monotonously regular in size. Another standard item is the diameter of a capillary, between five and ten micrometers in all sorts of animals—an earthworm, an octopus, a fish, a person. Yet another is the diameter of the finest branches of the internal air-transporting pipes, the tracheae, of insects—Pickard (1974) noted that whatever the overall size of the creature, the tracheae are about 0.2 micrometers in diameter.

Constancy of size ought to direct our attention to the existence of some underlying physical constraint. The present chapter is in large part a look at a basic physical phenomenon that might provide the constraint underlying these peculiar regularities—peculiar in the context of the spectacular diversity of size and structure in living systems.

Besides movement en masse, fluids engage in an entirely and fundamentally different kind of movement. This second sort is inseparable from the behavior of molecules as individuals; it is, in fact, one of the few phenomena in which the particulate character of matter has practical consequences in our macroscopic world. In gases and liquids, individual molecules wander around aimlessly at rates determined by their temperatures. (Or, from another viewpoint, their temperatures are determined by the rates at which they wander.) Collisions are frequent, with the changes in direction we would expect from our earlier discussion of momentum. The only peculiarity about the collisions is that they are fully "elastic"—not only is momentum conserved, but the molecules as a group go, on the average, as fast after colliding as before, losing no energy overall. The wandering is, of course, entirely a result of the collisions since, by Newton's first law, a molecule would not otherwise change direction.

A very important variable in this random motion is the average distance a molecule goes between collisions—it's called the "mean free path" and depends on the size of the particle and the nature of the medium in which it is moving. The denser the molecules are packed, the more frequent the collisions, so a molecule has a longer mean free path in a gas than in a liquid. Thus an oxygen molecule dissolved in water has a very much shorter mean free path than the same molecule in air—about 0.3 versus 90 nanometers, a factor of 300. Molecules in crystalline solids vibrate in place rather than wander. But some loose molecules do wander around even in solids, with important consequences.

These collisions of wandering particles provide an agency through which momentum is shared around and therefore a molecular explanation of the viscosity of gases. Gas molecules bang into the walls of their containers, and the interaction provides a good quantitative explanation for pressures in gases. And the collisions have an even more powerful consequence, at least for our understanding of physical science. Small particles in a liquid under a microscope are seen to be in a sort of continuous jiggling motion—"Brownian motion"—first noticed with pollen particles in water by Robert Brown in 1827. If the motion of the visible particles is interpreted as being due to hits by individual invisible molecules, then it is possible, as shown by Einstein early in this century, to calculate the size of a molecule. Pais (1982), in his splendid account of Einstein's science, reminds us that even that recently there remained the suspicion that molecules might be just the polite fiction of a formal chemistry. But arriving at the same size for a species of molecule from quite different physical starting points and by different lines of reasoning, as Einstein did, provided exceedingly persuasive evidence for their reality.

The random movement of molecules in liquids or gases is a process that inevitably disorganizes things, a point that will take on great importance in Chapter 15.

DIFFUSION

The macroscopic consequence of this aimless wandering is called "diffusion" and is a subject of daunting mathematical elaboration. You may be surprised to learn that equations can describe with enviable precision the results of a process in which we can say with certainty nothing about what any molecule is doing at any point in time. But making usefully precise statements about statistical aggregates certainly pervades social science—perhaps you just thought physical scientists were above such crudeness.

In reality, very few areas of science are uncontaminated by the pseudo-certainty of statistical conclusions describing individually uncertain events. What saves the physical sciences from obvious sloppiness is merely the smallness and abundance of molecules relative to us; Schrö-dinger (1944) makes just this point in the essay "What is Life?" One's relative uncertainty decreases with the number of entities under consideration, following what's called the "square root of n rule." Flip a coin once—the uncertainty is relatively large. Flip it a hundred times; the outcome will more likely than not fall between 45 and 55 heads. The range, 10, is equal to the square root of the number of trials, 100, and brackets a region of more confident expectation. The most likely outcome, a measure of what we think of as uncertainty, is 10% of the number of trials. Flip the coin a million times; the number of heads will probably be somewhere between 499,500 and 500,500. The range, 1000, is still equal to the square root of the number of trials but is now only 1/10 of 1% of that number. Worry about 10^{20} or more molecules, and the aggregate is awesomely lawful.

The random walk

Fortunately there's an intuitive model from which we can develop a decently quantitative notion of diffusion. Consider an individual at a point in a state of ideal inebriation—a drunk at a lamppost. He or she takes a step away from that comfortable support, stops, gets thoroughly disoriented, takes another step in a direction chosen randomly (that is to say, not chosen at all!), stops, goes another step in another random direction, etc. (We assume the steps occur at uniform intervals of time and that they are of equal lengths.) What happens over a longer time? Does the individual move away from the light? Nothing says he or she must. If an individual does move away, does this happen at a uniform rate or does the rate of movement depend on the distance away from the light already achieved?

While it might be easy to find volunteers for such experimentation, one does anticipate difficulty obtaining funds for the requisite chemicals. But since the drunk is only a model we might as well use a more socially acceptable one. Handiest is a random-number generator caged within a computer. With it drunks can be as ideal and as numerous as necessary, they can walk very rapidly, the random walks can be in one, two, or three dimensions, and the Volstead Act is not violated. Figure 8.1 gives two

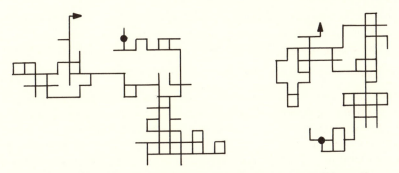

FIGURE 8.1. Two random walks done on my home computer. The walker has a choice of only four directions; nonetheless the distance achieved is, on the average, very nearly proportional to the square root of the number of "steps." Circles mark the starts, arrows the ends.

typical paths. (The outline of a program for doing such a simulation on a microcomputer is given in Appendix 2—the lack of standardization, especially of graphic displays, precludes greater specificity.) Two interesting results emerge from such simulations. First, the drunks do indeed wander away from the light. One may occasionally meander back, but with time returns become less and less frequent. The average position in space of the whole befuddlement is still the lightpost, but the average distance between drunks and post increases with time. Second, while that average distance increases, it does not do so in proportion to the time that has elapsed. Queerly, the drunks achieve a distance from the light (*on the average*) that is proportional not to the time but to the *square root of the time* they've spent. With the passage of time, their "progress" gets slower and slower. The same result occurs whether they keep their feet on the ground (a two-dimensional simulation) or (in a more surrealistic three-dimensional wandering) move away from a point in space.

If distance is proportional to the square root of time, then the square of distance is proportional to time. For time we can substitute distance over speed, so now we see (with a little algebraic juggling) that speed is inversely proportional to distance. If the verbiage seems too quick and slick, try the sequence on paper. Put back in terms of molecules or particles, *the rate of movement of particles on the average away from a point as a result of a random walk will be inversely proportional to the distance from that point.*

If you want a nonstatistical and macroscopic model for diffusion, just imagine a traveler connected to home by an infinitely extensible elastic band. The further the person goes, the slower is still further progress. The analogy is flawed, but it may still be helpful. The crucial point is the

161

peculiar dependence of distance on time. If it takes one second for a given amount of a substance to diffuse a given distance then it will take 100 seconds to go only 10 times as far.

Dealing with diffusion

We have here a truly odd sort of rate—to use yet another analogy, it is as if a trucking company, upon doubling the distance traveled with each load, needs not twice the number of trucks to carry the same total cargo in the same time, but four times the number. It is a world in which the speed limit is inversely proportional to the length of the journey. So we shouldn't be surprised to require similarly odd dimensions and units. Consider an ordinary rate, the kind measured by a speedometer, for which distance equals the product of rate and time. Rate or speed is thus distance divided by time, with units of miles per hour, meters per second, etc. For diffusion, though, distance *squared* equals the product of rate and time, so rate must have dimensions of distance squared divided by time, or L^2T^{-1}. Now this kind of rate is not our old, familiar, and comfortable distance per time or speed. Howard Berg, in fact, recommends that we not even use the term diffusion *rate* lest we be confused with ordinary rates; the term "diffusion coefficient" is, in fact, the normal designation.

The matter of units is comparatively ordinary. The SI unit for the diffusion coefficient is the square meter per second but is rarely used. The alternative $cm^2 \cdot s^{-1}$ is nearly universal. To get the SI unit, one just divides the conventional unit by 10,000. (Remember, if there are 100 centimeters in a meter, there are 100 squared = 10,000 square centimeters in a square meter.)

There's an entirely different way of viewing diffusion—a more formal but less explanatory approach. It leads more easily to an equation from which other equations that apply directly to practical situations can be derived. Consider a slab of material through which something is to diffuse (Figure 8.2). (The arrangement is conceptually simple if even less practical than that of the two-plate model for viscosity in Figure 6.2.) How much will diffuse through per unit time? According to what's known as Fick's law, an intuitively reasonable relationship, the mass per time diffusing will be proportional to the product of (1) the concentration gradient across the slab, that is, the concentration difference across the slab divided by its thickness, (2) the area of the slab, and (3) the diffusion coefficient of the material diffusing through the material of the slab:

$$m/t = (m_1/V_1 - m_2/V_2)(DS/x).$$

Using C for concentration (mass per unit volume) and recognizing that a

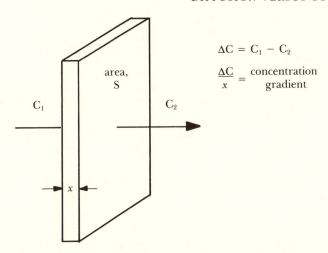

$$\Delta C = C_1 - C_2$$

$$\frac{\Delta C}{x} = \text{concentration gradient}$$

FIGURE 8.2. Steady diffusion down a constant concentration gradient showing the relevant variables for Fick's law.

minus sign ought to appear since material is moving down the gradient, Fick's law can be more tersely written as

$$m/t = -DS(\Delta C/x). \tag{8.1}$$

The skeptical reader will raise the objection that the equation above predicts a decrease in transport proportional to distance, not distance squared—doubling x only halves m/t. True—but the equation presupposes a rather different situation from the one considered earlier. Instead of decreasing in steepness with increasing distance from a point source, this gradient is constant in steepness across the slab—any locally greater gradient would cause a more rapid movement of whatever is diffusing, which would then restore the constancy of the gradient. The equation also presumes that the gradient does not change over time—in some way material is continuously supplied to the side having the higher concentration and removed from the side with the lower concentration, so neither concentration changes with time. So it's not inconsistent with the wandering drunks described earlier but just applies to a simpler set of circumstances. For more complex situations, a general introduction and references to specialized sources are given by Berg (1983).

The same skeptic may also ask how the molecules "know" in which direction Mr. Fick requires that they diffuse—the drunks wandered randomly, but here we suddenly assert that the process has directionality, that substances diffuse from regions of higher to ones of lower concentration. There is no chicanery or paradox—diffusion really occurs in both (really, all) directions. The initial random walk led to an overall dispersion—on the average, the distance between drunks and post increased with time. Dispersion resulted because the drunks were initially more concentrated in a particular place, so more were available to move away from than toward that place. Thus there was a *net* movement from higher to lower concentration, just as with diffusive transport across the slab.

The definition of the diffusion coefficient, using Figure 8.2 and equation 8.1, is quite analogous to the definition of quite a different physical property, "thermal conductivity." It turns out that heat transfer by conduction (Fourier's law) gets just the same formal treatment as mass transfer by diffusion, with temperature gradient replacing concentration gradient and a quantity called thermal conductivity replacing the diffusion coefficient. The equivalence, incidentally, is more than a mathematical convenience. Situations involving diffusion can be replaced for experimental purposes by models involving heat conduction, a handy thing since we can usually measure temperature more easily than concentration of one substance in another. I recently took part in a study (Hunter and Vogel 1986) that used this devious approach. We asked whether the spinning of snail embryos in their jelly mass might have the effect of reducing the distance over which diffusion had to occur and thus improving their supply of oxygen. The results, by the way, were positive, but not strongly so.

And some real values

Even if you have no occasion to do specific calculations involving diffusive transport, it's of interest to examine the relative values of diffusion coefficients, a few of which are given in Table 8.1. "Molecular weight" refers to the mass of a molecule relative to the mass of a standard unit, approximately a hydrogen atom.

TABLE 8.1. VARIOUS DIFFUSION COEFFICIENTS, D, IN CM^2·S^{-1}

Substance	Medium	Molecular Weight	D
Tobacco mosaic virus	water	40,000,000	0.03×10^{-6}
Human serum albumin	water	69,000	0.61×10^{-6}
Inulin	water	5,500	1.5×10^{-6}
Sucrose	water	342	5.2×10^{-6}
Glycine	water	75	10.0×10^{-6}
Oxygen	water	32	18.0×10^{-6}
Oxygen	air	32	0.2

Two things are immediately clear, and both have important implications for the functioning of organisms. First, large molecules diffuse more slowly than do small ones, although the diffusion coefficient isn't quite inversely proportional to molecular weight. Second, diffusion rates for gases or volatile liquids in air are *far* faster than in liquids—the roughly ten-thousandfold difference for oxygen between air and water is typical. It is one of several physical advantages of a terrestrial rather than an aquatic life.

Purely diffusive transport is still slower in living biological material than in water. Schmidt-Nielsen (1979) cites diffusion coefficients for oxygen in muscle and connective tissue (steak and gristle) as roughly one-third to one-half their values in water. And for diffusion within individual cells compared with water, data collected by Mastro and Keith (1984) show that large molecules have disproportionately low diffusion coefficients—coefficients for protein molecules may be more than a hundred-fold lower in cells than in water. The internal architecture of cells seems to slow macromolecular movement disproportionately. Indeed, the loss of much of this structure on the death of a cell is marked by an increase in Brownian motion that is quite conspicuous in time-lapse movies.

Diffusion versus convection

What must be emphasized in any discussion of diffusion is how strongly its effectiveness depends on the size of the system in question. Purcell (1977) gave an amusing but important illustration of the phenomenon by asking whether a bacterium ought to swim as it feeds on dissolved small molecules in the broth in which it lives. His answer is that swimming isn't worth the bother—food diffuses to the bacterium faster than the microorganism can possibly move. It lives in a world analogous to that of a cow in a bovine nirvana—a pasture of grass growing so fast that the cow has no need to walk as it grazes. Bacteria do swim, though, and Purcell noted that there is some profit in explorations that sample more concentrated or salubrious broth—new and perhaps greener pastures.

Closer to home, the delay in transmission of a nerve impulse from nerve cell to nerve cell is only about a millisecond—transmission depends on diffusion of a so-called transmitter substance across the synaptic gap, but the gap is only about 20 nanometers across. But for long distances the times involved in diffusive processes get devastatingly long. Schmidt-Nielsen (1979) did the following sobering calculation. If it takes a hundredth of a second for diffusion of a quantity of a substance such as oxygen between a typical cell and a capillary, then diffusion of the same quantity over a distance of 1 millimeter will take 100 seconds, and diffu-

165

sion of the same substance between a lung and a hand or foot will take no less than *3 years*.

Diffusion proceeds much faster in air than in water or tissue, which is to say that it is effective in air over substantially longer distances. But even in air diffusion is a slow process for what we regard as ordinary distances. A demonstration commonly done in high school physics classes purportedly shows the operation of diffusion—a jar of perfume is opened, and over time people increasingly distant from the jar report detection of the odor. The demonstration, however, is almost entirely a fraud. As a practical matter one can't get sufficiently low wind in a room to observe diffusive transport—merely the presence of warm bodies gives sufficient inhomogeneity to air temperature with resulting pressure differences that produce imperceptible but quite measurable wind. Most of the phenomena we casually attribute to molecular diffusion, such as the spread of the smell of coffee from kitchen bedroomward or of cooking fish toward the family felidae, are in fact cases of bulk flow of fluid, properly called "convection," upon which we've just spent two chapters.

Failure to distinguish properly between convective bulk flow and diffusion is no recent wrongheadedness. As Tabor (1979) pointed out, Lucretius (ca. 95–55 B.C.) thought he had discovered what we know as Brownian motion—dancing of particles due to hits from atoms, he said— but the movement of dust particles in a sunbeam are really caused by air currents. It's practically impossible to give a macroscopic demonstration of diffusion in a gas, and it demands some unusual precautions to do so even in a liquid—you have to create a stable density gradient through deliberate control of temperature or concentration gradients in order to prevent convection. The simplest alternative involves using some substance such as agar or household gelatin to prevent convection—a percent or two of gelling agent should make little difference to the diffusion coefficients. Visually complete equilibration of colorant in even a small beaker of gelled water will take weeks or months. Specific instructions are given in Appendix 2.

WHY SOME THINGS REMAIN SMALL

At the start of the chapter the remarkable constancy of the size of certain small biological items was remarked upon, with the comment that constancy pointed a finger toward some constraint outside the domain of matters manipulable by natural selection. Having established the character of diffusion, the factor toward which the finger was pointed, let's return to those small items.

First, the tiniest insect tracheae—Pickard (1974) commented not only on their constant diameters but on the relation of that diameter, 0.2 micrometers, to the mean free path of an oxygen molecule in air, about 0.09 micrometers. If such a molecule moves through a large container of air, its collisions will most often be with other gas molecules and only very rarely with the walls of the container. Thus the rate at which the molecule drifts around will be to all intents and purposes unaffected by the presence or geometry of the walls. But if the container becomes very small, of the same magnitude as the mean free path of the diffusing molecule, then a substantial fraction of the collisions will be with the walls, effectively shortening the average distance between collisions. As a result the ordinarily constant diffusion coefficient will appear to decrease. Thus, diffusive transport will get disproportionately slow as pipe diameter approaches the ordinary mean free path. In short, it would be a losing proposition for an insect to subdivide and proliferate its air pipes further than it does—it's already nudging up against the more restrictive physics imposed by frequent wall collisions. And the restriction has nothing whatsoever to do with the size of the insect, its respiratory rate, or much else besides mean free path.

And capillaries. For transport of liquids in pipes the problem of wall collisions has no practical significance. Mean free paths are so short and diffusion coefficients so low that (as far as I know) organisms simply never rely on diffusion to carry anything lengthwise in small liquid-filled pipes. Instead they use bulk flow, whose limitations are consequences of viscosity and that horrible *fourth* power of the pipe radius in the Hagen-Poiseuille equation (6.10). The constant size of capillaries may be in part a concomitant of the resulting cost of laminar flow in small pipes. Flow may be slowest in the capillaries of our circulatory systems—although the total cross-sectional area of our capillaries (about 0.2 m^2) may be greater than the cross section of the wider blood vessels (Chapter 2), a very large portion of the work done by the heart goes to overcome the pressure drop in the capillaries. Capillary size may also be related to cell size—circulating cells in a vascular system must obviously be able to fit through the tiniest pipes. On the other hand, for best exchange across walls (and to take advantage of the cells to get rid of the parabolic profile of speeds) the cells should just be able to squeeze through, as indeed our blood cells do through our capillaries.

And alveoli of lungs. A capillary is about 8 micrometers in diameter; the sites of gas exchange of the lungs, the alveoli, are each between about 150 and 300 micrometers in diameter (depending on the degree of inflation). In a capillary, oxygen is dissolved in liquid or chemically combined; in an alveolus oxygen is in the gaseous state. We saw earlier that the diffusion coefficients for oxygen were vastly different in air and water. One

167

can argue that capillary diameter "sets" alveolar size—an alveolus "ought" to be larger than a capillary by the square root (100) of the ratio (10,000) between the two diffusion coefficients. Getting much smaller than this amounts to shortening a time for diffusive transfer in alveoli that is already less than the time for transfer across the capillaries—"diminishing returns" is the common phrase. Taylor and Weibel (1981) suggested that a principle, which they term "symmorphosis," underlies the design of these complex systems. The idea is similar to that of the deacon's shay of Oliver Wendell Holmes—the best design is one in which no one part is any better than it need be, and that evolution will tend to generate such arrangements.

The smallness of cells in particular

More impressive in its ubiquity than anything about tracheae, capillaries, or alveoli is the constancy of cell size. Indeed, the very existence of cells represents a notable element of standardization in nature's design. It isn't immediately obvious why, as macroscopic life forms have evolved, the cellular level of organization has been retained—why, in other words, a major increase in size has inevitably involved proliferation and integration of cells rather than their enlargement. Exceptions to the multicellular arrangement of macroscopic organisms are so few and feeble that we rarely wonder about the peculiarity of multicellularity, of this federal union of semiautonomous units. But peculiar it truly is—a person has about 10^{14} cells, each with the full set of instructions, not only for its own activities but for those of all the cells of all the stages of his or her personal history.

A look at the small and constant size of cells might begin with a comparison of intracellular and intercellular communication—information transfer, in short. All protein is synthesized within cells according to the instructions encoded in the nucleic acid, DNA, even if the trigger for synthesis may come from elsewhere or if the protein is subsequently exported. Coordination of larger spatial and temporal matters is handled by nervous and endocrine systems. Despite the enormous complexity of the biggest brains, undoubtedly the greatest fraction by far of the information transferred within any organism is that involved in protein synthesis. Proteins, with their specific sequences of amino acids are just so complex, so diverse, and so continuously produced by each cell.

From this viewpoint, organisms operate as if they were very strange computers. The main information store is duplicated a vast number of times (more, in fact, than 10^{14} times, since many genes are present in multiple copies in each cell)—storage "core" must be very cheap, relatively speaking. But most of the information, transferred from nucleus to cytoplasm as messenger RNA, is never moved more than perhaps 5 micrometers. Somehow the system has evolved as if transmission lines are extremely costly, as if minimization of the distance the information involved in protein synthesis must be moved has been accorded a very high priority. But this isn't especially surprising if, as seems the case, the agency for this information transfer is the physical process of diffusion.

Perhaps the limitations of a diffusive transport system can be best illustrated with a simple (mathematical, although I hesitate to admit it) model. Consider some items that we'll call, for no good reason, "cells" and that vary only in size. Assume (1) that these cells are spherical, (2) that in each cell information is supplied by a central core itself having negligible volume, (3) that it takes a given rate of receipt of information to keep a unit volume of cell doing what needs to be done, and (4) that information moves radially outward following the rules governing diffusion. We divide each cell into very tiny elements of volume (thin concentric shells for convenience); for each element we figure the *square* of the distance to it from the center (shell radius) and multiply that by the volume of the element. We then add up these products—the result for each cell is a number proportional to the rate at which the core must supply information under the conditions specified—call it "demand." Demand might then be divided by the volume of the cell on the reasonable assumption that increasing size would also increase the productivity of the core—the resulting number is then "demand/supply."

Justification for setting the supply of information proportional to a cell's volume comes from some real observations. Where cells of a type (nucleated red blood cells of nonmammalian vertebrates, for instance) span some range of size, the DNA content turns out to be proportional to cell volume (Ycas et al. 1965).

Table 8.2 gives the outcome of the simulation for a doubling, a redoubling, and an order of magnitude increase in the diameter of a cell; the numbers have been divided by the values for a cell of unit diameter to convert them to relative factors of increase.

Doubling the diameter increases the volume eightfold; obviously at

TABLE 8.2. DIFFUSIVE TRANSPORT FROM THE CENTER OF A "CELL"

Diameter	Volume	Demand	Demand/Supply
1	1	1	1
2	8	32	4
4	64	1024	16
10	1000	100,000	100

least eight times as much information must be supplied. But distance from core increases simultaneously. Were times for transport directly proportional to distance (as in "ordinary" conveyances) then sixteen times the information rate would be needed. If transport time is proportional to the *square* of distance, then no less than thirty-two times the information sending rate is required to keep life creeping in its accustomed petty pace. Thirty-two is four times greater than eight—the larger cell's core must disgorge data four times more rapidly than the cell half its diameter, *even relative to its own octupled volume*. And that's just for a doubling, a trivial change in the size range of organisms. For a tenfold increase in diameter, the information supply must increase by 10^5, a hundred-thousandfold, or a hundredfold more drastically than the increase in volume. If, by the way, calculations involving spherical shells are not your pleasure, you can get very nearly the same numbers by using a pair of cubical cells—try cubes with sides two units and four units long and treat each cubic unit as an element.

The point of all these numbers is that cells, since they are internally integrated through diffusive transport, face a world in which any increase in size exacts a severe penalty in the times needed to carry out many of their activities. What seems to be the case is that the amount of information involved in protein synthesis is so vast that no organism has ever devised a way to disseminate it other than by simple diffusion. For the large but lesser rate of information transfer involved in all intercellular activities, specialized transmission conduits are employed—impulses in nerves, endocrine signals in bloodstreams. These conduits transmit at much higher speeds for macroscopic distances but have relatively limited capacity.

Parenthetically (at least to the theme of the present book) one can ask why cells are not smaller. For bacterial cells, Morowitz and Tourtellotte (1962) have argued persuasively that it takes a certain minimal amount of machinery to achieve the status of a reasonably autonomous living unit, one capable of growth and reproduction in a cell-free medium, and that bacterial cells are not much bigger than this minimum. The cells of animals are about an order of magnitude larger than bacterial cells but have some additional functional complications, so the same sorts of arguments, appropriately adjusted, ought to apply. Alternatively, one can apply the converse of Schrödinger's argument about why atoms are small—a cell has to be of a certain size before the number of entities of which it is composed becomes large enough to give adequate statistical precision to the operation of its parts.

In short, cells seem to be only a little bigger than some vaguely known but very potent lower limit; they cannot be smaller, and being larger exacts severe penalties in the efficiency with which they operate.

CONVECTION AND FLUID TRANSPORT SYSTEMS

It has been about a billion years since organisms were limited to cellular dimensions. Clearly these limitations of diffusion do not apply with quite the same inescapable tyranny above the cellular level of organization. The explanation isn't especially mysterious—most cells dwell in the domain of diffusion; most multicellular organisms escape into convective country in which a different set of laws is in force. The correlations aren't quite as rigid as all that—no "law" says that convection can't happen within cells, and so it does in large cells. Plant cells commonly use internal bulk flow in a process called "cyclosis," and long nerve cells have a specific system for bulk transport of material down their axons. And there's no rule requiring convection as a precondition of multicellularity. Very flat or threadlike organisms, essentially those whose cellular proliferation has occurred in one or two but not three dimensions, can be quite large without internal bulk transport—the algae, especially, include a wealth of such cases. Incidentally, functionally flat organisms may not be overtly planar—large anemones and jellyfish are bulked up with nonliving "mesoglea," and most of their cells are not more than one or two cell diameters from the environment.

In rationalizing the small size of cells, we have made the basic argument for the existence of all fluid transport systems in organisms. Whether they transport air, water, blood, or any other fluid or slurry, whether they are unidirectional or recirculating, they boil down to devices to circumvent the limitations of diffusion. A final diffusive step always remains at the cellular sites of metabolic processes, but an amazing diversity of pipes and pumps have been contrived to move fluid and suspended solid material between these sites and the environment. A few years ago Michael LaBarbera and I (LaBarbera and Vogel 1982) attempted to write down a few general rules underlying the design of these fluid transport systems—the material that follows draws heavily on that paper.

Diffusive exchange inward from the periphery shares the same limitation as export from a core; perhaps the situation ought not be quite so severely size-limited, inasmuch as net inward movement encounters a somewhat less extreme geometrical problem—times should vary with distance only to the first rather than the second power. Thus relative demands in Table 8.2 should be "only" 1, 16, 256, and 10,000. In fact, the

171

situation is precisely as bad—supply is now a function of an area (outer surface) instead of volume, so demand should be divided by the square, not the cube of the diameter. Thus the final column, demand/supply, is left unchanged. In effect, the demand-to-supply ratio is proportional to the square of the diameter—with increase in size, the demand still badly outstrips the supply.

What is the simplest convective scheme imaginable? Say the internal material of the system is merely stirred (Figure 8.3)—approximately what happens in cyclosis in a plant cell. The factor for distance from periphery to center drops out—once material gets inside it's effectively disseminated. The demand-to-supply ratio is now proportional to the first, not the second, power of diameter, so the final column in Table 8.2 should now have figures of 1, 2, 4, and 10. The difference can be easily demonstrated by looking at the scaling factors for heat loss as a function of size in two series of beakers, the first filled with water and a little agar to prevent convection, the second filled with water only and free to generate its own thermal convection (for procedural details, see Vogel and Wainwright, 1969, pp. 18–19).

The solution isn't complete, but there's a world of difference between exponents 2 and 1. And that reduction over diffusive migration is what internal fluid transport systems accomplish, usually with quite a lot more machinery than the scheme in Figure 8.3. We can, in fact, use the phenomenon as the basis of a functional definition of an internal fluid transport system: *any system in which internal convection reduces the effective distance between two parts of an organism or between the organism and the external environment.* And we can state our first rule for the design of organisms: *(1) While diffusion is always used for short-distance transport, it is augmented by bulk flow for any long-distance transport.* While the compensation isn't complete, neither is the story. Its other pieces made their appearances in Chapter 2—large organisms proliferate surface area disproportionately, getting less spherical with size, and live more leisurely lives, with metabolic rates increasing less rapidly than volume.

What else can we say in general about these fluid transport systems? Recall the Hagen-Poiseuille equation (6.10). It's disproportionately expensive to move fluid through narrow pipes—for a given flow (volume per unit time), the pressure drop per unit length is inversely proportional to pipe radius to the *fourth* power. But narrow pipes are better for exchange across their walls—they have more wall area relative to their volume. So these characteristics conflict; the practical result is an evasion rather than a compromise. What happens can be stated as another general rule: *(2) Transport systems use both wide and narrow pipes, narrow ones at sites of exchange and wide ones for moving fluid from one exchange site to another.*

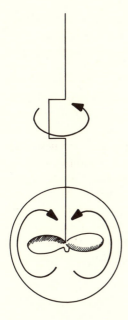

FIGURE 8.3. The minimalist's circulatory system—
turning the crank turns a propeller, causing convective transport within a "cell."

One example has already been given, contrasting the diameters of aortae
(about 26 millimeters) and capillaries (about 8 micrometers)—a 3200-fold
difference. Another is the comparison of the main tracheal trunk with
the smallest bronchi in our lungs—18 millimeters versus 0.45 millimeters,
or a 40-fold difference. Among other consequences, this wide range
makes it very difficult to give accurate pictorial representations of biological fluid transport systems!

But these constrictions at and near exchange sites raise another problem as they solve the previous one. If a pipe is narrowed, the speed of
flow through it increases in inverse proportion to its cross-sectional area,
by the now familiar principle of continuity (equation 7.2). This turns out
to be as unfortunate as it could be. In the narrow pipes, slow speed is
desirable both to keep down the cost of transport and to provide some
reasonable time for exchange processes. In the wide pipes, by contrast,
high speeds would not be uneconomical and would have the virtue of
minimizing both transit time and the mass of fluid temporarily tied up,
passive and unused, in transit. The universal solution to these conflicting
requirements is obvious when we recognize that the principle of continuity needn't apply to the cross section of each individual pipe but to the
aggregate cross section of pipes running parallel to each other. If flow
goes from a wide pipe through a branching manifold into a set of narrow
pipes, and the total cross-sectional area of the narrow pipes is actually

173

larger than that of the original wide one, then the speed of flow will decrease rather than increase (Figure 8.4). We have, in the resolution of the paradox, another general rule: (3) *Total cross-sectional area of the narrower pipes always exceeds that of the wider pipes so the speed of flow in the narrower pipes is less than that in the wider ones.*

Again, a few examples. That of our circulatory system was mentioned in Chapter 2—0.21 m² of capillary cross section versus 0 .000530 m² of aorta, with a concomitant 400-fold decrease in velocity of flow. In our lungs, the comparison of trachea (0.00025 m²) and final bronchi (0.17 m²) gives a 700-fold speed decrease. Only one system of pipes is known in which aggregate cross section remains unchanged as the pipes branch—the air-carrying tracheal system of some insect larvae investigated years ago by Krogh (1941). But he, one of the greatest physiologists, stressed that these pipes had no bulk flow within them but were entirely diffusive in operation. The exception is nice to have since it shows that the increase in area is not just a geometrical consequence of practical branching systems.

At this point we run out of universal features. There is, though, at least one other rule underlying the design of these fluid transport systems. Any system of pipes requires a pump of some sort to offset the viscous friction of flowing fluid. How might an appropriate pump be arranged? Here the physics, mainly the principle of continuity, is almost entirely permissive; and here, not surprisingly, the biological diversity bursts forth. Amidst all the detail, two general kinds of pumps are recognizable (Figure 8.5). One, which we might call a "macropump," is exemplified by the familiar muscle-and-valve pulsing device, four of which make up a

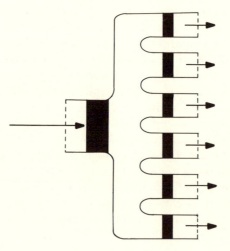

FIGURE 8.4. In a branching array of pipes the speed of flow, here proportional to the length of the arrows, is inversely proportional to the *total* cross-sectional area at any stage of the array. Thus, individual narrow pipes may have slower flow than individual wide ones.

(a)   (b)

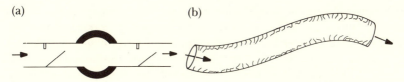

FIGURE 8.5. Two schemes for pumping fluid through a pipe. (a) Contraction of a muscular chamber flanked by one-way valves gives unidirectional, if pulsatile, flow. (b) The coordinated beating of cilia lining the walls of the pipe pushes fluid along.

mammalian heart. Just squeeze the bag repeatedly and it produces a pulsating but unidirectional flow. The pump is typically located in the largest pipes and generates the fastest flow in the system. Contrasting is a "micropump," usually the coordinated beating of a coating of cilia or flagella on some flat surface or lining some pipe or chamber. Examples are the ciliated gills of clams and the flagellated chambers of sponges. (We have cilia lining some of our respiratory passages, but rather than pump air they merely push around surface mucus.) These micropumps are capable of producing only low speeds, but the system as a whole need not be so limited—a collecting manifold can act as a nozzle and generate quite considerable velocities. In the case of sponges, the increase, to $0.2 \ \mathrm{m \cdot s^{-1}}$, may be as much as 10,000-fold.

Both types of pumps occur; which is preferable? The choice may simply be a matter of economics. Cilia and flagella operate at very low Reynolds numbers and consequently are limited to low efficiencies. Recovery strokes have an appreciable fraction of the drag of power strokes, and velocity gradients have to be severe for them to protrude into the free stream—points made earlier. Thus, where sizes overlap, creatures that swim with muscle-driven appendages can go much faster than those propelled by cilia. The question then becomes, why ever use cilia or flagella? One possible answer can be stated as another general rule, but one more of guidance than decree: (4) *High-speed pumps in the large pipes are preferable to low-speed pumps in the small pipes; the latter are used mainly when they confer some other functional advantage.*

So we find no higher animals with cilia lining their capillaries that eliminate the need for hearts. Clams and sponges, of course, use their pumps also as filters—another function is involved. And, it must be admitted, there are minor ciliary pumps here and there among muscle-equipped creatures. The major apparent exception to the rule is the pump used to raise water up a tree. As explained earlier, it is a solar-powered evaporative scheme, with water coming up to take the place of water that has evaporated across the interfaces in the cell walls of the leaves. Located at

175

the point of lowest speed and greatest area, this mechanism is clearly a micropump. But the argument against micropumps was economic, and, with direct solar power as an energy source, that argument doesn't work.

The discussion in this chapter built upon the nature and distance limitations of diffusion to rationalize both the cellular arrangement of macroorganisms and the existence of systems within them using forced convection. It then went to the physical imperatives underpinning forced convection and, in the process, provided an application for some material developed earlier. More importantly, it constitutes a cogent case for the pervasive operation of the constraints imposed by the physical world in which organisms find themselves, together with the wide exploitation of the opportunities provided by that world—the theme of the book as presented with more bravado than backing back at the beginning.

A matter of materials

"If it is love that makes the world go round, then it
is surely mucus and slime which facilitate its
translational motion."

Roger H. Pain

SOLID materials are at once more familiar intuitively and more diverse
and complex than either gases or liquids. To remind ourselves: gases
resist only compression, liquids resist tension as well as compression, but
solids resist three different kinds of stress—compression, tension, and
shear. Again, we recall that gases and liquids, collectively fluids, resist *rate
of* shear, while solids resist shear stress per se.

The next few chapters will focus on solids. We'll include a great diversity of solid materials, ranging from the very soft (but still elastic) mucuses to the very hard (but still deformable) dental enamels and corals.
The degree of "solidity," in the vernacular sense, is oddly enough definitionally irrelevant—an ordinary word has again been bent to an unfamiliar task.

THE SORTS OF SOLIDS

Aside from their underlying physical differences, fluids and solids play
different roles in natural design. While nature may capitalize ingeniously
on the diverse properties of the air and water she encounters, solids are
with few exceptions deliberately synthesized with properties tailored species to species and even tissue to tissue. As a consequence, the diversity
of solid materials in biological use vastly exceeds that of fluids and we
have to face the great bugbear associated with any kind of diversity, that
of classification. For solid materials it isn't even clear upon what we
should base a set of rational and useful categories.

The most straightforward classification is a hierarchical one based on
chemical composition. Thus, we can recognize some general categories,
such as polysaccharides and proteins, as well as subsidiary and somewhat
less heterogeneous categories, such as the particular proteins silk and collagen. And then silk from silkmoths and silk from spiders have major
differences. Moreover chemical composition differs among the silks of
different silkmoths and even among the silks used in different parts of

even a single spider's orb web. So the hierarchy has quite a few levels and gets relatively complicated even if the scheme is fundamentally logical.

A classification of materials based on chemical composition turns out to correspond only sporadically to their mechanical properties. We might instead divide solids by their mechanical behavior and biological roles. The commonest version of such a scheme, and the one we'll use here, considers three categories—*tensile*, *pliant*, and *rigid* (I think the arrangement was first introduced in a most useful book by Wainwright et al., 1976). Its drawbacks will become apparent as we proceed. A few other ways of dividing materials might be borne in mind as well. One useful distinction is between *simple materials*—homogeneous accumulations of only one chemical species (with perhaps some water of hydration)—and *composite materials*—nonhomogeneous combinations in which the mechanical properties reflect the interactions of two or more simple materials. Another distinction is between *isotropic* materials, in which mechanical properties are quite indifferent to the direction in which the material is aggressed upon, and *anisotropic* materials, in which direction makes a distinct difference.

On such matters I cannot go further without announcing my enthusiasm for two books by J. E. Gordon—*The New Science of Strong Materials* (1976) and *Structures* (1978). Not only are they germane to the next few chapters, but I know of no better writing for nonspecialists about any area of contemporary science.

Tensile materials

These function by resisting being pulled upon—that is, *tensile* stress—they're basically biological ropes. At least four biochemically distinct kinds are common among organisms.

(1) The protein *silk* is made in profusion by a variety of arthropods, of which spiders and silkmoths are simply the best known silk spinners; it serves a variety of functions. In fairly pure form it resists tension in the webs of spiders; combined with a glue into a composite material it makes a tough cocoon for moths. It's used in a fine meshwork in the catch-nets of caddisfly larvae, mentioned earlier, as the world's smallest silk stockings. We use the silkmoth version by unwinding the thread of the cocoons of specially bred strains of *Bombyx mori* or by chopping the cocoons and dissolving the glue of some species of *Antheraea* ("wild" or "raw" silk).

(2) Another protein, *collagen*, occurs in fairly pure form in our tendons, the ropes connecting muscles to bones. In combination with other materials, collagen forms a major component of composite materials with a wide range of properties and functions—skin, bone, muscle, and arte-

rial wall are just a few. It's present in almost all phyla of animals but is apparently absent in plants. In general, the price of edible muscle—red meat particularly—is inversely related to its content of collagen. It is, in fact, the tough stuff in several senses of the word "tough," but as any competent cook knows, long treatment with heat or mild acid has a decently destructive action. The ballistae of the ancient Greeks and Romans used collagen in the form of cow tendons to store energy (about which more later), but its excellent mechanical properties were offset by its cost and susceptibility to rot and decay in a warm climate. Of course the military mind has never been notably fastidious about matters of economics.

(3) A polymeric sugar (or polysaccharide), *cellulose*, is the basic fibrous, tensile material of plants. The plants don't seem to use it in any unadulterated state to withstand tensile forces—again it performs its mechanical function mainly as part (often the main part) of composites, usually as the thickened walls of individual cells. But cellulose does occur in fairly pure form in external, drag-increasing fibers of wind-dispersed seeds such as those of cotton. We make extensive use of its ability to withstand tension, collecting it from various sources, spinning it into ropes, and often weaving the latter into fabrics—cotton, linen, sisal, etc. We even extract it from wood, dissolve it, and reconstitute the fibers by forcing the material through tiny holes to make the semisynthetic, rayon. Cellulose is very rare in animals despite the wide occurrence of the underlying monomer, glucose; it does occur as a structural material in one group— the ascidians, or tunicates ("sea squirts").

All organisms use proteins as chemical catalysts—enzymes—and use polysaccharides for energy storage—starch and glycogen are the commonest examples. But there seems to be a major difference between plants and animals in the choice of structural materials—plants have a predilection for polysaccharide and animals for protein.

(4) The nearest thing to an exception to the above generalization is a modified polymeric sugar, *chitin*. After collagen, it's the most widely used tensile stuff in animals; as a pure tension-resisting material it connects muscles to exoskeletons in insects, crustaceans, spiders, etc. in the form of "apodemes," their equivalent of tendons. (There's a piece of decent size in the claw of the Maine lobster, *Homarus*, if you want an excuse to look into one; it's thermally stable, so the animal can be killed by immersion for a few minutes in boiling water.) Chitin occurs here and there in the animal kingdom and occasionally among plants. Its main use and most common appearance is as a major part of the composite exoskeletons of insects and other arthropods. There doesn't seem to be any substantial technological application of this fine material.

Pliant materials

All materials deform under stress to some extent—no material is perfectly rigid. In materials specifically designated as pliant, though, deformation is crucial to functioning. Where they differ among themselves is in the degree of their squidginess and the extent to which and speed with which they recover their original forms upon removal of a stress.

(1) So-called natural rubber really isn't—the native substance of rubber-yielding plants is a liquid sap. Three (at present count) functional, mechanical rubbers do exist in nature; all are animal proteins. By a rubber we mean a material that can be rather drastically distorted in one or more directions yet remain capable of rapidly and forcefully returning to its undeformed shape.

Perhaps the best of these *protein rubbers*, the one that returns the largest fraction of the imposed work following removal of the stress, is *resilin*. It's located in little pads or tendons at the wing hinges of insects where it has the very obvious function of decelerating a wing at the end of, say, the downstroke and then using the work thus done on it to accelerate the wing at the start of the succeeding upstroke.

Another is *abductin*, found in the shell-opening ligaments of bivalve mollusks. The two valves ("shells") of a bivalve are closed by one or two adductor muscles (the edible part of a scallop is one of these); in closing, the muscles compress the resilient ligament which expands and reopens the shell if the muscles relax. Interestingly, the fraction of the work recovered is greater in scallops, which can swim by repeatedly clapping their valves together, than in more sedentary species.

Our own rubber, *elastin*, occurs mainly as a component of two composites, skin and arterial wall. The nearest thing to pure elastin is the nuchal ligament of large grazing mammals. It runs from a ridge on the rear of the skull back along the top of the neck to the thoracic vertebrae, and seems to act as a shock absorber and support for the head. Dissected from a lamb neck (the local butcher can get you one of these), it is distinctly rubbery and responds quite differently to stretching than the collagenous tendons from, say, a lamb shank.

One might expect some similarity in composition among these three proteins, but in amino acid content they are about as different from each other as proteins can be. At a higher level of organization, though, strong similarities appear—pulling on them amounts to pulling on molecular entanglements rather than on chemical bonds, which is why they are stretchier than the tensile materials.

(2) The other pliant materials are pliant only as complete composites and not as single chemical compounds; these *pliant composites* are a varia-

ble bunch both chemically and physically. At one extreme they grade off into true fluids, retaining only a little resistance to shear—the various mucuses that line pipes, coat surfaces, and even provide traction for snails and slugs, as well as odd materials such as the synovial fluid that lubricates our joints and the Wharton's jelly of the umbilical cord. Then there's the jelly of jellyfish and sea anemones ("mesoglea"), made up of sparse collagen fibers in a complex gel matrix. The cartilage of our ears, noses, and intervertebral discs is similar but stiffer—more collagen fibers and less gel. Skin and arterial wall are more complex still—pliancy is crucial, and its decrease with age helps neither appearance nor function.

Rigid materials

Rigid materials resist stresses without undergoing much deformation. None is a simple chemical compound—all are composite materials. There's another general difference between pliant and rigid materials besides the obvious one of stiffness. Most of the pliant materials are to a substantial extent "viscoelastic," which is to say that their behavior is somewhat time-dependent in the manner of a fluid. A prolonged load or a load beyond some transition value results in something akin to flow—a permanent or semipermanent deformation. The rigid materials rarely show much trace of viscosity—whatever their other complexities, they are lawfully solid in a more important way than mere hardness.

What we're calling rigid materials includes many very familiar biological items. There's *arthropod cuticle*—a composite of chitin fibers in some proteinaceous material, with the addition of calcium carbonate salt in the larger crustaceans. In many instances the fibers are arranged in sheets, each with a specific orientation, rather like plywood. There's *bone*, a composite of collagen fibers (about 50% of the volume), some other proteinaceous material, and calcium phosphate salt. In most mammals, the collagen is in neat layers (lamellae)—again the material is anisotropic, its mechanical response varying with the orientation of a stress. There's *keratin*, a composite of microfibrils in a matrix, both proteins; it's what hair, horn, and bird feathers are made of. We make use of it as tensile material in fabrics such as wool and mohair, but it's closer to a rigid material in nature. Horn is, in fact, excellent in resistance to compression, as the old Asiatic bowmakers knew.

And we cannot forget *wood*, the cell walls of plants—cellulose microfibrils specifically oriented in an unstructured matrix mainly of a substance called "lignin." The stuff is certainly the main biologically derived structural material in human technology; but, one hastens to add, the properties of fresh, tree-supporting wood are substantially different from those of the dried and sliced products of sawmills. Every woodcarver,

181

woodcutter, and woodsplitter is intimate with the anisotropy of the material as well as its variation from species to species.

There are yet other rigid materials, which Wainwright et al. (1976) refer to as "stony materials" and Vincent (1982a) calls "biological ceramics." These are distinguished by being very heavily mineralized, with more mineral (some inorganic salt) than organic matter. Some are tiny, such as the spicules of even very large sponges—mainly either calcium carbonate or silica (roughly, glass). Others occur in massive pieces, such as the skeletons of the stony corals and the shells of most mollusks, both made of calcium carbonate with a small amount of organic matter. Stony materials occur here and there in the vertebrates—the eggshells of birds (mostly calcium carbonate) and the dentine and enamel (mainly calcium phosphate) of our teeth are a few.

STILL another possible division of solids is tripartite—materials, structures, and structural systems. Steel is a material, a beam is a structure, and a bridge is a system. An imperfect distinction, perhaps, but useful. In nature, matters are messier since no functional benefit presumably rewards an organism for paying attention to our tidy groupings. Bone is made of several materials in an orderly array; but is bone therefore a structure or merely a composite material; or do we call bone generically a material but a particular bone, with shape, a structure? What matters, really, is realization of the large elements of awkwardness, arbitrariness, and verbal convention. Having admitted the implicit sloppiness, we've nonetheless begun with materials; with the aid of a set of properties that describe the mechanical behavior of the materials, we'll build upward toward structures and systems.

THE PROPERTIES OF SOLIDS

For fluids we needed only a few properties—density, viscosity, surface tension, compressibility. Not only do solids resist stresses such as torsion and shear to which fluids are indifferent (Chapter 5), but they resist even a single stress in very much more complex ways. Even worse, they fail in a variety of ways—they may crack, crush, or otherwise collapse. So we need a complex of quantities to describe the behavior of a solid—several elastic moduli, one for each sort of loading, as well as strength, extensibility, toughness, resilience, and so forth.

Perhaps the place to start is with the most straightforward abuse of a material—pulling on a sample at opposite ends to apply a tensile stress. The Procrustean testing apparatus is called a "tensometer" and may be either handmade or commercial—materials testing is standard practice among mechanical engineers. One slowly and steadily stretches a material while simultaneously measuring both the stretch (strain) and the stress with which the material resists extension. The procedure sounds just as routine as weighing out a pound of coffee, but it's all too commonly one of the most frustrating jobs in the whole masochistic world of experimental science. To test a metal or ceramic, the engineer fabricates a sample of a deliberately convenient shape—large ends for attachment, gentle taper to the thin part that will take the most stress, and so forth. The biomechanic is determined to pull on a specimen of some anisotropic material shaped by its own devious ends, which furthermore is typically wet and slimy, awkwardly small, rapidly deteriorating, and hard to separate from the rest of the organism. There's an extensive and profane oral tradition concerned just with the design and use of grips for biomaterials. Finally, the biological data are only roughly reproducible—results vary rather a lot between samples and usually depend more than we'd like on the details of the testing procedures.

The outcome of a successful set of tensile tests is a so-called stress-strain curve (Figure 9.1). Neither variable, though, comes directly from the tensometer. To get stress (with dimensions of force over area), one divides force by the cross-sectional area of the sample. To get strain, or fractional extension (dimensionless as noted back in Chapter 4), one di-

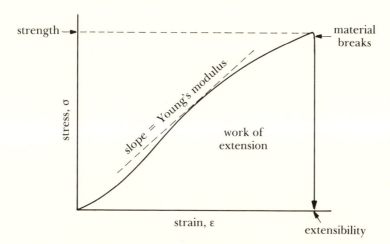

FIGURE 9.1. A stress-strain curve and the material properties it yields. The slope is the stiffness, or Young's modulus; the maximum height is the strength; the maximum horizontal distance is the extensibility; and the area under the curve is the work of extension, or strain energy storage.

vides extension by original length.[1] As we'll see, quite a few bioportentous properties can be derived from such a stress-strain plot.

(1) The relation of stress to strain is the first of the material properties of concern here. It is, simply, the change in stress divided by the corresponding change in strain, the slope of the graph, which indicates how much the material deforms for a given level of abuse. It is called, variously, the "stiffness," the "elastic modulus," or, best, *Young's modulus of elasticity* (Table 9.1). Note carefully that the steeper the slope, the stiffer the material. So the term "elasticity" is perhaps best avoided—its everyday use is precisely the opposite of what is meant here, which is taking the matter of redefinition a bit far. Note also that Young's modulus depends on how much the material has been stretched, except for materials for which the graph forms a straight line. Such materials are termed "Hookean" (after Robert Hooke, remembered also for irritating Isaac Newton); the best examples are metals (including springs). Ordinarily, biological materials should be presumed to be non-Hookean—thus, cited values presume some "ordinary" degree of strain, all too often left unspecified. Incidentally, since strain is dimensionless, Young's modulus must have the same dimensions as stress—force divided by area.

Several points deserve comment. The value for elastin is typical of rubbers (resilin, abductin, and butyl rubber are only two to four times stiffer); notice the two-thousandfold difference separating it from collagen.

TABLE 9.1. EXAMPLES OF YOUNG'S MODULUS OF ELASTICITY

Sea anemone mesoglea	0.001 MPa
Elastin (nuchal ligament)	1.0
Collagen (tendon)	2000
Silk	4000
Pine timber (with grain)	10,000
Chitin (locust apodeme)	12,000
Teeth (dentine)	15,000
Bone	17,000
Mollusk shell (nacre)	30,000
Teeth (enamel)	78,000
Steel	200,000

(Data from Waters 1980; Alexander 1983; Dimery et al. 1985)

---

[1] For extensions greater than about 0.1 or 10% there is a slightly more involved way to get so-called true stress, described in Wainwright et al. (1976) or Alexander (1983).

That's the quantitative view of the comparison of the nuchal ligament with a tendon from a lamb. Notice also that bones, shell, chitin, and wood stand up to tension very well—their values are only about an order of magnitude below that of steel.

(2) If the specimen in the tensometer is stretched to the point of breakage, two further numbers are obtained. One is how much stress a material can take before it fractures, the maximum height of the graph; it's called the *tensile strength* (the second number will come along shortly). Like stress and Young's modulus it has dimensions of force over area. We now have three items with the same dimensions, each a measure of something quite distinct. For each, the SI unit is the pascal, or newton per square meter, a very small unit—roughly, says a colleague, the stress of an apple atop a garbage can lid—so one often encounters megapascals (MPa), or meganewtons per square meter. Data for the strengths of a few materials are given in Table 9.2; it is most important that you maintain a decent mental distance between stiffness and strength—some small bit of ribaldry can be contrived if need be.

The low value for arterial wall is interesting—presumably the occurrence of ruptured aneurysms has not been a major selective agent in mammalian evolution! Notice, also, the low value for ceramic building materials. They're an order of magnitude better in pure compression (than in tension), which is how we use them (and why no one bothers to lay bricks with a decent glue). The low value (plus a thick callus) makes it possible to break a brick with a chop of the hand—you just have to

TABLE 9.2. EXAMPLES OF TENSILE STRENGTHS
OF MATERIALS

Arterial wall	2.0 MPa
Pine wood, across grain (dry)	4.0
Brick, concrete	5.0
Teeth (enamel)	35
Teeth (dentine)	50
Pine wood, with grain (dry)	100
Collagen (tendon)	100
Bone	200
Keratin (human hair)	200
Steel ("mild")	400
Nylon thread	1000
Silk (spider)	2000

(Data from Gordon 1978; Currey 1984; Craig 1987)

arrange a little unsupported area under the center of the brick so the bottom is put in tension. I once did this in a large hydraulic press; I screwed up my courage, shielded my face, and felt quite silly as brick after brick broke with only a tiny click at a very low load. Anything with a strength at or above 100 MPa has to be considered a good tensile material—wood with the grain and collagen have about this value. Nylon is outstanding; you may have noticed that, compared to other cordage, nylon ropes seem to have very generous load ratings relative to their diameters. And spider silk is superb—one can only wonder why, if one kind of creature can make a protein this good, the others, with the same synthetic machinery, don't do as well.

A slight permutation of a tensometer can be used to measure compressive stiffness and strength. The test is not as generally applicable as a tensile test for reasons that will be apparent when we take up beams and columns in the next chapter—but, if a short enough specimen is used, a measure of material properties related to crushing (in contrast to bending and buckling) can be obtained. Wood, tested axially (along the grain), has a compressive strength of about 30 MPa, bone (lengthwise) about 200, mollusk shell and tooth dentine about 300, and tooth enamel about 400.

(3) The other number that emerges directly from a graph of stretching to breakage is the breaking strain or *extensibility*; it's the maximum horizontal extent of the plotted line. Be reminded that extensibility is not elasticity. Extensibility is maximum strain, so, like strain, it's dimensionless—distance extended divided by original length. This quantity (Table 9.3 gives some values) is less often quoted than are Young's modulus and strength for reasons related, I'd guess, to our peculiar technological predilections; as we'll see, nature seems to care more about it.

In practice, extensibility varies more-or-less inversely with the stiffness

TABLE 9.3. EXAMPLES OF EXTENSIBILITIES OF MATERIALS

Mollusk shell (nacre)	0.006
Bone (yield or giving strain)	0.006
Bone (ultimate, breaking strain)	0.02
Collagen (rat tail tendon)	0.1
Steels	0.1–0.35
Keratin (wool)	0.5
Elastin (nuchal ligament)	1.2
Spider silk	0.3–10
Pregnant locust intersegmental membrane	15

(Data from Eshbach 1952; Vincent 1982a; Currey 1984)

of a material and so at least follows the commonplace notion of elasticity. As one might expect, nuchal ligament stretches further than tendon. The current record holder, the intersegmental membrane of the pregnant female locust, achieved its renown in an investigation by Julian Vincent (1975); it's made of about 12% protein, 12% chitin, and the rest water and undergoes something called "stress-softening." Mother locust, a creature of dry habitats, stretches these membranes between her abdominal segments to get her eggs about 8 cm underground—deep enough so the desiccated eggs have a reliable source of water. We presume that the eggs are kept fairly dry before being expelled so the locust can hold a large number and still fly.

(4) It takes a certain force or stress to stretch something. It also takes a certain amount of work, but the two are quite separate notions, and materials that can take the most stress (the strongest) may not take the most work. Indeed, many very strong materials are fragile in a hard-to-pin-down manner; their practical strength seems different from that suggested by the data produced by the tensometer. As it happens, the everyday notion of strength corresponds more nearly to the work of extension needed to deform an object—force times distance—than to force or force per unit area.

Work, though, isn't an adequately general measure since it depends on the amount of material being tested. What we'd really prefer is a material property, which can be simply obtained by dividing the work of extension by the volume of the sample of material. In this nice, orderly world, it turns out that work per volume (joules per cubic meter in SI units) is dimensionally equivalent to force per area (again!). Force per area (or work per volume) is what one gets by multiplying together the dimensions of the axes of the plot of stress vs. strain. Figure 9.2 contrasts the very different stress-strain plots of two materials of the same strength. The one with the higher curve absorbs more work; most often it is said

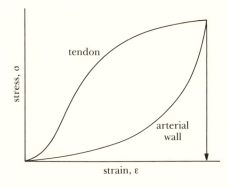

FIGURE 9.2. Two materials of the same strength and extensibility but with very different values of work of extension. While the values of strength and extensibility have been set equal to emphasize the contrast, the shapes of the curves realistically represent tendon and arterial wall.

CHAPTER 9

to have the greater *work of extension* or *strain energy storage*. As noted back in Chapter 4, the value of the property is equal to an area under the stress-strain curve, so this, too, comes off the same graph as the previous properties.

Table 9.4 gives this "strain energy storage," or "work of extension," (or in some accounts "toughness" or "resilience," names used here for other variables) for a variety of materials under, unfortunately, not exactly comparable test conditions. The main uncertainty is, of course, just how far to extend a material in a test—to the elastic limit (the maximum distance from which it will snap back), to some safe working limit before fracture, or to the breaking point itself.

The inclusion of work normalized to mass as well as volume in the table is an attempt to find a more reasonable basis for comparing natural and technological materials. Thus, collagen is only a little better than steel on a volumetric basis but far surpasses it relative to mass. Natural rubber is a peculiar case—it's very extensible but not at all strong; the area under its stress-strain curve, though, gives some idea why it is useful for powering model airplanes! Spider silk is out of sight, about which more later.

(5) And a final property from these stress-strain graphs. Say one stretches a material and then allows it to recover elastically its original length while measuring its stress throughout the test. A real material will not recover along quite the same stress-strain curve that described the stretch; instead it will trace a curve somewhat below that along which it ascended (Figure 9.3). Therefore, the area under the curve representing the stretch will be greater than that representing recovery. What this means is that not all the work put into stretch comes out again during relaxation in mechanical form.

TABLE 9.4. STRAIN ENERGY STORAGE, OR WORK OF EXTENSION, AT PARTICULAR STRAIN LEVELS

	Work/Vol. (MJ·m³)	Work/Mass (J·kg⁻¹)	Strain (%)	Stress (MPa)
Yew wood	0.5	900	0.9	120
Spring steel	1.0	130	0.3	700
Keratin (horn)	1.8	1500	4.0	90
Collagen (tendon)	2.8	2500	8.0	70
Bone	3.0	1500	2.0	150
Rubber ("natural")	10.0	8000	300.	7
Spider silk	200	160,000	30	1400

(Data from Gordon 1978; Currey 1984; Wainwright et al. 1976)

188

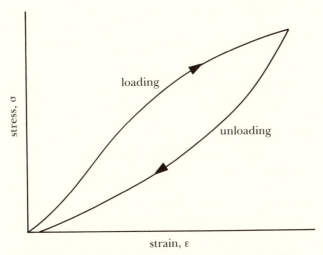

FIGURE 9.3. A pair of stress-strain curves for a material
that is first loaded and then unloaded; the noncoincidence
of the two curves tells us that the material isn't especially
resilient. Note also that this particular material doesn't
recover completely from the load but is permanently
deformed—as is common, it's a little bit "viscoelastic."

We'll call the fraction of the work that comes out relative to that which
goes in the *resilience*; it's the ratio of the area under the recovery curve
to the area under the loading curve, given as a percentage, which must
be below one hundred.[2] ("Resilience" has also been used simply for the
maximum possible strain energy storage before breaking, equivalent to
work to fracture—Gordon, 1976, uses this alternative definition. To com-
plicate things further, the phenomenon to which we are applying the
present measure is also described by variables called, variously, elastic
efficiency, hysteresis, energy loss per cycle, rebound resilience, and coefficient of restitution.) A few values are given in Table 9.5; again they're
just for rough comparisons, since all vary with state of hydration, temperature, and so forth.

The datum for collagen refers to operation at much lower strains than
those for the other materials; it's a much stiffer material, but is very resilient under appropriate conditions. As we saw, spider silk absorbs a lot
of energy, but the energy doesn't come out again in the form of mechanical work, appearing instead as heat. Resilin is as good a rubber by this
criterion as the best synthetic stuff; Vincent (1982a) commented that the

[2] Except for active muscle, of course, which is doing rather than absorbing work.

TABLE 9.5. THE RESILIENCE OF SOME BIOLOGICAL
MATERIALS (%)

Spider silk	35
Elastin (nuchal ligament)	76
Abductin (sessile bivalves)	80
Abductin (scallops)	91
Collagen (tendon)	93
Resilin	97

(Data from Wainwright et al. 1976; Denny 1980; Vincent 1982a)

efficiency may be important, among other things, to avoid local heating in the intense activity of flight. This business of heating will reappear in a decent context in Chapter 15; for now just feel the tires of a car after rapid driving even in cold weather—the heat from noncoincidence of loading and unloading curves will be immediately apparent.

(6) One particularly important property doesn't come directly from a stress-strain curve. It takes a certain stress to break something, what we call its strength. Similarly, it takes a certain amount of work per unit cross-sectional area to break the item—often called the *work of fracture* or, as I'll do here, *toughness*. (Again usage is inconsistent; toughness is often equated with strain energy storage or work to fracture. In practice the two meanings can be distinguished by their dimensions—work per volume for work to fracture or strain energy storage; work per area for toughness.) The idea that it takes work to make new surface appeared earlier. Surface tension (Chapter 5) has dimensions of force over distance; that's equivalent to work over area—it takes work to make new surface in a liquid whose bits prefer a cozy communality, and solids just behave in an analogous manner.

Table 9.6 gives data for toughness. You may note some similarities in the relative rankings of materials for strain energy storage and for toughness—the terminological confusion is mildly ameliorated by a certain practical congruence. But the differences between strength and toughness deserve attention. A tough material is one across which cracks do not easily propagate so, where the two properties diverge, it is the toughness to which we ought to pay most attention. As a general rule, the more jagged the break, the more work it took to make it, and the tougher is (or was) the material. A brick not only breaks at a low tensile stress but requires little work to make the break when it occurs. Glass has a high tensile strength, but a crack, once started, similarly propagates with very little work. Mild structural steel has far less tensile strength than so-called

TABLE 9.6. TOUGHNESS, OR WORK OF FRACTURE

Cement, brick, stone	3–40 J·m^{-2}
Wood, with grain	150
Mollusk shell, parallel to surface	150
Tooth enamel	200
Tooth dentine	550
Pig aorta	1000
Sea anemone mesoglea	1200
Insect cuticle	1500
Mollusk shell, crosswise	1650
Bone (cow leg)	1700
Steel, high tensile	10,000
Wood, across grain	12,000
Rabbit skin	20,000
Mild steels	100,000–1,000,000
Fiberglass	1,000,000

(Data from Gordon 1978; Vincent 1982a; Alexander 1983)

high tension steel, but the former may be safer inasmuch as it is the tougher and thus less apt to fail catastrophically. Toughness is the inverse of our ordinary notion of brittleness—cloth is fairly tough; glass is quite brittle.

So we apply a stress and are rewarded with a strain and eventually a fracture, and we thereby encounter this complex spectrum of properties—(1) Young's modulus (stiffness), (2) strength (breaking stress), (3) extensibility (breaking strain), (4) work of extension (strain energy storage), (5) resilience, and (6) toughness (work of fracture). Note that numbers 1, 4, and 5 refer to how an intact material responds to stresses, while 2, 3, and 6 refer specifically to actual fracture.

WHICH PROPERTIES ARE MOST IMPORTANT WHEN

The specific synthesis of solid materials by organisms has a result more consequential than mere proliferation of kinds of materials. It's a matter we might simply call "tuning"—the harmonization by natural selection of the properties of materials with their applications. At one level the subject is almost trivial—the availability of steel cable permits a different kind of bridge from one made of bricks; the notion of a rubber can opener is not a great idea, and a soft, sway-backed engine block is strictly an Oldenburg, not an Oldsmobile. For a given task some materials are more appropriate than others. But beneath that bald statement lurk substantial biological issues. For one thing, the materials are of interest in themselves—for instance, just what are the limits to the mechanical versatility of protein? Or what are the practical trade-offs among various properties, that is, what deficiencies must be suffered in one property in order to achieve a high value of another?

But the business of tuning provides a scheme whereby investigation of the properties of materials can give special insight into the tasks to which they're put—we assume good "design" and use the assumption as a tool to ask what really matters in an adaptive sense in the operation of an organism. Here comparisons are exceedingly helpful—we can usefully ask how material properties change from species to species, from habitat to habitat, from place to place within a single organism; we can look for similarities of mechanical adjustments among very distantly related organisms subjected to similar environmental changes; we can investigate differences in properties of materials that interact harmoniously in some common function.

Consider our teeth (Waters, 1980, has a good discussion of their properties), crucial to feeding us, subject to stresses and abrasions over long periods, and (as we know too well from the disproportionate professional attention they require) none too good. Enamel, the outer layer on teeth (Figure 9.4), is very stiff and resistant to compression—handy for biting and chewing—but takes tension poorly and is exceedingly brittle—stiff-

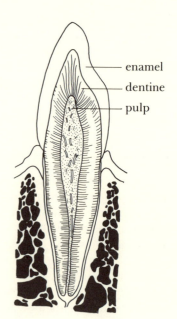

enamel

dentine

pulp

FIGURE 9.4. A lengthwise section through a human tooth. Flanked by gum (above) and bone (below), the tooth has three main layers—enamel, dentine, and pulp—going from the hardest on the outside to the softest in the middle.

ness and toughness seem in practice as antithetical here as in the various commercial steels. Dentine, the next layer in (and the bulk of the hard stuff in teeth), is less stiff than enamel but not quite so brittle. An off-axis load might safely develop some substantial tensile moment in the dentine—harder dentine would most likely lead to increased functional fragility. Perhaps there's a general principle involved—limit the hardest material to small pieces at the sites of wear, as we do in making carbide-tipped drills and saws. Teeth do crack without enormous provocation—their poor behavior in actual fracture is an unpleasant converse of the hardness needed for trituration. Trade-offs seem unavoidable, although it's no easy thing at this stage of our understanding to give them proper quantitative expression.

Wood is poor stuff when loaded across the grain, no surprise and a convenience for splitting it lengthwise. It's nicely tough when loaded lengthwise—not only can we use all sorts of intrusive fasteners without initiating fracture, but a tree can survive a crosswise axe stroke without cracking in the next storm. Indeed, the sawyer must cut almost through the trunk before the tree topples. Wood is relatively weak in compression, at least as compared with bones or teeth, but one can argue that the compressive loads on a tree trunk are not all that great—they're distributed over a wide cross section. Probably much of the wood is almost pure ballast, providing weight to offset any tendency toward upset; concomitantly, most trees are quite stiff, so the weight stays usefully centered and the center of gravity does not drift around (as we'll elaborate upon in Chapter 14). While there's a wealth of information on commercial timber, the analysis of Jeronimidis (1980) is one of the few that focus on functional, fresh wood.

Bone, although totally different in composition from wood, isn't all that different mechanically. It's mostly a little less anisotropic but similarly stiff. The stiffness appears at first glance to have a similar significance—holding a massive body off the substratum. In fact, impact loading is commonly more critical than steady gravitational loading. Furthermore, muscles act mainly by pulling on bones, using them as levers. And an overly flexible femur would be as unhandy as a rubber axe handle or pry bar, whatever its strength or toughness. Currey (1984), in fact, argued that properties associated with fracture are not really central—survival means avoiding fracture, so the niceties of failure are adaptively irrelevant.

The use of collagen for storage and release of work in the ballistae (Figure 9.5) of the ancients has already been brought up; Gordon (1978) has a good discussion of the matter. Ballistae were probably the ultimate machines of classical civilization, products refined over centuries by the

193

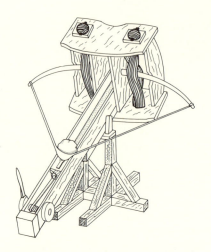

FIGURE 9.5. A somewhat imaginative reconstruction of a ballista. Energy was stored in the two vertical bundles of tendon through which the arms pass. Note that pulling back on the arms stretches the tendon-cord with great force but only by a small amount. The winch in particular must have been much more elaborate in reality.

military research establishments of the Mediterranean powers—their history and associated lore are collected in Marsden (1969, 1971) along with translations of the surviving ancient artillery manuals. As can be seen from Tables 9.4 and 9.5, collagen is excellent stuff for the purpose, high both in strain energy storage (especially relative to mass) and resilience, and surpassed only by rubber, of which the ancients knew nothing. Bronze springs were tried—after all, collagen is perishable and expensive—but metals then available just didn't give anything like the same performance on a weight-for-weight basis.

Throwing a 40-kilogram projectile 400 meters is no small achievement, but a smaller mass would be much less effective against stone or masonry fortifications, and a shorter throw distance would put the artillery within range (about 300 meters) of enemy archers. These ballistae were far more powerful than the early cannon of a millenium or so later. My guess (I can't find mention of the idea) as to their real limitation has to do with their prime movers. Collagen only stores work; it may be reusable whereas gunpowder is not, but the artillerymen had to crank in all the work expended in tossing each shot. According to my crude calculations based on human metabolic capacity, this Herculean task must have taken some time and restricted the shooting rate to once every few minutes. With cannon it must have been a matter of better killing through chemistry.

What about the biological role of collagenous tendons? They usually run between the ends of a muscle and its attachments to bones, so the force *and shortening* of the muscle is transmitted to the bones. Work is stored—in a running person or hopping kangaroo about 40% of the work done on a leg in landing is recovered as it pushes off again (Alexander 1983)—and the mass of tendon relative to that of the whole animal is impressively low. The inextensibility may initially seem odd, but shortening must be transmitted. Thus, a conventional rubbery material, as used in the hinges of insect wings and bivalve shells, would do no better in tendon than in bone—both must be adequately stiff, the former to withstand tension and the latter compression.

Certainly the most extraordinary material among those tabulated here is spider silk (that of silkmoths is substantially less extreme)—it has the greatest tensile strength, astonishing extensibility, and by far the greatest strain energy storage. It should be admitted that the figures are samples of diverse data—the tuning of silks to particular tasks is particularly extensive. The task of the silk in an orb web, though, is itself extraordinary (see Denny 1980). The fibers must be thin to minimize visibility. They must use a minimum of material—protein is valuable stuff, and it takes a lot of thread to make a web. If a large prey enters, it should be caught or at least not destroy the web—extensibility can multiply the effect of strength by what we might call the "tar-baby effect"—if you can't stretch something *far* enough to stress it much, what does your own strength matter? And high extensibility may permit a spider to lower itself on a line only barely strong enough to support the animal's weight—impact loading can be largely ignored if the structure just gives upon impact. The poor showing in one property, silk's lack of resilience, in fact may be a virtue. The work of prey against web has to be dissipated somehow—the device is, after all, a catch-net and not a trampoline!

One last story. On the rocky shores of the Pacific northwest a large alga, the bull kelp *Nereocystis*, grows as long as 10 meters (Figure 12.1). It's attached to a submerged rock by a holdfast and has a long cylindrical "stipe" at the far end of which are a float and long thin blades. Loading is almost purely tensile, but Koehl and Wainwright (1977) found that the tensile strength of stipe is indecently low—orders of magnitude below tendons, cuticle, or steel. Strain energy storage was entirely respectable, though, since a stipe could withstand extensions of 30–45%. They suggested that the material of the stipe withstands breakage by stretching, secure in the knowledge that no single wave lasts long enough for the stipe to reach breaking strain.

CRACKS AND COMPOSITES

It was casually put on record that most pliant materials and all rigid materials in organisms turn out to be composites of several components. And the toughness, or work of fracture, of a material proved to be related to the surface created as it broke. We're now in a position to make something of these facts. First, an observation. Take a sheet of aluminum foil, fold it once, and make a small slit at right angles through the fold. Then pull lengthwise on the fold—with very little applied force the foil fails at the slit. Now, with a paper punch, make a round hole at the far end of a similar slit and pull again. This *further* removal of foil should go far to foil the facility of the foil to fail.

What has happened gets us back to the difference between force and stress—the latter, force per unit area, is what matters most in the real world. The sharp end of the slit (or almost any crack) acts as a force concentrator and thus as a local stress increaser (Figure 9.6). As you keep pulling, of course, the crack keeps advancing, so there's always a sharp edge. Always—unless the crack runs into a smooth-edged void, which may just stop it. Start a crack in glass, and the glass breaks easily. Learning the lesson, you round the corners of the portholes and hatchways on a ship, or of the window and door cutouts on an airplane, to avoid locally high stresses. Round the edges of the grooves between the teeth on a gear. Put lots of little spaces (voids) in bone, in the tiny hard ossicles of a starfish, or in coral, but never let them have sharp edges. Fracture depends not just on the material but on its arrangement. The notion isn't new—I'm told that in 1545 Roger Ascham, tutor to Queen Elizabeth I, advised bowyers to prick small holes at the ends of compression fractures on the bellies of longbows.

But there's an even better way to keep cracks from propagating disastrously. If a crack runs through a little fiber of stiff material and then reaches a very unstiff (compliant) component, the latter will give a little, accommodate the disturbance, and reduce the stress concentration at the tip of the crack (Figure 9.7). Result—the crack stops here. While the material as a whole, the composite, may be less strong than the material of the stiff component, it now takes far more work to get the whole to the breaking point and still more work actually to fracture it—work of extension and toughness have been increased. Several tricks help here. First, the stiff stuff should be divided into small pieces. A crack must have at

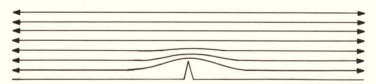

FIGURE 9.6. The force-concentrating effect of a crack. One can draw so-called force trajectories, analogs of streamlines. At all points on such a line force runs tangent to the line. Where lines are closest together, forces are most concentrated and stresses greatest.

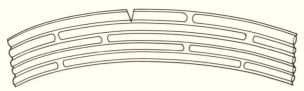

FIGURE 9.7. A composite material of parallel cylindrical fibers glued together. Bending initiates a crack,
but the stress concentration at the tip of the crack is
reduced when the tip hits a soft glue layer, so it doesn't
continue into the next fiber.

least a certain minimum size before it will propagate decently, and keeping the stiff material smaller than this critical crack length makes best use
of it. Second, a goodly *difference* in properties between components
makes it harder for a crack to make the transition from one component
to another. Third, making a lot of internal contact surface between components maximizes the chance of a crack hitting such an interface. And,
in addition, orienting the components so a crack must run a jagged
course increases the distance the crack must travel and thus the work it
has to do in cleaving bonded material to make surface.

Except when we build things of wood, our technology makes only limited use of composites (and when we talk of them, they're always referred
to as "advanced" composites!). Part of the explanation is the trouble it
takes to fabricate them, and part is related to the excellence of those
ubiquitous pure materials (pure in the present sense, even if alloys),
metals. But the difference between a (pure) glass rod and a fiberglass
(composite) rod is enormous—far more than one would reasonably attribute to a small fraction of relatively soft epoxy glue. Organisms don't
use pure metals and inevitably keep their stiff material divided into small
pieces in composites. Even coral, mollusk shell, and teeth are composites.
The sulfurous smell when the teeth are drilled is the smell of burning
protein—enamel may be only 2% organic (and is really brittle stuff), but
dentine is fully 29% organic, the organic fraction being mostly protein
(Waters 1980).

The egg shells of birds are mechanically impressive devices, surprisingly resistant to external loading; Vincent (1982a), though, complained
about how little we understand them, muttering at "half-boiled notions"
in the literature. The shells are mostly mineral but have a critical 2 to 4%
of organic matter, making them composites. Still, cracks can propagate,
which the chick puts to advantage to get out—before pushing, it pecks
around a circle so it can then break the egg along the dotted line.

Equally impressive is the composite character and consequent resist-

197

ance to crack propagation of the leaves of grasses, also investigated by Vincent (1982b). If a grass leaf is slit or notched, it does tear more easily but only (and fairly precisely) in proportion to its reduced cross section—there's just no sign of any significant stress concentration. Do your worst to a grass leaf—it just doesn't go along with attempts to tear it crosswise. This phenomenon has major implications for the grazing techniques of grass-eaters—grass has to be either sliced almost completely across or actually torn up. A cow wraps its tongue around some grass and pulls; it doesn't take the time and trouble to chew the truly tough stuff. A lot of ecology and evolution must depend on such mechanical events.

OTHER STRESSES, OTHER STRAINS

Thus far we have defined all but one material property (resilience) in terms of response to a steady tensile load. Compression was mentioned only briefly, with regard to crushing. What about bending, shear, and torsion? All are important modes of loading, but we much less commonly make tables of basic stress-strain data for them. For one thing, most of the tests are a bit trickier than for other mechanical properties. More important, perhaps, is the role of specimen geometry. The results of pulling on something are nearly indifferent to its shape and depend almost entirely on the area of its cross section. By contrast, for compression we find that, except for simple crushing, the modes of distortion and failure—Euler and local buckling—are highly dependent on shape. So we enter a world in which we have to consider structures as opposed to materials.

On the other hand, tensile tests prove more useful than you might at first expect. It turns out that failure in loading modes other than tension quite often ends up being attributable to failure in tension. Thus, a compressed cylinder typically bends before breaking; one side of the bend is in tension even though the ends of the cylinder as a whole are being pushed toward each other. Similarly, local tensile loading is an inevitable component of overall loads described as shearing and torsional.

If you stretch a rubber band and then monitor the stress while holding the band at its new length, you'll see the stress gradually drop a little for a few minutes. Something in the rubber has "given." Time-dependence has not been of concern in the discussion thus far. But whether we talk of "give," "creep," "yield," or "stress-relaxation," we enter a world of time-dependent material properties, a world in which for all properties we must specify the time course over which the load is applied. In short, many real materials are appreciably "viscoelastic"—they have some of the viscosity characteristic of fluids and so care how *fast* as well as how *far* they're stressed. Of soft materials, a rubber band is minimally viscoelastic—most biological materials are much more so. Indeed, one can state as a loose generalization that, among living things, the lower the Young's modulus the more viscoelastic the material is likely to be, with the jellies and mucuses showing the most complexly time-dependent behavior. For the most part these complexities will escape our close attention; but take note—they're real and important. A sea anemone stands up to a single wave but deflects in a tidal current that imposes the same drag (Koehl 1977). And a bone can take sustained loads that would crack it if applied in an impact (Currey 1984). And you're shorter before bedtime than upon arising—try a pair of measurements if you're skeptical. The cartilage of your intervertebral discs gradually compresses under sustained load; since the compression is noticeable relative to your overall height, it's really substantial relative to the thickness of the discs. There's a long-term creep as well, described by that closet biomechanic T.S. Eliot: "I grow old . . . I grow old . . . I shall wear the bottoms of my trousers rolled."

For some viscoelastic materials the presence of sharp yield points at which they begin to flow is crucial to their functions. Probably the most extreme example is the pedal (foot) mucus on which a slug locomotes (Denny and Gosline 1980). A bit of mucus resists shear, so a bit of slug can get a purchase and push forward against it. But after a certain amount of shear, the mucus suddenly becomes a viscous fluid, and the bit of slug slides forward, with purchase provided by adjacent, less stressed bits. Left unstressed briefly, the mucus "heals" back into a very soft solid and can again be pushed against. The process can be repeated many times; indeed it must—there are quite a few waves of transition beneath a given slug at a given time.

A totally different kind of yield underlies the operation of the pit of an ant lion larva. An unbound pile of solid particles will form a slope no steeper than some "angle of repose," a phenomenon of great interest to both highway engineers and designers of automatic parts feeding and packaging equipment. Lucas (1982) showed that the insect carefully maintains the greatest possible slope; the addition of a wandering ant precipitates a miniature avalanche, with a large and efficient set of jaws barely buried at the bottom of the pit.

Where a complex material property exists, it seems as if some species will make some special capital out of it. I suggest, quite seriously, that a good way to make a neat discovery is to begin, not with your favorite organism, but with some physical property or phenomenon, and then ask how organisms might take advantage of it. The people I've mentioned in the preceding paragraphs—Vincent, Koehl, Denny, Lucas—appear to have done just that; none defines his or her field in terms of the particular kind of organism whose special feature is described here.

For ancient technology, at least, biological materials were crucial. Stone was about the only nonbiological thing in common use, and horn, tendon, various woods, and fibers were certainly more important than minor bits of metal and ceramic. But the *biological* role of biomaterials received very little attention even before the great metallic revolution of the nineteenth century and the plastic revolution of the twentieth. Only within the past few decades has more than an occasional offbeat biologist done more than speculate about the subject. But the immediate past has seen a welcome surge of activity—the contemporaneity of the references here isn't just the scientist's usual prejudicial focus on the most recent work.

CHAPTER 10

Arranging structures

> Q. "What's the difference between girder and
>   joist?"
> A. "One wrote *Faust*, the other *Ulysses*."
>
> Heard on BBC program, "My Word"

I N EXPLORING materials ranging from the enamel of teeth to the
mucus of slugs we found that their properties "made sense" relative
to their uses by organisms. To get such nice coincidence between materials and applications, though, only very simple modes of loading were
discussed—tensile loads and, almost parenthetically, crushing loads.
These loads share a unique characteristic—as stresses they elicit strains
that depend on the material at hand but almost not at all on how that
material is arranged. Only the cross-sectional area of the sample matters
and is automatically taken care of in the definition of stress as force per
unit area. So we can specify, for instance, a value of the stiffness, E, that
characterizes a particular material and that shifts only slightly with the
degree of strain.

By contrast, responses to other modes of loading—compression other
than simple crushing, bending, shear, torsion—depend on something
else in addition to the properties of the material. That other thing is the
arrangement of the material, a topic touched previously only when we considered composites. The need for an additional factor is, I think, easiest
to see if we consider what happens when a reasonably stiff thing is bent.
Bend two rods of the same material and of equal cross-sectional area (not
counting empty space), one solid and the other a hollow (but not excessively thin-walled) tube—the first will deform much more easily than the
second. Their tensile strengths and tensile stiffnesses are the same—
there's no percentage in making ropes hollow—but the tube is clearly
more resistant to bending. Obviously we need some quantity beyond the
material properties of the last chapter to describe this effect of *shape*, a
quantity that will now be introduced.

FLEXURAL STIFFNESS, BEAMS, AND COLUMNS

By "flexural stiffness" we mean resistance to (or stiffness against) bending, a property for bending analogous to Young's modulus of elasticity
for tensile loads. The measure turns out to be the simple product of two

factors, one reflecting the material present and the other its arrangement. The first is nothing more nor less than Young's modulus itself. The second is given the name "second moment of area," which amounts to an explicit definition for those familiar with such matters; it's designated I. So flexural stiffness (sometimes called "bending modulus") is just EI; it has no common symbol of its own, and only the I needs explanation at this point.

"Second moment of area" and "neutral plane"

If you were to bend an infinitely thin rod or sheet, nothing would be either compressed or extended—the process would be unadulterated bending. But as soon as an object has some finite thickness, then bending has the effect of stretching the surface on the outside of the curve and compressing the surface on the inside. Still, if one side is stretched and the other squidged there must be some plane running down the middle that is completely unstressed in either tension or compression. It's called the "neutral plane," and, as ought to be obvious from Figure 10.1, the stresses increase with distance to either side of that plane. Not that the neutral plane is truly unstressed—it's quite clearly sheared by the opposing stresses to either side. Thus two apposed strips of wood bend less readily if glued together than if unattached—the glue joint bears the shear stress and prevents slipping at the surface of contact. Still, material near the neutral plane can contribute little to the object's antipathy toward bends, so the further a bit of material is from the neutral plane, the greater can be its contribution. Clearly, the way to make a stiff object is to move the material away from the neutral plane—that's why a tube is stiffer than a rod with the same cross-sectional area of material, and that's why the joist under a floor is made higher than it is thick.

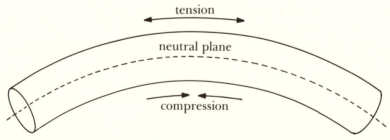

FIGURE 10.1. When a rod or tube is bent, one side is subjected to compressive stress and the other to tensile stress. Somewhere between is a "neutral plane" that experiences neither. If the bent object is axisymmetrical, and bending can be in any direction, we can recognize a "neutral axis" instead of a plane.

In fact, material contributes to the I of flexural stiffness in more than direct proportion to its distance from the neutral plane—the tube is disproportionately stiffer than the rod, with two factors combining in its favor. The matter is perhaps best explained using a rather contrived model. Consider a structure, a kind of beam, that protrudes outward from the wall to which it is attached—two easily bent (but nonsagging) struts, each of which can change length by stretching or compressing a spring (Figure 10.2). The springs are linearly elastic—they obey Hooke's law in either tension or compression—and each strut is the same distance, d, from the neutral plane. The free end of the beam is depressed a distance y by a weight F_w that simultaneously generates a force F_s stretching the top spring a distance Δl. We need worry only about the top spring since its extension is just mirrored by the compression in the bottom one.

Now the stretch of the top spring will be proportional to the product of its distance from the neutral plane and the deflection of the whole beam:

$$\Delta l \propto dy.$$

By Hooke's law, the stretch of the spring is proportional to the force exerted on it, so

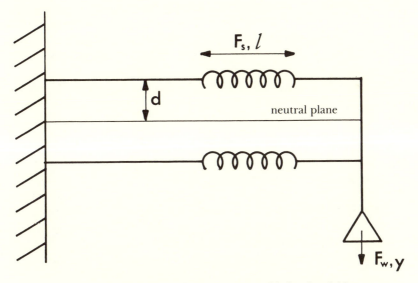

FIGURE 10.2. A "beam" composed of two inextensible but bendable struts, each with an inserted spring. The right angle attachments of either end of the struts are assumed fixed. Thus a weight at the end bends the beam by stretching the top spring and compressing the bottom one.

$F_s \propto \Delta l.$

Combining these, we see that the deflection of the whole beam will be proportional to the force on the spring divided by the distance from the neutral plane:

$y \propto F_s/d.$

At the same time, by the notion of moments of forces, as introduced in Chapter 2, the force on the spring will be proportional to the weight on the beam divided by that same distance from the neutral plane—think of the neutral plane as the pivot and the vertical end piece of the beam as the board of a seesaw:

$F_s \propto F_w/d.$

What we're looking for, I, is something proportional to how much force it takes to get a given deflection of the beam—but that's now easily obtained by few substitutions from these relationships:

$F_w/y \propto d^2.$

In that exponent of 2 lies the disproportionate stiffness of the thicker beam. Put in verbal form, the material of an object being bent resists stress in proportion to its distance from the neutral plane. At the same time, the applied force causes an effective force on the material in inverse proportion to that distance from the neutral plane—the more distant material is both more effective in resisting stress and less stressed by a given weight.

To get the overall second moment of area, I, one takes each little element of area of a cross section of the object being bent and multiplies its area by the *square* of its distance from the neutral plane. But just where is the neutral plane? Imagine a cross section of the object transformed into a board of uniform thickness that preserves the shape of that cross section—the neutral plane is the hinge line about which the board balances. If bending can occur in any direction, then the neutral plane is replaced by a line, the "neutral axis," running the length of that plane. For symmetrical forms such as cylindrical rods or tubes, the neutral axis is typically the center line. For an object of some less simple cross section this neutral axis corresponds to the "centroid" of the section. The latter is similar to the center of gravity (Chapter 2)—the centroid is the center of gravity of a cutout of the section using a material of uniform thickness.

How does this second moment of area (and hence flexural stiffness) vary with the cross-sectional size and geometry of an object? Consider a set of cylinders, including both hollow and solid ones, of a given kind of material, in three simple situations. First, assume a given amount of material per unit length of cylinder, that is, a constant cross-sectional area, the situation with which we began. I (and hence EI) will be proportional (very nearly) to the *square of the average radius*—as the radius is increased, the material is being used with increasing effectiveness. Second, assume a tube with a given thickness of wall. Now the cross-sectional area of material also will increase in proportion to the average radius, so EI will be proportional not to the square but to the *cube of the radius*. Third, assume isometry, where shape is unchanged as size increases so wall thickness also increases in proportion to the radius. Cross-sectional area will now be proportional to the square of the average radius, and EI will be proportional to the *fourth power of the radius*. Flexural stiffness thus increases even more drastically than does weight! Unfortunately for big sorts such as we, this does not make creatures relatively sturdier as they get bigger—bones get longer as well as fatter with increases in size, so the big beast is, as pointed out back in Chapter 3, worse off than the small one.

The situation can be summarized by citing a pair of useful formulas for the second moment of area. For a hollow tube of outer radius r_o and inner radius r_i, or a solid tube for which the latter is zero,

$$I = \pi (r_o^4 - r_i^4)/4. \tag{10.1}$$

For a solid rectangular beam of height (sometimes called "depth") h and width w, loaded in the height plane,

$$I = wh^3/12. \tag{10.2}$$

I, quite obviously, has dimensions of length to the fourth power; EI is thus force times the square of length or, in SI units, newton-square meters (not "per" anything).

The basis for the preference of both nature and human technology for hollow tubes over solid rods ought to be quite obvious, whether you prefer to think of bicycle frames or bamboo poles or the exoskeletons of insects or our own long bones. We use the spaces in such bones for generating red blood cells; birds fill them mainly with air (presumably in the interest of weight economy); insects put their muscles and everything else on the inside; aircraft may run electrical or control cables through them or use them for fuel storage—but all these are secondary uses of structures made hollow for mechanical reasons.

We now have a quantity, EI, or flexural stiffness, that gives the resistance to bending of a *structure*. A minor caution, though, is necessary. For this quantity, as determined from data for E and calculations of I, to correspond to practical reality, quite a few conditions must be met. The material must be homogeneous and isotropic. It must be linearly elastic, which means that its stress-strain curve must be uncurved. It must deform equally under the same tensile and compressive loads. And its shape must not change appreciably as it is loaded. Metals, as we use them in ordinary structures, pretty well meet these conditions. For biological materials, the conditions are often much less reasonable and frequently quite disabling. So what do we do? We retain the quantity EI, but we resort to measuring it in some sort of bending apparatus rather than calculating it from its components.

In fact, EI turns out to be one of the easiest of all mechanical properties to measure—you can easily improvise a testing apparatus with a pair of supports (saw horses will do), weights (a bucket to which water can be added), and a tape measure. The specimen, some elongated object, is laid across the supports and the bending stress applied to the middle (perhaps with a rope). For small deflections, the deformation downward in the middle is given by Gordon (1978) (and standard textbooks of structural engineering) as

$$y = Fl^3/48EI \qquad (10.3)$$

where F is the weight (not mass) applied and l is the distance between the supports. The formula is worth examining as well as using—it states that the deflection of such a beam under constant load is proportional to the cube of its length, making a long beam a bad business. If, as it often will, the load itself increases with length, then the situation is even worse. A similar formula that assumes a distributed rather than point load encapsulates the main challenge in building long bridges.

Beams

Having established a measure of stiffness in bending we can move to somewhat more specific structures. First, then, beams. By a beam we usually mean some more-or-less horizontal girder subjected to up-and-down or side-to-side loads (downward and gravitational most often) that tend to bend it. The joists beneath a floor or flat roof are typical beams; so is the roadway of most bridges; sometimes too is the entire bridge. And so is the protruding neck of a bird or quadrupedal mammal, the petiole that holds a leaf away from a branch, a horizontal branch itself, the vertebral column between forelegs and hindlegs of a quadruped, the tubular abdomen of an insect, and so forth. A beam may be supported every so often along its length (a "continuous beam"), at its two ends (a "fixed beam"), or it may protrude outward from a single support (a "cantilever beam")—the latter was, of course, the version we used to rationalize the behavior of I.

Having begun with them, it's simplest to continue (to stay on the beam, one might say) cantillating about cantilevers. Consider, then, a cantilever beam whose length is decently greater than its vertical thickness (Figure 10.2 again), with the load concentrated at the unsupported end. (A very short, thick cantilever isn't strictly a beam in that its primary deformation will be a downward shear rather than a bend—you can persuade yourself of this with a simple sketch.) As already mentioned, both tensile and compressive forces will increase with distance up and down from the neutral plane. In addition, though, the forces will steadily increase as one moves inward toward the support from the load. In effect, the loading force acts as a turning moment (clockwise in the figure) with the attachment point of the beam as the pivot—the beam can be viewed as one big lever or half of a seesaw. Closer to the pivot the deflection will be less while the force will be more—the product of force and distance has to remain constant. So what should be the shape of a good cantilever beam loaded at its outer end?

The main thing is that material ought to be concentrated where it is most needed, which is to say where the stresses are greatest. If the deflecting moment decreases with increasing distance from the support, then to make best use of material the beam ought to taper from its support to a point where the load is attached (Figure 10.3a). If the beam weighs much less than the load, the taper should be linear. The arrangement is frequently seen on cantilever supports for hanging loads in construction machinery.

Some degree of taper, though, is virtually universal for long necks and upheld tails, for archy's long, thin cockroach antennae as well as for the cat's whiskers of mehitabel (Marquis 1927). Such beams are appreciably

(a)   (b)

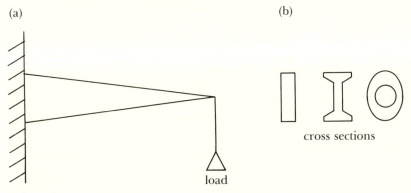

cross sections

FIGURE 10.3. (a) The optimum shape for a weightless beam loaded only at its end. (b) Various cross sections that do well at resisting bending caused by vertical forces.

207

self-loaded—they have appreciable weight. But not much has really changed, even if the beam is the entire weight. In this case, the best lengthwise taper is no longer a straight line but rather a concave curve so that more material is nearer the base, and the region near the tip (now supporting less) is thinner. One biological case might be a long horizontal branch of a tree, assuming that gravitational loading is the most important factor in determining its design. But I haven't seen an appropriate set of measurements from which to judge its taper, and it is all too likely that this view of its mechanical problems is excessively simplistic.

In cross section (Figure 10.3b) the beam should be high but not wide. A timber beam is arranged in just such a manner—the only common exceptions are diving boards and leaf springs, where substantial deformation is positively essential. One can do even better—material can be put mainly at top and bottom—after all, material on the neutral axis takes no tensile or compressive stress and only a little shear. If the material withstands tension and compression with equal stiffness, the same amount ought to be used top and bottom. We've now invented that most basic structural member, the "I-beam," with the serifs on the printed "I" representing heavy top and bottom flanges. Of course, some material is still needed between top and bottom flanges to prevent any shearing strain that would bring the two closer together and ruin everything.

An I-beam makes good use of metals, but biological materials are commonly more specialized. So the beam may be present, but the "I" is obscured, with tension-resisting stuff concentrated at the top (tendons and other fibrous materials) and stuff that resists compression (bone and so forth) at the bottom. Consider a leaf on a petiole (Figure 10.4). The greatest forces on a petiole probably occur as it is pulled by the drag of the leaf in a wind storm, but these forces are purely tensile and thus easy to resist. Without wind, the petiole is a beam, and it uses thick-walled, liquid-filled cells along its bottom and long cells with lengthwise fibers along the top. The leaf itself is as much a cantilever beam as any springboard. Veins protrude downward to get some height ("depth") to the beam and to continue the compression-resisting material of the petiole. The blade is always at the top—a flat sheet can take tension but is almost as bad in compression as a rope. An insect's wing is likewise made up of a membrane and veins but is stressed about equally on up- and downstrokes; concomitantly, and in contrast with leaves, its veins protrude equally above and below the membrane.

And what if the beam is subjected to loads from all directions, not just downward? One obvious solution is to use a double-I girder, one with a cross section in the form of a Maltese cross. But that shape just begs to

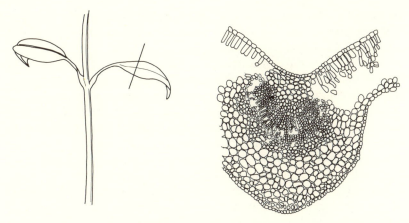

Figure 10.4. A leaf and its petiole act as a cantilever beam. The upper part is loaded in tension and the lower in compression. At right is a cross section through the midrib of a leaf—the large cells near the bottom are nearly spherical and liquid-filled; the smaller ones further up are elongate and fibrous.

have its corners connected for a little more stability; with corners connected the internal webs that connect opposite flanges become unnecessary, and we have reinvented something long known as a "box girder" or "box beam." Connecting the corners has the incidental benefit of increasing resistance to torsional loads. Robert Stephenson used just such a beam in his Britannia Bridge of 1850—trains ran through the center of a horizontal iron box (Billington 1983). One can do even better—stresses will concentrate at the corners of the box, which therefore need reinforcement and which have no specific function anyway. With more corners, each one can form a more obtuse angle and thus be less stress-concentrating; the ultimate beam for withstanding forces from all directions is clearly a hollow circular cylindrical tube. We've thus brought the business of beams back to where we began, to our argument (and others will come) for the functional utility of hollow cylindrical shapes.

Trusses

After talk about beams, the subject of trusses holds few additional complications—a truss is just a structural framework that functions as a beam, a set of interconnected elements that do the task we previously assigned to a single piece. Again it may be supported at several points or at opposite ends, or it may extend as a cantilever outward from a support. Note, though, that a bridge may be built as two cantilever trusses reaching toward each other so the result looks superficially continuous. Or a bridge may be designed to work as a pair of cantilevers during construction and as a fixed beam thereafter.

The obvious feature of a truss is the arrangement of its elements as sets of triangles. In a well-designed truss, each element is called on to resist only compression or tension under the intended loads; for trusses of metal the two sorts of elements may look quite the same. Nineteenth-century engineers tried, it seems, every conceivable kind of truss, and the arrangements—Pratt, Warren, etc.—still bear the names of these early explorers. Analogous trusses are present but relatively uncommon in living systems. D'Arcy Thompson (1942) pointed out that the bracing of the metacarpal bone of a vulture's wing takes the form of a Warren truss; a similar arrangement is found in the arms of the free-swimming larvae of the sand dollar *Dendraster* (Emlet 1982; Figure 10.5).

The rarity of natural trusses similar to those of engineers almost certainly traces in large measure to a basic difference between biological materials and metals, repeatedly noted here—biological materials are not equally resistant to compression and tension. Thus an "ordinary" truss is used only when the direction of loading varies and its elements must withstand both compression and tension. If the predominant direction of loading is certain, then nature goes a step further in specialization and uses different materials to withstand the different stresses. A mammalian backbone, as D'Arcy Thompson eloquently argued, is a truss working as a beam (Figure 10.6); the neck and thoracic regions are best analyzed as paired cantilevers extending each way from the forelegs. Tensile forces are concentrated at the top, where they are resisted by muscles and ten-

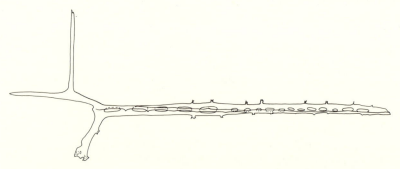

FIGURE 10.5. In some kinds of sand dollar larvae the spicules that stiffen the arms are "fenestrated," forming a truss instead of a simple beam. In this one the truss has three longitudinal members; the structure is about 0.3 mm long.

FIGURE 10.6. The head and cervical vertebrae of a mammal—the latter are the compression-resisting elements of a cantilever. The minimal tension-resisting components have been drawn in with dashed lines; the real array of muscles and tendons is more complex.

dons running between the vertical struts ("neural spines") of the vertebrae; compressive forces are concentrated at the bottom, and there the vertebrae are appressed, with alternating bony "centra" and cartilaginous intervertebral discs. And the neural spines and diagonal tensile elements run crosswise to each other, the former taking compression and the latter tension. The use of muscles as one form of tensile element gives options quite beyond any technological analog—such a living truss permits moment to moment adjustments of both shape and stiffness.

Columns

Columns are not much more than vertical beams subjected to downward loads. But they have several special points of interest. The initial effect of a weight on a column is purely compressive, and the analysis is just that of a material rather than a structure. Increase the weight, though, and complications ensue—that is, unless the column is so short that it fails by simple crushing. The complications result from the fact that, one way or another, the column bends before it collapses. There are two distinct ways in which bending ("buckling," in the trade) can happen; which one occurs depends more than anything else on the relative thickness of the walls of the column.

The first form of buckling, "Euler" (pronounced "oiler") buckling, involves a smooth bend, with compressive forces concentrated on the concave side and tensile forces on the convex side. By this point, it shouldn't seem strange to find that there are tensile forces generated from an overall compressive load, and that the initial failure in some materials may be tensile. Which is another reason the tensile properties in the last chapter have wide application. Buckling is a form of degenerative collapse in that the more a column buckles the more the buckling will progress under the same load. So what matters is the critical force that initiates the process. Fortunately, a decent formula for the critical force in Euler buckling is available (Wainwright et al. 1976; Gordon 1978; and other sources):

$$F_E = n\pi^2 EI/l^2, \tag{10.4}$$

where l is just the length of the column and n is a dimensionless coefficient that depends on how the ends of the column are restrained. If both ends are merely pinned—that is, free to bend at the support, n is 1; if one end is firmly fixed (inserted into the support or firmly braced or rooted), n is 2; if both ends are firmly fixed, then n is 4. The condition of having one fixed end is of particular biological interest—it's the situation of long, slender plant stems such as those of dandelions, grass, bamboo, etc.

Equation 10.4 has major implications for the scaling of skeletons; it was tacitly invoked in Chapter 3. Young's modulus, E, is a material property with no intrinsic geometric character; we'll ignore it for the moment. The second moment of area, I, is proportional to the fourth power of a linear dimension such as radius. So for an isometric series of designs, the critical Euler buckling force is proportional to length to the fourth power over length to the second power and thus to the square of a linear dimension—it takes a much larger force to buckle a larger column. But a second-power relationship isn't really as good as it sounds, since loads are mass-related and increase as the third rather than the second power of lengths. In short, bigger is weaker, as Galileo first pointed out—the leg bones of large animals must be disproportionately thick. The other size-related implication results from the square of length in the denominator—for a constant diameter, the longer column is dramatically weaker. If this seems odd, try to push together the ends of several different lengths of thin wooden dowling.

In one case, at least, Euler buckling may be a Good Thing, deliberately employed as a primitive kind of joint. A tall sea anemone, *Metridium*, has an area of its columnar body just beneath the crown of tentacles that is narrower than anywhere else. The material properties of the stalk don't vary, but when a gentle water current is present, the stalk bends at this narrow region, and the tentacles are exposed broadside to the flow in the best position for feeding on suspended matter (Koehl 1977). It doesn't take much narrowing to concentrate the bending—I, as equation 10.1 shows, is severely dependent on the radius.

The second form of buckling is familiar to anyone who has squashed an empty beer can; it's called "local buckling." Unsurprisingly, it is mainly a problem for hollow columns and beams with very thin walls—it's the main practical reason why the advantage of tubular over solid cylinders can't be pushed indefinitely to very thin, wide tubes. A formula is available for the critical local buckling force for a thin-walled cylinder in compression, a more rough-and-ready empirical affair than the one for Euler buckling:

$$F_L = K\pi t^2 E. \tag{10.5}$$

Here t is the wall thickness, the difference between outer and inner radii. K is a semiempirical constant, varying between 0.5 and 0.8 depending on the extent to which the cylinder has imperfections at which buckling might be initiated (Wainwright et al. 1976). Notice that neither radius per se nor second moment of area make any difference to the initiation of local buckling.

Local buckling does occur in biological columns—it's certainly involved in the "lodging" of slender crop plants in wind storms, and it can be deliberately induced in any dandelion stem. But I can't point to any specifically mechanical investigations relating environmental forces, natural structures, and the propensity to buckle this way.

BRACED FRAMEWORKS

Trusses are only one of a general class of structures sometimes referred to as "braced frameworks"—systems of struts or, if compression- and tension-resisting elements are distinct, of struts and cables. In such systems, by definition, loads are accommodated by changes in the lengths of individual members rather than by changes in their geometric inter-relationships. A very well-developed theoretical branch of engineering is devoted to the design of stress-resisting frameworks using minimal amounts of material; a good reference is Parkes (1965). About the only place where the results appear congruent with biological design is in the internal bracing (the so-called trabeculae) of bones. D'Arcy Thompson (1942) first noted the similarity between a cross section of the head of a femur and the head of a theoretical crane (derrick, not bird), and Currey (1984) also gave the matter some attention. Figure 10.7 gives an example of the phenomenon. It might be mentioned as an item of biological hubris that if a bone is broken and reset in a different orientation, or if the stresses on a bone change, the trabeculae are reorganized to adjust the stiffening to the new situation.

Why, beyond a few obvious trusses and the trabeculae of bones, are braced frameworks in the classical sense of the engineers so notably rare in organisms? Wainwright et al. (1976) suggested (and I quite agree) that such frameworks accomplish something organisms ordinarily avoid— making structures of very great stiffness; there is lurking here that general biological preference for strength over stiffness, for accepting rather

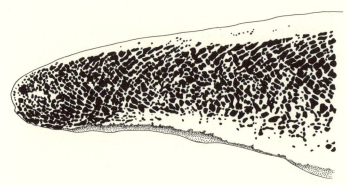

FIGURE 10.7. A cross section of part of the ischium of a horse; this is the rearmost bone of the hip, loaded by leg muscles pulling downward on its free end. The dark parts are voids in the solid; notice the generally diagonal arrangement of the trabeculae. You can make a similar preparation by grinding down lengthwise the upper part of a well-dried lamb or pig femur.

than preventing deformation. For instance, geodesic domes are impressively efficient but stiff structures; in nature they seem to appear only as the shells of viruses (and there more likely as a matter of informational simplicity than for any mechanical advantage).

To support or distort

The subject of braced frameworks, though, is still of biological interest. It's worth inquiring about what distinguishes nondistortable, or "statically determined," structures from distortable ones, officially known as "mechanisms" (an extraordinary use of a most ordinary word!). The test of whether a structure is statically determined, and thus a braced framework, is whether knowing the length of each element gives you an unambiguous picture of the shape of an array. Figure 10.8a quite obviously shows a mechanism in which the struts can rotate at the pins with respect to each other. In Figure 10.8b, by contrast, even though the joints may allow free rotation, the shape of the array is nonetheless fixed—it is statically determined. The distinction between these simple cases rests on the different consequences of three versus four or more joints—all of which is a fancied-up way of saying that while triangles are intrinsically rigid, quadrilaterals are not.

Three points are worth emphasizing. First, rules are available for determining from the number of joints and struts whether more elaborate structures are statically determined or flexible, although we won't get into the details here. Second, neither is superior to the other in any sim-

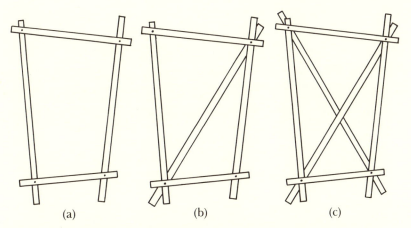

FIGURE 10.8. Three arrays of pin-jointed struts. (a) is a "mechanism"—its shape is indeterminate. (b) is a statically determined structure—it has a definite shape. (c) is also statically determined but redundant—any single strut can be removed without changing its shape.

ple sense—the two just play very different roles in mechanical designs. The skull and jaw of a fish or a snake, with their complex mechanisms, serve one function, while the trusses already mentioned do something quite different. Finally, any mechanism can be made statically determinate by sufficient reinforcing of the joints. The result is almost always a bad design, involving stress concentrations and excessive material.

One of the peculiarities of statically determined structures is that a minimal structure is usually the best one, not only for simple material economy but also for both ease of construction and effective operation. We normally regard redundancy as righteousness—as I write I save my prose on two different discs and am comforted by knowing that I have access to a second word processor. But consider a nonredundant design to which a redundant element is added (Figure 10.8c). The nonredundant design is tolerant of minor variations in the length of each strut—it can shift its geometry by rotation at the joints so the pieces will fit. One class I taught built a geodesic jungle gym of 120 struts joined by single screws at the apices. We put it together without tightening the screws and found the last piece apparently 5 cm short. We gave the structure a vigorous shake, and the final hole dropped onto the final screw as if, one might say, they had been made for each other. We hadn't really made such precisely measured struts—the structure was nonredundant. Any *additional* strut would have had to be made to fit with real care and precision.

215

And the resulting redundant structure would most likely have been weaker, not stronger. Look again at Figure 10.8c—strengthening with the final strut is a delusion unless the struts are very carefully measured or very elastic—the last strut may carry all the load, it may carry none, or the load may be partitioned in some way that depends (if there's any stretch) in a nearly unpredictable way on its own magnitude. But (and perhaps this is the lesson to us from organisms) these problems of redundancy and load sharing are ones about which only the stiff need be uptight. The automatic remodeling of the trabeculae of bones in the face of changing loads may be remarkable; as remarkable, for organisms, is the necessity of remodeling only in a few unusually stiff structures.

Spicular reinforcement—grit in the Jello

Should a braced framework or a mechanism that acts in a supporting role be made of a few large struts or of a lot of little ones? The problem is the now familiar one of scaling, in particular of the scaling of the tensile and compressive elements of a framework. The scaling of tensile elements is simple—Young's modulus, the stiffness, is independent of the length of the element. So for an element of given cross section and material, longer is not weaker. For compressive elements long enough to encounter Euler buckling, the situation is not so nice. Again, for a given material and cross section (constant E and I), the critical force is, by equation 10.4, inversely proportional to the square of length—longer is weaker, drastically so. The implication is that quite different guidelines apply to the best forms of tension and compression elements—the latter ought to be kept short, but for the former length matters little.

R. Buckminster Fuller (1962) recognized this distinction as well as the economy of using tensile elements—where buckling can't happen, less material is needed to resist a given load. He coined the term "tensegrity" to describe structures in which the tension elements form a continuous lattice into which short, isolated compression elements fit—"islands of compression in a sea of tension" is his metaphorical phrase. While the resulting structures are esthetically satisfying (there's a neat mast by Kenneth Snelson outside the Hirshhorn Museum in Washington, D.C.), their disadvantage is that they aren't particularly stiff without causing huge tensile stresses in the cables. But if, as we've argued, nature isn't usually as interested in stiffness as we are, then such latticeworks of tiny struts should abound in nature. Which they don't, especially. The main limitation of the scheme, I'd guess, has to do with motility. Concomitant with self-propulsion is usually a set of muscles pulling on solid elements. The smaller the compression elements, the greater the number of muscles needed to work them; indeed, starfish, with very short ossicles, have quite a fabulous number of muscles (Eylers 1976).

There is, though, a group of animals that use very short struts (called "spicules")—typically a few tenths of a millimeter in length—and yet achieve heights of up to two meters and withstand severe environmental forces. These are sponges, with variously shaped spicules of brittle material laced together by a form of collagen. Their spicular skeleton is (as are sponges generally) all too often dismissed as primitive, but for sessile animals nearly devoid of contractile tissue it's an eminently sensible arrangement. Indeed, even the largest sponges devote only a few percent of their wet weight to supportive material, both spicules and collagen together. (Bath sponges are aberrant sorts with no spicules and extra collagen instead.)

At one stage I had the idea that the spicules and collagen of sponges ought to be Fuller-on-earth, with geodesic lattices, tensegrities, octet trusses, and all of his imaginative and efficient structures. I looked at quite a number of sponge skeletons without finding much coincidence with Fuller's designs; I was about ready to admit that sponges were truly primitive in the ordinary sense. But then in Bermuda, on the edge of the ocean, I found the skeleton of a tiny sponge that (being my own) I could abuse (Figure 10.9). I compressed it nearly flat (between fist and desk) whereupon it promptly sprang back to its normal shape. Examination revealed a three-dimensional array of mechanisms without a single statically determined set of spicules. Sponges seem to be consummate specialists in accepting deformation—tiny, brittle spicules, too short and isolated for bending to make trouble, laced together with as good a set of inextensible but bendable straps as nature has devised—one is tempted to call them "unbraced frameworks."

If mechanisms, rather than statically determined structures, are ac-

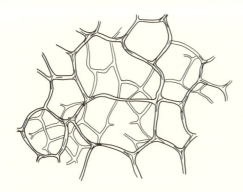

FIGURE 10.9. A few millimeters of the skeleton of a small sponge—tiny, stiff spicules are embedded in an easily bent collagenous meshwork.

ceptable, then yet another possibility arises. At a crude level of analysis, a statically determined structure is something specific while mechanisms are everything else. Putting small pieces of brittle material into a pliant matrix gives a composite called a "filled polymer"—it amounts to a kind of random array of mechanisms. Koehl (1982) looked into the extent to which the connective tissues of animals that had embedded spicules behaved like proper filled polymers—the use of embedded spicules is fairly widespread, not just in sponges, but in some coelenterates, echinoderms, mollusks (the chitons), arthropods (stalked barnacles), and ascidians. She took isolated animal spicules of various kinds and concentrations, embedded them in gelatin (raspberry flavored), and performed various abusive manipulations on the products. Since the normal function of spicules is to stiffen tissue (we're still considering relatively un-stiff structures), she used that as a criterion of effectiveness. Even a relatively small proportion of spicules dramatically increases stiffness; more spicules, or more elongate or irregularly shaped spicules, give more stiffness, and small spicules are more effective than large ones for a given added mass. What seems to matter is the area of contact between spicules and matrix, not unlike other composites. So whether the spicules are in a specific framework or in a random array, roughly the same rules seem to apply.

INTERNALLY PRESSURIZED STRUCTURES

Now for quite a different kind of structure. Balloons are familiar to everyone—for children they are perhaps the first evidence that air is a real substance, something enough of which can transform a limp piece of rubber into a large, well-shaped object (or lack of which can do the opposite, as found by Piglet in *Winnie the Pooh*). We use inflated tires, inflated blimps, and sometimes even inflated buildings—all are structures best described as "internally pressurized." All carry to an extreme the use of separate materials to withstand tension and compression. Tension is entirely resisted by the outer membrane, which may, as in a tire, be quite a thick and complex piece of material. Compression is taken by the internal fluid, in these cases air or helium, both gases. The gases are lightweight, cheap, if air, and usefully buoyant, if helium, and (perhaps a mixed blessing) somewhat compressible.

Many organisms carry internal bags of air such as lungs and seed pods. Only rarely, though, can the air be regarded as a structural material—the floats of some large algae and of the Portuguese man-of-war (a colonial coelenterate) are the best natural analogs of blimps. Life is wet, and the common working fluid of nature's internally pressurized systems is water. It's functionally incompressible and, while dense, is no denser than the other materials of organisms. For aquatic organisms, it's neutrally buoyant. It's also cheap—its acquisition entails none of the inevitable metabolic costs of making protein or polysaccharide.

So—how to make a structure based on an internally pressurized system, an aquatic blimp, so to speak? The overall shape will be cylindrical, either circular or elliptical in cross section, or spherical, or ellipsoidal— for reasons that will get lots of attention in the next chapter corners are hard to arrange and functionally awkward in such systems. Since these structures are substantially deformable, either by environmental or internal activities, we ought to consider both shape and the possibilities for change of shape. Considering mainly cylinders for now, at least three general arrangements suggest themselves. First, the system might have constant volume, so any increase in length is matched by a compensatory decrease in cross-sectional area. This most obvious constraint (essentially just an incompressible system) is central to the functioning of a peculiar group of structures collectively named by Kier and Smith (1985) "muscular hydrostats," to which we'll return in Chapter 13. Here some of the situations will involve constancy of volume; others will not. Second, such a system might have constant surface area, an outer membrane which can be changed in shape but not size. As far as I know, no biological case of the latter has been described.

Consequences of fixing fiber length

The third possibility is both the least obvious and the most common. *A system might be arranged with some tension-resisting fibrous component in the membrane where the lengths of the fibers are fixed.* (It may take you a moment's thought or some cross-hatching on paper to persuade yourself that the second and third possibilities are actually distinct.) It's the one we'll explore here, and its behavior will take a bit of explaining.

For a start, there are two fundamentally different ways in which fibers might reasonably be arranged in a surface membrane. Either some might run lengthwise and some circumferentially, as in the sleeve of a shirt. Or they might run helically, with separate left- and right-hand helices extending down the length of a cylinder, as in a good-quality necktie or a dress cut "on the bias." The two possibilities turn out to determine thoroughly different responses to the various stresses the structures might encounter (Figure 10.10); the contrast is clearly evident using balloons inflated in tubes of loosely woven fabric (Appendix 2). The model with longitudinal and circumferential fiber reinforcement does not deform in response to pushes or pulls—compression or tension—on the whole cylinder; for a push the circumferential fibers prevent expansion in girth, and for a pull the longitudinal fibers prevent increase in length. It does

219

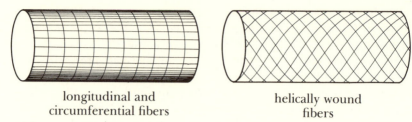

longitudinal and
circumferential fibers

helically wound
fibers

FIGURE 10.10. Two contrasting ways to arrange fibrous reinforcement in the membrane of an internally pressurized cylinder—fibers may run lengthwise and circumferentially, or in right- and left-hand helices.

deform in response to a bending stress, but it very easily goes into local buckling—it kinks. Only twisting—torsional stress—elicits a smooth deformation. By contrast, the helically reinforced model lengthens and shortens smoothly as it is pulled and pushed. It bends about as easily as the other but has much less tendency to buckle locally. Only torsional stresses are not accommodated with a noticeable deformation—the two sets of helical fibers resist twists in either direction.

Only one of the two reinforcing schemes occurs in nature. All known cases of internally pressurized, water-filled cylinders (often termed "hydrostatic skeletons" or "hydroskeletons") have helical reinforcing fibers. And this particular arrangement is no rare thing. It occurs in the stems of young herbaceous (nonwoody) plants such as sunflowers; it provides a wrapping for flatworms (platyhelminths and nemerteans), roundworms (nematodes), and segmented worms (annelids); it supports the tiny tube-feet of echinoderms such as starfish; it stiffens the body wall of sea anemones; it determines the response to muscle contraction in the outer mantle of squids; it's a major functional component of shark skin; and it even provides intermittent stiffening for the penises of many (some have bones) male mammals including those of our own species. The material of the fibers varies widely, and the functions of these hydroskeletons are even more diverse, but the wrapping is always helical.

The relationship between the structure and the behavior of these hydrostatic beams and columns is perhaps best developed with a hypothetical and idealized mathematical model expressed as a graph (Figure 10.11).[1] "Fiber angle" is the angle between the reinforcing fibers and the long axis of the cylinder. If the cylinder is stretched out, the fiber angle

---

[1] I've replotted the graph given elsewhere to put length rather than fiber angle on the x-axis; the result is a little like a mirror image of the usual version, which I have always found confusing.

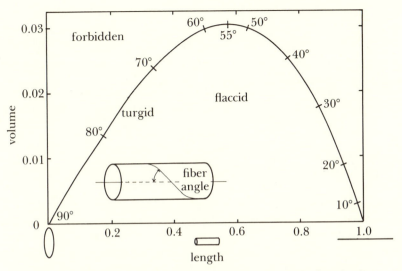

FIGURE 10.11. The changes in volume and fiber angle as the length of a helically wound cylinder is varied. The length of each fiber (and thus of the fully extended cylinder) has been arbitrarily set at 1.0. Beneath the curve the cylinder isn't fully inflated and so isn't cylindrical; above the curve it would be overinflated and exceed its intrinsic volume, an impossible condition.

decreases—the fibers run more nearly lengthwise—if it is compressed, the angle increases. The graph represents the results of taking a cylinder and either compressing it into a disk (moving left) or stretching it into a line (moving right). At full extension or compression, of course, the cylinder has a volume of zero, so maximum volume must be reached somewhere in between. "In between" turns out not to be something ordinary, such as 45° or 60°, but an angle just a little less than 55°.

What happens in real organisms?

With hydroskeletons we can't simply fall back on our usual use of the products of engineers for guidance to the practical possibilities for natural design—the scheme is more nearly unique to living structures than anything we've previously considered. At this point we just have an abstract graph; still, implicit in the graph is a surprising diversity of biological arrangements. A brief description of several follows.

Herbaceous stems. If you make a tiny lengthwise slit in the outer layer

221

of a young sunflower stem, the cut edges pull apart. You've relieved some of the tension in the outer layer, so it shrinks a little. At the same time you've relieved some compression in the core, but, being mostly water, it doesn't expand noticeably. The sunflower amounts to a flagpole built of water—no other compression-resisting material is present in any great quantity until woody fibers begin to be laid down. Without muscles or motility the system doesn't "do" anything as do the animal schemes we'll come to—it's a constant volume arrangement with, presumably (unfortunately I don't have figures), a fiber angle in the outer layer ("collenchyma") somewhere around 55°.

The compressive core is of especial interest, at least to me. In a freshman botany course I dutifully memorized the "fact" that the inner, thin-walled cells ("parenchyma") of such a stem had a particular shape— a so-called orthotetrakaidecahedron, a fourteen-sided solid with eight six-edged faces and six four-edged faces (Figure 10.12). Only a few years later did I read D'Arcy Thompson (1942) and get an idea why that shape might occur. If a set of distortable spheres of equal size are squeezed together so as to completely fill some large volume, each will (ideally at least) take on this particular shape; it's the shape that permits each to expose the minimum surface area. So the peculiar shape of these cells is indirect evidence that they are being squeezed and thus that they function as a compression-resisting core. Only thin cell walls are needed— there ought to be no particular pressure difference between individual cells, and most of the motile animal systems lack any partitions at all. Still, these flimsy-looking cells have a crucial supportive function. The fourteen-sided shape, incidentally, is something of an idealization—the real cells are actually a bit variable. Wainwright et al. (1976) mentioned this system, using the *Nasturtium* petiole as an example.

Limp worms. The first to recognize the importance of fiber arrangement in hydroskeletons seems to have been Cowey (1952), in a nemertean worm—a long, flat, and unsegmented creature; a more general treatment followed a few years later (Clark and Cowey 1958). These worms are normally severely flattened cylinders; the fibers in their surface layers lie at angles to their long axis somewhere in the 40° to 70° range. Since they are not circular cylinders, their contained volume is less than it might be, and they live beneath the curve of Figure 10.11 in the region labeled "flaccid." If such a worm contracts longitudinal muscles, it will get shorter, moving to the left along a horizontal line. Sufficient contraction will bring it up against the curve, where it finds itself circular in section and stiffer ("turgid") to boot. If the worm contracts circumferential muscles, it gets longer, moving to the right, but again it gets more nearly

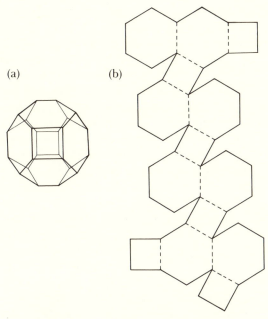

(a)   (b)

FIGURE 10.12. (a) An orthotetrakaidecahedron.
The thin lines represent the surface facing away
from the viewer. (b) Plane facets (partially joined)
of the same solid—if you make a few of these solids with paper and tape, you can demonstrate
that they pack together without leaving voids
between.

circular and turgid. Not all of these sorts of worms can change shape
enough to hit the limiting line, but they all use this curious scheme in
which shape and stiffness are predictably interrelated. Internal pressures, though, are never very high in these creatures—even when actively
locomoting, they are fairly limp.

Some stiff worms. Nematodes, or roundworms, contrast sharply with
the various sorts of flatworms. They have a strong external cuticle, a normally round cross section, and only longitudinal muscles. A grotesquely
large species, the intestinal parasite *Ascaris*, was studied by Harris and
Crofton (1957). *Ascaris* has a normal fiber angle of about 75°; since it is
unflattened it lives just on the curve of Figure 10.11, part way down the
left-hand slope. Contraction of its muscles can only shorten it further.
But the worm can shorten only if it decreases in volume to move down

223

the slope, which it can't, or if its fibers stretch, which they do only a little. What mainly happens is that it gets much stiffer, generating internal pressures up to 30 kPa—around a third of atmospheric. The worms do get shorter, but only by about 10%. They can bend by contracting muscles on one side only, and this, too, increases internal pressure. Circumferential muscles are quite superfluous—the resilience of the cuticle antagonizes the action of the longitudinals. High internal pressure does create some awkwardness—eating cannot be done simultaneously with locomotion, and the digestive system is equipped with a sturdy-looking structure that seems to act as a one-way, antivomit valve. The arrangement permits nematodes some unpleasant features of lifestyle such as burrowing through our flesh.

The squid's mantle. How a squid refills its mantle cavity was mentioned in Chapter 7; another piece of the story of jet propulsion by the squid fits here. A squid's mantle lacks longitudinal muscles but is well equipped with circumferential ones. Collagen fibers in the mantle make an angle of about 25° with the squid's long axis, so the system is on the right-hand slope of the curve of Figure 10.11 (Ward and Wainwright 1972). Contraction of circumferential muscles tends to extend the mantle, but, with such a low fiber angle, extension implies a reduction in volume. Here the volume does indeed reduce and very forcefully too—water squirts out the siphon and, to conserve momentum, causes the squid to accelerate in the opposite direction.

Shark skin. It came as more of a surprise than it should have when Wainwright et al. (1978) showed that sharks utilized a hydroskeleton. Sharks, after all, do have a conventional skeleton, albeit somewhat less calcified than that of most other vertebrates. Shark skin is sturdier stuff than fish skins in general (it has been made into leather on occasion) and has the crossed helical fiber array diagnostic of hydrostatic arrangements. The system in sharks, though, is more complex than those described previously—muscles attach directly to the skin which thus acts as both an external pressure-resisting membrane and an external whole-body tendon. Fiber angles, not unreasonably, vary with location on the fish—the unstretchy skin must transmit the forces generated by the body musculature back toward the tail. During locomotion, the pressure inside the body of a shark rises to as much as 200 kPa—twice atmospheric and as high as that inside an automobile tire. Sharks, it seems, are just shark-shaped balloons with teeth.

CHAPTER 11

Insinuations about curves

"And so no force, however great,
    Can stretch a cord, however fine,
    Into a horizontal line
That shall be absolutely straight."

William Whewell, *Elementary Treatise
on Mechanics*, 1819—inadvertent
poetry originally printed as prose.

As SHOULD be evident from the last chapter, flat surfaces are rather flimsy affairs unless very thick or elaborately braced. Little has been said, though, about the positive advantages of spheres, ellipsoids, cylinders, shells—the deserved praises of bulges and sags. That's the present task, admittedly a bit awkward in that trigonometry, the mathematics of curves and angles, will for the most part be avoided.

GOOD WORDS FOR SAGGING

We'll approach the epitome of curvature, spherical shells, stepwise through sagging wires, beams, catenary suspensions, and cylinders. Consider, first, a bird perched on the middle of a wire extending between two fixed poles. Under the weight of the bird, the wire sags downward; if the wire has a negligible mass of its own then Figure 11.1a adequately represents the situation.

We ask what may initially seem an odd question—how does the tension in the wire vary, not with the weight of the bird, but with the amount by which the wire sags downward from the horizontal? One's first quick presumption is that only the weight of the bird and not the sag of the wire matters, especially if the wire is itself weightless. But try a real version of the situation—hold an end of a thread in each hand, put a weight (a pair of scissors or something) in the middle, and pull the ends of the thread apart. The force each hand must exert is just half the applied weight when the thread runs vertically; it then increases as the sag is gradually reduced. However you try, I (and the late Mr. Whewell, quoted above) promise that you won't get to zero sag, a horizontal thread—the thread will break first.

Neither the explanation for the sag nor prediction of the tension in the wire due to the bird is especially difficult. Remember that a wire can re-

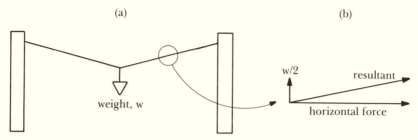

(a)   (b)

weight, w

w/2   resultant

horizontal force

FIGURE 11.1. (a) A weight makes a wire sag. (b) Both the resultant force in the wire and the horizontal force that each attachment must counteract may exceed the magnitude of the weight.

sist only tensile forces, and these must run along its length. So we know the direction of the force in each leg of the wire—it is the direction the wire takes going away from the bird. We also know the magnitude of the vertical component of the force in each wire—it must be equal to half the weight of the bird, since the downward force of the bird must be exactly balanced by the two upward forces (Figure 11.1b). What we're interested in is the resultant tensile force in each half of the wire—we can get it by much the same procedure used to resolve an aerodynamic resultant into lift and drag in Chapter 7, the vector analysis introduced in Chapter 2. Here again the direction of each resultant sets the ratio of its two components—half the weight (upward) and the horizontal force (outward). When the wire is nearly horizontal, the outward force obviously gets much larger than the upward one. But, since each upward force must remain equal to half the weight of the bird, the outward horizontal forces become far larger than that weight. Since a horizontal force has no component in the vertical direction, it's quite impossible for a horizontal thread to support any weight—the force in the thread would be infinite. Sag is mandatory, and the more sag, the less force required! The lesson is crudely learned when one erects a wire fence or puts up strings for climbing bean plants or hangs clothes on clotheslines.

To put some numbers on the argument, consider a bird weighing two newtons (about eight ounces) in the middle of a wire between a pair of poles two meters apart. If the wire stretches enough so each side descends at 45° to the horizon, the vertical sag is fully a meter, and the horizontal and vertical components are equal for each string—application of the Pythagorean theorem (Appendix 1) gives a force in each string of 1.4 N. Thus 2.0 N of bird is exerting a total force of 2.8 N, with the increase caused by the fact that the wires don't run vertically. If the string stretches only enough so the sag is 10 cm, a similar calculation gives a force in each string of about 10 N; if the string sags only 1 cm, then the force rises to about 100 N; for a sag of 1 mm, the force in each string reaches the impressive value of 1000 N, or 225 pounds—all from a modest bird!

Now let's get rid of that weightless wire and substitute a realistically massive chain, which for present purposes does nothing but support its own weight. Obviously it must sag, and sag it does into a smooth curve, as shown in Figure 11.2. What we have is equivalent to a weightless wire with a standard weight hanging from each element of its length. The curve of its sag turns out to be neither an arc of a circle nor a parabola nor a hyperbola but something called a "catenary"—we'll have no particular use for catenaries, but they certainly matter to people concerned with the sag of electrical cables or the construction of suspension bridges. Life itself is like a catenary chain—you can't relax without sagging, and being uptight and high-strung is inevitably fraught with tension.

How then can one extend a nonsaggy beam from one support to another if the beam has any appreciable weight? There's no contradiction—what happens is that the depth of the beam merely hides the sag. A pair of weightless cantilever beams on which the only load is at their point of contact is really little different from our bird-on-a-wire—a pair of wires (taking tension) is braced by a pair of so-called jibs that take compression (Figure 11.3). The arrangement isn't just hypothetical—the jib on a sailboat works in precisely this way. A sail can resist only tension, so it must be held clear of the mast by a compression element. And our collarbones,

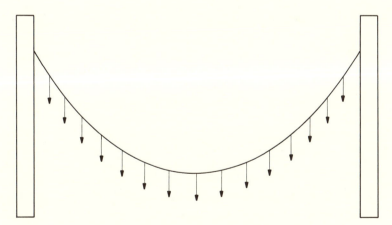

FIGURE 11.2. The catenary curve assumed by a flexible chain extending between two supports and stressed downward either by its own weight alone or by a constant force per unit length of chain.

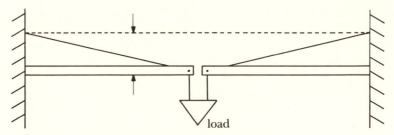

FIGURE 11.3. A pair of jibs, each with a bracing wire, support a
weight. They're equivalent to a single solid beam, shown by the dashed
line and the arrows, or to a pair of cantilevers.

the clavicles, are just such jibs. A weight (arm plus any load) is held away
from the body's midline by this simple cantilever of clavicle plus the tendons and muscles running from the neck across the top of the shoulder.

Next step—we drop gravity (a note of levity) from immediate consideration. Let's close the ends of the chain into a loop with some force
pushing equally outward on all links (Figure 11.4). The result approxi-
mates a circle, and the messy math of the catenaries is (and will remain)
gone. What we'd like to know about this circle is the relationship between
the outward force and the force pulling the links apart. Recall the bird
on the wire, where the downward force was constant but the tensile
forces in the wire depended on the angle of sag; the outward force here
(again assumed constant) is analogous to that downward force. Consider
any joined pair of links—the force pulling them apart is obviously proportional to the outward force and also depends, as previously, on the
angle between the links—the sharper the angle, the less the force. What
this means is that the fewer the links, the less the force pulling each pair
apart! So you have a smart idea—make a chain with only four or five very
long thin links. That would be fine, except that the individual links will
bend if the outward force isn't located solely at the junction points.

It turns out that the analysis doesn't depend on having individual links
of any appreciable size—it works as well for any object that can be
formed into a circle and pushed uniformly outward. There are now no
more links, so there is no angle between them. But the curvature of the
circle is quite analogous to the angle between the links; it's variable and
depends on the circle's size. We specify the curvature of either a circle or
a piece of one (an arc) with the so-called *radius of curvature*, simply the
radius of the circle whether complete or only a short arc. Logically, but
still oddly, a short radius of curvature means a sharp curve—a small circle or a very curved arc—while a long radius implies a gentle curve. A
small radius of curvature is like a chain of few links—a given outward

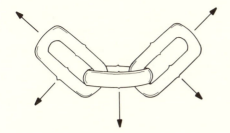

FIGURE 11.4. Outward forces on a loop of chain generate circumferential forces that pull each link against the adjacent ones.

force per unit of circumference will produce only a small tensile force around the circumference of the circle. Conversely, a large circle will have a large tensile force running around its circumference for the same outward force per unit of circumference. The point is important—the conversion of outward force to circumferential tension depends on the size of the circle.

LAPLACE'S LAW

We now have to move out of the plane of the paper and consider proper three-dimensional reality. The transition is easiest if we move from the circle to a cylinder that runs in and out of the plane of the page—the circle is then just a typical cross section of the cylinder. The only other alterations are in our variables. "Outward force per unit circumference" is now "outward force per unit circumference and per unit length of cylinder," or force per unit area, or net *pressure*. And tensile force now becomes "circumferential force per unit length of cylinder," commonly called just *tension*. Try to avoid confusing tensile force with tension; the former is a force, the latter a force per unit distance. And keep both distinct from tensile stress—force divided by cross-sectional area.

The rules

What we'd like to have is a formula describing the relationship between the net pressure within a cylinder and the tension in its walls. This isn't trivial—it's tension that splits cylinders, not pressure inside per se. The wall of a cylinder "knows" only tension as we apply internal pressure. Notice again that tension and pressure have different dimensions—tension is force divided by length, while pressure is force divided by area. The apparent complication turns out to be the key to the formula. We know that a given pressure will make a greater tension in a big cylinder than in a small one, this from the earlier talk of links and chains; and bigness is easiest to specify by giving the radius of curvature, a length.

The three variables can be arranged in only one dimensionally correct fashion—pressure must be proportional to tension divided by radius of curvature. All we lack now is the (dimensionless) constant of proportionality. And that turns out to be the simplest possible constant, 1. In short (using T for tension here) the relationship is

$$p = T/r \text{ or } T = pr. \tag{11.1}$$

This equation is sometimes referred to as "Laplace's law" after the French mathematician and astronomer, Pierre Simon Laplace (1749–1827). There's a more explicit way to derive the law. Consider a cylinder split lengthwise into halves (Figure 11.5). The halves are being pushed apart by pressure, acting in opposite directions on an effective area equal to that of the split surface—the length, l, times twice the radius, $2r$—so the outward force will be $2prl$. But the cylinder is held together by the tensile force at the cut top and bottom edges; the latter will be $2Tl$. Equate the two forces, rearrange, and you have equation 11.1.

The situation for a sphere is similar. Dimensional reasoning comes up with the same proportionality; it misses the constant, which in this instance is 2—reasonable since spheres curve in two directions while cylinders curve in only one. Thus

$$p = 2T/r \text{ or } T = pr/2. \tag{11.2}$$

Again one can do a more explicit derivation; this one I'll leave to the reader.

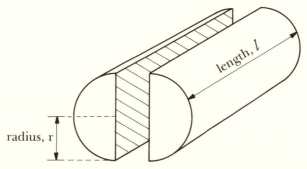

FIGURE 11.5. A piece of a hollow, thin-walled cylinder of length l and radius r, split lengthwise. Internal pressure pushes radially outward; for the split shown, it acts as if pushing on the plane marked with diagonal lines—thus over an area of $2rl$. Wall tension opposes such a split along the length $2l$ of the cut wall.

Consequences—cylinders

The implications and applications are at once profound and ordinary. The skinny tires of a racing bicycle are normally inflated to pressures around 700 kPa; the tires of an automobile take only about 200 kPa—yet the latter must withstand greater tensions in their walls, which is why their walls must be considerably thicker. In my laboratory, flow tanks[1] for subjecting organisms and models to currents are built of plastic sewer pipe. We're very much aware that the cost of pipe, reflecting the amount of plastic, goes up with the *square* of the diameter. Larger piping has not only a greater circumference, which increases in proportion to the diameter, but also needs thicker walls to withstand a given pressure; under the rule of Laplace's law, thickness must also increase in proportion to the diameter. This matter of economics (or economics of matter) led me to build my first big flow tank with a return channel (beneath the experimental trough) of plywood instead of plastic pipe, and the use of plywood led to adoption of a square cross section for that channel. Laplace then hit from a different angle, and I learned the hard way why pipes should be round, not square. A square pipe has flat walls; these have an infinitely large radius of curvature (or, strictly, a very large radius, since the walls have some curve-concealing thickness). According to equation 11.1, an infinite radius implies that *any* internal pressure (such as that of my 3-foot depth of water—10 kPa) will generate an infinite tension in the walls. In fact, the walls bulged a bit, greatly stressing the corner joints and generating recurrent and persistent leaks. Since then, we've always managed to acquire decently behaved cylindrical piping for the return channel.

It should come as no surprise, then, that very high pressures should be recorded in small nematode worms. Nor should it be surprising that the relatively thin walls (around 20 μm) of the xylem elements (with radii of about 0.1 mm) in the trunks of trees can withstand negative pressures (recall Chapter 5) of the order of 2 MPa or twenty times atmospheric. Here wall tension will be negative as well, with failure a matter of buckling collapse rather than explosion, but the same relationship between pressure and tension must apply.

Blood vessels are cylindrical, withstand internal pressure, and (as rationalized in Chapter 8) vary greatly in diameter, so the operation of Laplace's law ought to be evident in their design. Our capillaries have an internal pressure averaging about a third that of the aorta—4 instead of 13 kPa—but (fortunately for diffusive transport) they take the pressure

---

[1] In a flow tank, the liquid analog of a wind tunnel, water is pumped around a circuit; the arrangement amounts to an inside-out motor boat.

with walls only a single cell layer thick. The tension in the walls drops not threefold, but 10,000-fold, as you can readily check by plugging the radii of aorta (13 mm) and capillaries (4 μm) into equation 11.1. Wall thickness drops from 2 mm to 1 μm—2000 fold (Burton 1954), so the tensile stress (tension divided by wall thickness) actually drops by 10,000/2000—a factor of 5. Thus the stress on the material of the aorta and a capillary wall is really almost the same.

The operation of Laplace's law is evident in another feature of the design of blood vessels. You may have noticed that cylindrical balloons inevitably inflate in a specific and peculiar way—one particular part inflates almost fully before much stretching happens anywhere else. The particular part varies, balloon to balloon, and its location may even be predetermined by a little preliminary manual stretching, but the phenomenon remains. It's counterintuitive—any initially inflated portion of the balloon is stiff and would seem likely to resist further inflation more effectively than would a limp, uninflated portion. But no—it inflates fully. What is happening is that the pressure, uniform inside the balloon, produces more tension in an inflated portion, with its larger radius of curvature, than in an uninflated portion. So the bigger gets bigger still. Still, the aneurysm doesn't burst—blame that and the practicality of cylindrical balloons on a peculiarity of rubber. Just before failure, the stiffness of rubber increases; the slope of its stress-strain curve gets a little steeper, and that's all it takes to allow the rest of the balloon to inflate.

The larger blood vessels have to avoid this premature, local inflation that characterizes balloons (or surgical rubber tubing or laboratory vacuum hose). They do so by an elaborate structural arrangement that gives them extremely nonlinear stress-strain curves, as mentioned earlier and illustrated in Figures 4.8 and 9.2—what happens at the extreme of stretch to commercial rubber here characterizes the whole plot. As the wall of a vessel is stretched, it resists, not in proportion to the degree of stretch but disproportionately strongly as the process proceeds. The structural basis for the phenomenon rests on the contrasting properties and different arrangements of the elastin and collagen fibers in blood vessel walls. Initial stretching works mainly against the un-stiff elastin, because the collagen fibers lie thoroughly kinked in the wall. Further stretching extends the collagen fibers until more and more of the them are straight and unkinked and therefore resist being pulled against. It is this gradual recruitment of very much less extensible fibers that gives the peculiar and utterly crucial curve to the plot of stress against strain.

One incidental note about stresses in the walls of cylinders. The stress of concern above was circumferential, a so-called hoop stress. Internal pressure in a cylinder generates tensile stress in the longitudinal direction as well. The longitudinal stress comes from a balance of forces equivalent to that used to derive Laplace's law for a sphere—pressure is trying to blow an end, circular in outline, off the cylinder. Tensile stress in any wall is, as mentioned, equal to tension divided by wall thickness. So we can rewrite equations 11.1 and 11.2 in terms of tensile stress, σ, using t for wall thickness. For the circumferential stress in a cylinder, we get

$$\sigma_c = pr/t. \tag{11.3}$$

For the wall stress in a sphere *or the longitudinal stress in a cylinder*, we get

$$\sigma_l = pr/2t. \tag{11.4}$$

Notice that the circumferential stress is exactly twice the longitudinal stress. Blow up a cylindrical balloon until it bursts—the slit at which failure occurs will run lengthwise. A lengthwise slit means that failure was circumferential—the balloon was blown apart radially rather than having an end blown off. A blood vessel must be (and is) more strongly braced against circumferential than against longitudinal failure. If a cylinder is helically wound with reinforcing fibers, they should run a bit more circumferentially than longitudinally—the stresses will be equally accommodated if the fiber angle is somewhat over 45°. More specifically, the best angle is a little less than 55°—familiar from Figure 10.11 as the angle that maximized volume, an attractive congruence. Gordon (1978) commented that a cooked sausage fails with a lengthwise slit. That certainly agrees with my experience, at least with sausage made with artificial and presumably isotropic casing. But I wonder about the character of failure of natural and perhaps fiber-wound casing.

Consequences—spheres

With its two-directional curvature, a sphere is automatically stiffer and stronger than a cylinder of the same radius of curvature. Or, to put it another way, a given pressure is half as effective at generating tension in the walls of a spherical container as it is in the case of a cylindrical container. That's, by the way, why there's no particular problem in putting hemispherical ends on a cylindrical tank and why the ends may have a larger radius of curvature (be flatter) than the side walls of the cylinder.

There is, though, an additional consequence of Laplace's law for spheres. The situation is best introduced by inflating balloons again. With the aid of a Y-tube and a few rubber bands, there's no great difficulty arranging so that two spherical balloons can be pressurized at once with their insides interconnected (Figure 11.6a). But when you blow, inevitably one balloon inflates fully (or even explodes, sometimes) before the other begins to fill—the setup is dramatically counterintuitive, with a soft

233

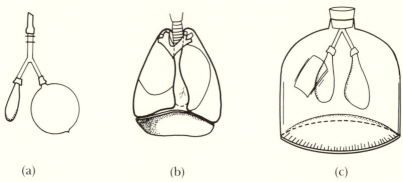

(a)   (b)   (c)

FIGURE 11.6. (a) A pair of balloons on a Y-tube—only one inflates. (b) A pair of lungs in a chest—both inflate because they're in separate compartments. (c) A common lung "model" of balloons in a bell jar with a rubber diaphragm on its bottom—pull downward on the diaphragm and it really works like (a) not (b).

balloon connected to a turgid one and yet with no possibility of different pressures inside the two. If the turgid balloon is squeezed sufficiently, it will begin to fill the limp one; at a certain point the squeeze becomes unnecessary and the (formerly) turgid balloon spontaneously completes the process, becoming fully flaccid and nearly empty. As with the cylinder, pressure generates more tension in the wall of a big curved surface than in the wall of a small one, but here we're looking at quite separate vessels.

An immediate biological application seems to have escaped the notice of the people who design laboratory material for introductory biology courses. Each of us is equipped with two lungs—subdivided balloons inflated through a Y-tube. Inhale, and *both* lungs inflate—something that should now seem less than automatic. The fact that the lungs are inflated by reducing the pressure outside them (dropping the diaphragm and expanding the rib cage as in Figure 2.2) rather than by increasing the pressure inside (oral pumping, as in frogs) is inconsequential. What makes both lungs fill is that they are inflated separately—each one is in its own compartment separated from the other by a sturdy wall, the mediastinum (Figure 11.6b). Separation is a good thing in another way as well— puncture of one lung isn't uncommon but the condition, "pneumothorax," also isn't life-threatening since only one lung collapses. The common (and commercial) model, with two balloons in a bell jar (Figure 11.6c), is really a fraud perpetrated on generations of innocent biology students. Its failure is routinely rationalized as a matter of nonuniformity of balloons. I leveled scorn at the situation to little effect almost twenty years ago (Vogel and Wainwright 1969)—another antecedent of the impulse to write this book!

The individual alveoli have somewhat the same problem as the pair of lungs—why doesn't one alveolus expand to the point of explosion (pneumothorax again) before the others begin to inflate? Moreover, here the balloons are quite clearly in the same compartment. Their particular trick is analogous to that used by blood vessels to distend uniformly. Lungs filled with air take more force to inflate than do lungs deliberately filled with a salt solution. With air inside, the outward pressure difference across the alveolar walls must exert its force against two resisting components of about the same magnitude, the elasticity of the tissue and the surface tension of the layer of water inside the alveoli. The latter opposes the formation of additional air-water interface as the the alveoli are expanded. The surface tension, though, is drastically reduced by a wetting agent secreted by cells in the alveolar walls. But, and here's the trick, the effectiveness of the wetting agent varies with its concentration, which in turn falls as the alveoli expands. Thus the force of surface tension rises sharply as an alveolus inflates, opposing further inflation. As a result of this wetting agent (or surfactant or detergent), the alveolar wall has a functionally curved stress-strain plot and the requisite nonlinear elasticity. "Hyaline membrane disease" of infants involves a failure to produce sufficient wetting agent (Clements 1962).

And Laplace's law explains why very small spheres with only thin walls can withstand enormous internal pressures—plant cells often have internal pressures of over a million pascals (ten times atmospheric pressure) with walls only a few micrometers thick. (Discussion of the origin of such high pressures from osmotically generated forces will wait for Chapter 13.) Nothing very special is happening, although cellulose fibers are certainly good at resisting tensile stresses (Table 9.2)—the cells are just small. Carpita (1985) tested the pressures a variety of thin-walled cells could withstand by subjecting them to sudden decompressions. He got values ranging from 3 MPa for cultured carrot cells to around 10 MPa for a bacterium and a green alga. Using the relationships given here as equations 11.3 and 11.4, he showed that these pressures corresponded to a narrow range of tensile strengths—450 to 840 MPa (less than twofold)—although the cell radii varied from 0.25 to 30 μm (120-fold) and their wall thicknesses varied from 0.003 to 0.1 μm (33-fold).

The basic idea behind Laplace's law also applies to large domes. Here the analog of pressure is again gravity, as with the catenary chain. The loading forces are inward instead of outward, so the dome must take negative tension, that is, a compressive force per unit distance. And the load is no longer exactly radial except at the peak, so the analysis gets less like that of a sphere as one moves down from the peak of the dome. These differences notwithstanding, we can easily see that a dome with a larger radius of curvature (bigger or flatter) is weaker—a given weight produces a greater compression in its skin. Thus one has to make it stronger; if this means that it gets proportionately heavier, then nothing is gained and one needs a different design or better materials. Domes do occur in nature—a human skull is one—but even after some heavy thinking I haven't come up with an example of a gravitationally loaded one.

Surface tension revisited

Although surface tension was introduced in Chapter 5, very little was said about curved surfaces—a large number of the consequences of surface tension were silently deferred. But surface tension is just a special kind of tension, so, with no substantial alterations, Laplace's law applies— to bubbles of air in water, to droplets of water in air, and to partial bubbles or curved interfaces. Only two aspects are unusual. First, the value of tension is generally a given—the surface tension of the particular interface. Second, something such as a soap bubble with air both inside and out, has two interfaces, and the pressure inside is therefore twice what equation 11.2 predicts. (Boys, 1902, wrote the classic account of the physics associated with soap bubbles; it's still informative and enjoyable.) By custom, one just replaces the T used earlier with γ, lower case gamma, as used much earlier for surface tension.

Bubbles

The immediate result of Laplace's law is that the pressure inside a bubble or droplet is greater than otherwise expected—surface tension compresses whatever is on the concave side of the interface, whether the inside is liquid or gaseous. As you'd expect from equation 11.2, the smaller the bubble or droplet, the greater the extra internal pressure. As a quick illustration of how surface tension becomes a big thing for small spheres, we can calculate the extra pressure (above atmospheric) inside some air bubbles in water. For bubbles of 1 mm, 10 μm, and 0.1 μm diameter the pressures come out (assuming a surface tension of 0.073 N·m^{-1}) to 290, 29,000, and 2,900,000 Pa, respectively. The first may be trivial, but the second is over a quarter of atmospheric pressure, and the third is nearly thirty atmospheres.

The consequence of an extra pressure of thirty atmospheres in a sphere of water may not amount to much, but for a sphere of air the effect could not be more drastic on account of Henry's law. Gases become increasingly soluble in liquids as the hydrostatic pressure is increased;

we're talking about a huge hydrostatic pressure, equivalent to what you'd encounter beneath about 300 meters, or 1000 feet, of water. And the gas in the tiny bubble can't be far from the interface, so the times involved in diffusion are negligible. Result—the gases of the air go into solution in the water (which, being at a pressure 30 atmospheres lower, can't be saturated with gas relative to the pressure in the bubble). The bubble gets still smaller, the pressure rises further, more gas goes into solution, and quick as a wink the bubble ceases to exist.

There's even a worse corollary to this argument. Assume a liquid that is supersaturated with gas—a state in which the gas would "prefer" (in a thermodynamic sense about which more in Chapter 15) to bubble out of the dissolved state. A bubble, though, must start small—it's hard to imagine enough gas molecules coming together at the same time to create a big bubble at one stroke. But a small bubble of gas will collapse in accord with the laws of Laplace and Henry—small, spherical bubbles are unlawful! So how can a bubble ever form even in a liquid supersaturated with gas—a freshly poured or slowly warming glass of beer, for example? The problem is not trivial, and certain aspects are still argued. Presumably this is what a physical scientist ponders while staring at a glass, deep in thought.

Still, bubbles do form, and *careful* observation of the effervescence of the beer in the glass reveals the basis of the best explanation. The answer, incidentally, came from biologists (Harvey et al. 1944)—gas bubbles let loose within organisms are not at all nice things. Notice (if you haven't) that bubbles always seem to form on the surface of the glass and more abundantly at certain particular sites from which they come off in a steady stream. A bubble-generating place has two important characteristics. First, it is hydrophobic—that is, gas spreads on the surface while water shrinks away from it. Second, it has a scratch or other irregularity. The combination permits formation of a local gas pocket whose boundary (gas-liquid interface) curves (or is at worst flat) inward toward the gas phase (Figure 11.7). There a bubble with a concave surface can grow until it reaches a critical size—large enough so that when it bulges out and rounds up into a properly convex sphere it will have a sufficient large radius of curvature so the excess inward pressure will not cause collapse.

If no hydrophobic surface is available, enormous degrees of supersaturation of gases in liquids can be attained by a variety of simple schemes; such a lack of appropriate surface seems to explain the great (and fortunate) reluctance of bubbles to form within cells and blood vessels. A more drastic example of what is possible when all surfaces are hydrophilic is provided by the water columns in the xylem vessels running up

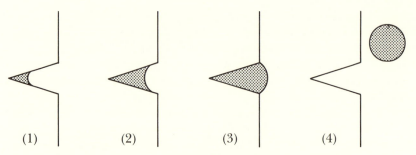

FIGURE 11.7. The formation of an air bubble (shaded) in a scratch on the wall of a water-filled glass container. At no point is a highly concave water surface present to apply a great squeeze.

the trunk of a tree. With megapascals of negative pressures, virtually *any* dissolved gas ensures supersaturation, yet bubbles do not form. It's a good thing, too—a tiny bubble would rupture a water column since any bubble is itself an appropriate surface for gas formation; once formed, bubbles grow almost explosively in a supersaturated liquid. But trees persist despite the seeming fragility of this most basic element of their physical existence. They can even (although we don't know how) refill vessels emptied by bubble formation due to an air or vacuum embolism. Nor is the question of initiation of bubble formation purely biological. Sudden and self-driven carbon dioxide outgassing from the supersaturated depths of Lake Nios seems to have been the basis of the disaster in central Africa in 1986, in which a gas cloud asphyxiated many people living near the lake.

Critical matters in our daily lives depend on the peculiarities of bubble formation and growth. Some truth lies in the rumor that champagne tastes better when imbibed from the best quality glassware—fewer scratches means less bubble formation, more persistent supersaturation, and release of more effervescence on the palate. Treatment with dishwasher detergent may help matters—the stuff is designed to leave a little residue to make glass less hydrophobic, so glasses will air dry without local deposits from streaks and bubbles of impure tap water. Plastic glasses are bad for champagne—most plastics are terribly hydrophobic. Drinking beer from an unglazed mug appreciably changes the taste—the irregular surface precipitates a great flurry of bubble formation, so the head is high while the beer is flat. Of course you may prefer beer this way, especially if you like to drink it very cold but would rather avoid the concomitant gassiness. Water boils more easily in an old kettle with scale inside its bottom—in a new, stainless steel kettle the water can get a bit above its normal boiling temperature and then suddenly break out into vapor. Vaporization then lowers the temperature, and the process may repeat in a series of pulsations. In chemistry labs, so-called boiling chips are often introduced into reaction vessels to provide enough rough surface for easy bubble formation so boiling occurs steadily and smoothly.

If many small bubbles coalesce into one large one, the total gas volume increases—for a thousand bubbles each 1 μm in diameter, the final volume will be almost triple that of the original, individual bubbles. Still, I haven't heard of any bit of biology being hung on the phenomenon. It may be important in some circumstance where ice melts—freezing commonly causes bubble formation because air is less soluble in ice than in cold water—during melting, bubbles might coalesce instead of redissolving.

Pores

Surface tension is the manifestation of a liquid's abhorrence of a surface. Water rises against gravity in a capillary tube rather than make a greater area of air-water interface at the top. Put another way, the liquid does its best to avoid interfaces—and under proper provocation it does other things besides squeezing tiny bubbles of air out of existence. With a sieve of sufficiently fine mesh you can bail water—the interfaces get caught in a meshwork through which either air or water alone can freely pass. If the phenomenon seems preposterous, consider something less hypothetical. About to drown, you take off your pants, knot the end of each leg, fill the pants with air, and hold the waist closed and downward. If the knots are tight enough you have a good float. But note—either air or water can pass freely through the pants; it's only the interface that seems to get caught. If the pants remain dry, the float won't last! Again, for an interface to get through a tiny pore, a radius of curvature has to get very small, and it therefore takes a lot of pressure to push the liquid through against surface tension. Lowering the surface tension of the water reduces the required pressure, which is one of the main advantages of laundering with detergent.

Further practical matters are germane here. Synthetic fabrics are said not to "breathe." In the damp southern summer that I annually endure, cotton is clearly preferable to polyester in any circumstance in which a person produces liquid perspiration. The trouble with most synthetics is that they are very hydrophobic, and when a liquid film bridges their pores they're about as vaporproof as a raincoat. The opposite side of the same coin is the treatment of tents and raincoats to make them hydrophobic so they pass air freely but not water even when the fabric feels wet as a result of stuck interfaces. (A truly waterproof and closed tent is not pleasant—the water vapor you lose from lungs and skin condenses on the cold walls and runs down as a liquid.) Leave a wet piece of absorbent cotton to dry, and it spontaneously contracts into a tight wad. The fibers get pulled inward by the receding interfaces—their hydrophilic character in this case makes matters no better.

A related problem must be dealt with by aquatic insects that carry underwater coatings of air in so-called plastrons; but, as pointed out by Thorp and Crisp (1947), they derive a special virtue from the situation. As mentioned earlier, the air is located between tiny hydrophobic hairs, with a complexly curved interface running from hair tip to hair tip—the interface looks like a tent root held up by lots of poles. Increasing depth increases the water pressure, which increases the inward curves of the interface. But surface tension opposes the consequent creation of extra surface, so the pressure in the entrapped air is lower than the local hydrostatic pressure outside. The lower pressure retards outward diffusion and prolongs the life of the air store. With two million hairs per square millimeter, the pressure differences attending curved interfaces between hairs can get quite large. In fact, the hairs are apparently just barely strong enough as columns to take the compression (they're but a hair's breadth away from collapse), according to observations made while pressures were artificially increased and calculations based on hair dimensions and the Young's modulus of chitin.

Plants again. Even the soil in a desert contains liquid water a little ways below the surface. It's called "capillary water" and is often thought of as very firmly stuck to soil particles. The binding, though, is as much physical as chemical—the water is in the soil interstices, caught in tiny recesses between soil crumbs where it has minimized its exposed interface with air (Rose 1966). Extracting the water involves subdividing surface and thus takes a very great pull, which eventually appears as an additional (negative) component of the pressure in the vessels running up a stem or trunk. The lowest (most negative) pressures known in plants occur in desert species, which must suck really hard on the ground to get any water out.

Of the several places where the resistance of an interface to passage through a pore is biologically relevant, one is overwhelmingly the most important. Earlier (Chapter 5) we used the internal coherence of water to explain how a column of sap can be held up in a tall tree by tension from the top rather than compression from the bottom. But these columns of sap in the xylem are in some very real way open at their tops— very large quantities of water evaporate ("transpire") from leaves. Why, then, isn't air sucked into the leaves, filling the entire system? Clearly, in some equally real way, the columns of sap are also closed! The explanation isn't especially mysterious (more mysterious is why the situation is so rarely mentioned in textbooks), and a good treatment is given by Nobel

(1974). The functional interfaces are in the interstices of the cell walls within the leaves, in their feltwork of cellulose fibers. The gaps between the fibers are about 10 nm across; making the rough-and-ready assumption that these gaps form cylindrical pores, it's no trouble at all to calculate (equation 11.2 again) that it would take an internal negative pressure of 29 MPa to suck air into the plant's vascular system. This magnitude is substantially greater than the negative pressures reported for sap within the leaves of the tallest trees or the most water-deprived desert plants.

CHAPTER 12

Systems of support

"An octopus once started a string quartet. But the
act fell flat on its face having been left without a leg
to stand on."

Anon.

IT IS AT the level of systems that the guiding hand of evolution is most
likely to be evident. While this is certainly the most complex level of
mechanical design in living things, it is the level at which we can inquire
most reasonably about the biological consequences of design. Success in
propagating progeny depends more immediately on good supportive
systems than it does on good materials; indeed, the idea of quality in a
material has meaning only in the context of the system in which it is used.
Evolution is sometimes referred to as a process of optimization; I agree
with the general notion, but I'm not enthusiastic about using the word
"optimization." Part of my hesitancy is linguistic—something can't be
more optimal just as it can't be *more* unique, so "to optimize" is synonymous with "to perfect." We mean, of course, "to improve" or "to make
more nearly optimal," but a battle cry that needs a footnote is too easily
misunderstood. Still, if natural selection doesn't commonly push organisms in the direction of some sort of functional improvement, then modern biology is in real trouble. Concomitantly, if our mechanical analyses
are inconsistent with nature's functional demands on organisms, then
we're probably barking up the wrong tree in these analyses.

SAFETY FACTORS IN SUPPORTIVE SYSTEMS

One of the variables in any process of practical (by which I exclude
strictly esthetic) design is what we call "safety factor." By this term we
usually mean the magnitude of the load that just causes failure of a device divided by the maximum load the device is expected to experience.
For instance, if a swing is normally used by no child heavier than 50 kg
but accepts 500 kg before failure, it has a safety factor of 10. The factor
must be evaluated for each structural component that bears a load, but it
is the supportive system as a whole that shoulders the final cost of failure.
Gordon (1978) prefaces a sobering chapter on accidents with the comment that factors of safety are often no more than factors of ignorance
and that engineering design is really applied theology. More recently

(and equally readably) Petroski (1985) devoted an entire book to the failures of mechanical designs.

And organisms do suffer mechanical failures in their supportive systems. Alexander (1981) and Currey (1984) have given considerable attention to the matter of safety factors, which turn out to be quite a lot more complicated in both theory and measurement than I would have naively guessed. Not only do modern, non-aged humans break bones and pull tendons loose in various athletic activities, but skeletal remains from members of earlier societies show that a large fraction of their total number suffered and recovered from fractures. Healed fractures are very common among baboons, gibbons, and macaques. And the data are likely to understate the incidence of breaks if a significant percentage had fatal consequences. Zebras and antelopes may break legs in racing ahead of predators. Hard corals and trees suffer breakage from their drag during storms. Even the crushing strength of different foraminifera—tiny, shelled protozoa—has been shown to correlate with the coarseness of the sediment in which they live (Wetmore 1987), so mechanical failure may be presumed to have been a selective factor.

How strong should a system or a component be? Whether we consider the technology of humans or of nature, two kinds of considerations bear on the question. First, there are uncertainties—uncertainty as to the maximal load borne during the lifetime of the system, uncertainty arising from inconsistency in the quality of materials, and uncertainty as to the correctness and completeness of the mechanical analysis. Second, there are costs—the incremental (marginal, if you've taken an economics course) cost of making the system a bit stronger, the incremental maintenance costs associated with more material (but perhaps a negative increment from a more lenient maintenance schedule), and the cost associated with failure. Let's look at these factors, one by one.

Uncertainties

An internally pressurized system, whether a tank of gas or the cell of a stem, is subjected to a fairly constant load. A bone in the hind leg of a kangaroo, or a strut on a short vehicular bridge, is subjected to very much more variable loads. The forearm of a brachiating ("Tarzan" locomoting) gibbon probably experiences still more variable loads. Clearly, maximal load is a complexly statistical notion. And materials vary in capability—for critical applications, samples of each batch are tested as they are used. I work in a building to which extra reinforcement was added when it was discovered that samples of concrete from batches already poured were below specifications. The Brooklyn Bridge has miles of sub-par cables within its main catenaries—a supplier faked some tensile tests,

243

but the fraud was uncovered in time to make adjustments. Much uncertainty attended nineteenth-century anticipatory tests—it was a period of use of brittle materials such as cast iron and of little understanding of the characteristics of stress concentration and crack propagation. But even proper material has an acceptable range of variation. Living materials must also be variable, although, as Currey (1984) pointed out, the evident variability in existing data is more likely a result of difficulties in testing than of real variability in the material. Uncertainty in the analysis, though, applies only to the engineer—nature neither analyzes nor tests but goes right to production and field use!

Costs

Foremost, of course, is the cost of failure. Is it the economic cost only of downtime and replacement of a transmission tower after an unusually large storm, or do human life and such economic imponderables have to be included? We accept cars that pose perils to existence quite unacceptable in airplanes in which large numbers of people travel at once. What increase in cost of material or fabrication results from a given reduction in the likelihood of failure? The answer involves an analysis of materials, design, the relationship of the magnitude of the load to the probability of failure, and the statistics of the uncertainty of the loading regime.

For organisms we at least have an applicable currency in which to measure costs and benefits—it's called "fitness" and is a measure of an individual's contribution to future generations through the production of reproductively competent progeny. Neither life, liberty, nor the good of spouse or species has relevance except as it contributes to fitness. The trouble is that the currency isn't a convertible one—I think no one has yet linked a nonpathological change of a specific amount in the strength of, say, the Achilles tendon with reproductive success. But fitness does at least provide a very good qualitative guide—we're not really surprised to find (as did Currey) that a short-lived, rapidly reproducing barnacle exhibits shoddier workmanship than a species with the opposite kind of life history.

And breakage is certainly common where the cost is clearly low. Some corals break readily in storms, but they also grow rapidly, and breakage is secondarily an asexual reproductive device. Trees often lose branches in a storm, but the tithe of Boreas is only rarely serious. Also, by reducing drag, loss of a branch must reduce the stress on the trunk and base and thereby lessen the chance of a really catastrophic failure. Thus, at least for sessile creatures, *local* failure may be a mark of a fit design. It's altogether too easy to stay within our own perspective as organisms seriously discommoded by the loss of more than a digit or two. Autotomy, the self-amputation of an injured appendage, is common among creatures from spiders (a leg or so) to lizards (the tail). Indeed, the tails of some lizards may function mainly as items with which to propitiate predators.

Estimates of specific safety factors for a few individual structures have been reported by Alexander (1981) and Currey (1984). In Chapter 5 mention was made of the incompressible gas-filled buoyancy chambers of cephalopod mollusks such as cuttlefish and the nautilus. These have been tested to find the pressures at which they implode; on the bases of the depths, and hence the pressures, at which the various species are known to live, safety factors of 1.3 to 1.4 have been calculated. Such constancy is remarkable—the normal maximal depths for the different species range over almost an order of magnitude. And equally remarkable are the low values—pressure at a given depth must represent an unusually invariable load.

Safety factors are higher for bones. Alexander (1981) gives values ranging from about 2 to 6, with slightly lower values for loading in tension than in compression where the same bone normally experiences both. (Human weight lifters push their safety factors down to unity, but their fanaticism must be maladaptive.) The uncertainties in the analyses are great, and the data don't have sufficient safety factors for drawing detailed conclusions. Currey (1984) noted that the cost incurred for an increase in skeletal mass will depend on how an animal spends its time. A marine mammal must offset the buoyancy from air in its lungs, so heavier bones may even be a benefit, not a cost. Extra mass will be bad, but not a major cost, for a slow runner on level surfaces. For fast runners, though, the work needed to accelerate and decelerate limbs becomes appreciable, and extra mass is a more serious impediment. It must be still more serious for animals whose whole-body acceleration is critical to getting food or being eaten. Fliers, of course, ought to be the most mass-sensitive creatures of all and should accept the lowest safety factors—they have to stay aloft as well as reciprocate their appendages. But the data to test Currey's attractive suppositions are still lacking.

A variety of estimates exists for the safety factor of tendons in vertebrates and of at least one insect apodeme. These are lower than for bones—mostly below 2.0. But three nearly independent explanations seem reasonable. First, these structures are loaded in tension; tensile loads are simpler than any others to withstand and less likely to be unpredictably increased by changes in the time course or direction of loading forces. Second, tensile loads are either applied by the animals themselves through muscles of definite strength or transmitted through muscles whose tendency to stretch provides protection against overloads. The third suggestion, Alexander's, is the most interesting. Tendon and

245

apodeme take little enough material—tension is, after all, the easy load. But these structures are also used to store work, so what matters is the area under the stress-strain curve. Using a structure very conservatively amounts to working low on the curve, where the increment of area under the curve is less for each unit of extension than it is on upper portions of the curve. Better strain energy storage is obtained if the material is stretched out to where the area under the curve increases more for each increment of extension; that means working the material less conservatively and living with a lower safety factor.

Staying in tune with the habitat

Safety factor, though, is only one of the subjects in which mechanical analyses touch areas of biology such as morphology, ecology, and evolution. To take another example, consider the tuning of design to the details of the mechanical characteristics of the environment. One can view this tuning through interspecific comparisons, comparisons of individuals of a single species living in different habitats, or comparison of organisms subjected to varied mechanical regimes during their development. Each of these approaches has been taken, using a wide variety of creatures from a wide variety of habitats. The reorganization of the trabeculae of long bones following an alteration in the direction of loading has already been mentioned. Trees do something analogous, adding wood in an asymmetrical pattern to offset a nonaxial load on a trunk or branch. Curiously, softwoods (gymnosperms) reinforce with "reaction wood" especially good at resisting compression, adding it on the side subjected to increased compression, while hardwoods (angiosperms) do the opposite, adding tensile reaction wood on the side where shortening will restore a tree's original posture.

This subject of tuning appears to be especially attractive to a younger generation of ecologically oriented and outdoors-tolerant investigators. Koehl's (1977) study of a pair of species of sea anemones of different structure and different habitats is thought of as an old one; she found (among other things, some already noted) that structural changes compensated for habitat changes with respect to some mechanical properties. The material of a tall anemone inhabiting sheltered bays experienced about the same stress as that of a low, wide anemone in tidal surge channels. Armstrong, for a species of large cabbagelike algae, showed not only that plants from a low-current habitat could not tolerate transplantation to an exposed locality, but that habitually exposed algae found a sheltered habitat inhospitable—they shed silt poorly and fared badly in competition with the local conspecifics.

The actual mechanics of adjustment to different habitats still holds surprises. Telewski looked at pines from windy and sheltered places and found he could duplicate their structural differences within a group of genetically homogeneous plants by subjecting some to periodic mechanical perturbation as seedlings. Stressed trees had a higher flexural stiffness than unstressed trees—one guesses that where winds are consistently present, a photosynthetic machine must maintain a good, sky-exposed orientation up to higher velocities. More interesting yet was his dissection of the changes in flexural stiffness EI into variations in E, Young's modulus (the material factor), and I, the second moment of area (the geometric factor). The wind-stressed or perturbed trees proved to have a *lower* E but more than made up for it with a higher I. Does this reflect some inescapable correlation of increased material stiffness with brittleness and functional fragility? Interestingly, Best found quite the same phenomenon when she compared sea pens (coelenterates) from bay bottoms having different speeds of tidal flow—faster flow was associated with higher flexural stiffness but lower material stiffness (Young's modulus). But Charters et al. (1969) found increases in both E and I in an erect marine alga in a more stressful environment, so there goes another fine generalization.

CATEGORIZING SUPPORTIVE SYSTEMS

Out of materials are made structures; out of structures are made supportive systems. The sequence is logical and helps organize one's thinking (or writing) about the mechanical designs of organisms, but ultimately it suffers from the inherent deficiencies of any scheme we impose upon nature. Just as the difference between materials and structures became rather arbitrary when composites came up, so the difference between structures and systems lacks a decent dichotomizing distinction. In introducing materials, quite a few pages back, the point was made that no selective advantage accrued to a creature for adhering to some logical arrangement that we might find convenient. The point still applies. Some subtle advantages may underlie the distinction between the simple spicules that stiffen the arms of the larva of one species of sand dollar and the fenestrated spicules of a closely related species. Quite irrelevant is the fact that a few perforations change a beam, which we consider a structure, into a truss, which we might prefer to regard as a supportive system.

Nonetheless, a degree of arbitrariness is tolerable in the interest of reducing one's cerebral jumble. And looking at structures as we've done has resulted in an unavoidably piecemeal and jumbled approach to systems. So at this point a classification ought to prove useful—providing, of course, that the reader recognizes this particular hierarchy as my scheme, not nature's.

First, though, a few parenthetical notes. (1) Where loading generates only simple stresses (e.g., tension or crushing compression), any distinction between structure and material becomes minor—in effect we jump directly from material to supportive systems. It's only with bending, shear, and torsional loads that the tripartite scheme of material, structure, and system really proves applicable. (2) Several types of supportive systems may coexist in an organism, either in different places or as systems and subsystems. Thus a strutted system may have hydrostatic elements—a person lifting a heavy weight tenses the walls of the entire torso and thereby assists the vertebral column with some internal pressure. Similarly, some supportive systems categorize less tidily and definitively than others.

Tensile systems

Figure 12.1 shows biological systems in tension. A modern technological example is a boat anchored in a current or a tethered balloon. Of course, humans have long made suspension bridges from vines—vines, after all, are normally loaded in tension, so they ought to provide useful sources of tensile material. In nature, tensile systems are reasonably common and certainly diverse in biological affinities. Among lower plants the

(a)   (b)

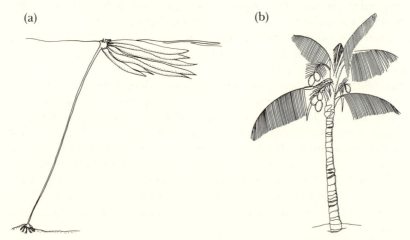

FIGURE 12.1. Tensile systems. (a) The bull kelp *Nereocystis* (an alga) is loaded in tension by almost any water movement. (b) The stems of the heavy fruits of coconut palms are loaded in tension by gravity.

stipes of some large algae are notable—in the Pacific northwest there are algae whose holdfasts are attached to rocks and whose fronds are on the surface; stipes of twenty or so feet long are loaded in tension by waves and tidal currents. Among the higher terrestrial plants are various epiphytes, tendrils, and dangling fruits. Mussels are attached to rocks entirely through the use of tough byssus threads (mainly collagen). Even the setae (hairs) by which a small lizard hangs beneath a rock or ceiling are proper tensile systems.

Tensile systems ought to scale in an unusual way, as first pointed out by Galileo. Length is independent of either load or diameter; thus the thickness of your fishing line doesn't have to be adjusted for the depth at which you hang the bait. Cross-sectional area, though, ought to be proportional to load, so diameter ought to scale with the square root (0.5 power) of load. Peterson et al. (1982) looked at several structures normally loaded in tension and found that, in practice, diameter scales with load to the 0.3 power; cross section thus goes up in proportion to load to the 0.6 power, not the 1.0 power that seems most reasonable.

Strutted systems, with or without external skins

Single or branched strut. We use branched struts in many forms—frame buildings (the collection of joists, studs, and rafters), bicycle frames, support towers for electrical transmission cables, and so forth. In nature we notice trees, branching corals, and other fairly stiff items (Figure 12.2). In these systems all the struts join in a common lattice, and no

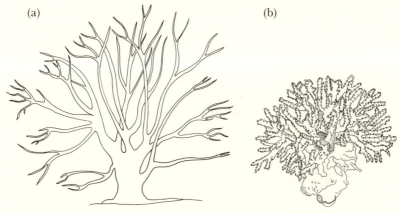

(a)　　　　　　　　　　　　　(b)

FIGURE 12.2. Two systems composed of branched struts. (a) The trunk and branches of a live oak from Yoknapatawpha County, and (b) a hard coral from some tropical reef.

249

motion is permissible at joints—we have simple, statically determined systems. If the system is branched, it's usually conspicuously equipped with lots of triangular elements, although some crude cases (frame houses) use an array of mechanisms braced against any possible deformations by a structural skin of plywood or something similar. In nature, more often than not, the branches of a system diverge without rejoining, although struts are sometimes joined into trusses—the arms of some starfish larvae and some bones in the wings of large birds have already been mentioned. Some cacti (such as the cholla) have supportive columns of branched and rejoining struts—trusses in the form of hollow cylinders with perforated walls.

Articulated strut. Articulated strut systems (Figure 12.3) still have a common lattice of compression-resisting elements, but motion is permissible at joints (articulations). We use them less frequently but do deliberately build versatile joints into many bridges, for example, so the resulting mechanisms can distort safely under changing wind loads, varied "live" or functional loads, or thermal size changes. In nature, the arrangement is very common—most components of vertebrate skeletons are best regarded as a set of articulated struts. Another example is the hard elements (ossicles) and their connections in echinoderms such as starfish.

Articulated strut systems combine nicely with muscles; sometimes, as

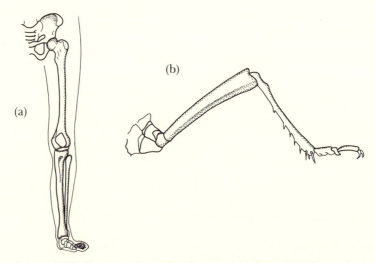

FIGURE 12.3. Systems of articulated struts. (a) The endoskeleton of a vertebrate leg, and (b) the exoskeleton of an insect's leg.

in insect skeletons, the muscles are on the inside, but the principle is the same. One of the neat features of these systems is that, while the stiffness of their materials is fixed (at least for short times), the stiffness of the system may be as capable of alteration as its shape. You drastically increase your stiffness when changing from a supine to an upright posture by appropriately increasing the tension of a very large number of muscles. But even tensile tissues other than muscle are sometimes capable of a relatively rapid change of properties in response to some chemical signal—these alterations have been studied extensively in echinoderms (Motokawa 1984). After all, a starfish undergoes an impressive transformation between being limp enough to crawl with its tube feet on an irregular substratum to being stiff enough so the same tube feet have adequate anchorage when pulling open the shell of a clam.

Dispersed strut. In our technology, dispersed strut systems are still less common than others and indeed quite unfamiliar to most of us. It's not even self-evident that decent support can be achieved in a system where struts aren't directly jointed with each other but instead connect through tensile elements. So-called tensegrity domes and towers have been noted earlier as dispersed strut systems but exist mainly as an art form. A few tents use discontinuous strutting, with the struts inserted into pockets of fabric like the battens of a sail—I'm surprised that more of them don't use this style of construction, which ought to be very weight-efficient and avoid the troubles of making rigid joints between short pieces of pole.

In nature the scheme is commoner but still far from widespread—the clearest example is the spicular skeleton of sponges, in which tiny rigid elements are laced together by collagen (Figure 10.9). And there are occasional forays in this direction among sea anemones (coelenterates) and sea cucumbers (echinoderms). It ought to be reemphasized that the arrangement is not intrinsically flawed in some way; the limitation is more likely to lie in problems of compatibility with attachments for muscles.

Nearly rigid surfaces

Planes. It's difficult to make a surface that will stay flat under appreciable stress, but we seem to like flat floors in our houses and on bridges and are willing to invest a lot of material to achieve the result. Astonishingly, we even make flat, horizontal roofs—they not only require sturdy supports but have to be waterproof rather than act as rain guides as do shingles, slate, and thatch. To get a flat surface, castings must be thick or the flat plate well braced and trussed beneath—heavy girders run up to our luxuriously flat floors, and strong joists, trusses, or flanges run just beneath them.

251

The main use of flat surfaces in nature is in photosynthetic structures, mainly leaves. These also are well braced beneath; most leaves seem to circumvent problems of loading perpendicular to their surfaces simply by flexing or reorienting in winds (Figure 7.9). Insect wings also have braced, flat surfaces—cylindrical cantilever beams ("veins") support a very thin membrane. A pound of fruit-fly wings laid end to end would stretch about 500 miles, a *very* low mass per unit length—a steel wire to go so far would have about the same diameter as a red blood cell. Yet in each second of flight the tip of a wing moves a total of several meters and reverses direction 400 times. Other paddles and fins are fairly flat as well, as are some feathers, the "book" gills of horseshoe crabs, and a scattering of other stiff structures. In all these cases, though, flatness is strongly desirable for functions other than support—from a mechanical viewpoint these systems, however impressive, are best regarded as necessary evils.

Hollow cylinders. The utility of tubes, mentioned earlier when we talked of flexural stiffness, has been evident to both nature and the engineers. We use them as subsystems in building bicycles and racing cars and as entire systems in so-called monocoque aircraft fuselages, cylindrical storage tanks, glass jars, and metal cans. Nature also uses them in diverse places—bamboo stems, vertebrate long bones, insect, spider and crustacean appendages, the wing veins of insects (just noted), the feather shafts of birds. Sometimes they contain the entire organism—lots of threadlike algae are arranged this way, although it seems unclear how much of their stiffness is due to fluid pressure rather than to being a tubular beam. Microtubules (Figure 13.4), stiffening elements in cells, are also hollow cylindrical beams—arranged as helices of two sorts of protein molecules.

My colleague, Wainwright (1988), argues forcefully that the cylinder is the basic shape in the design of organisms. Indeed, there are lots more cylinders than just the hollow ones categorized here—solid struts and internally pressurized containers appear elsewhere in this classification.

Spherical and ellipsoidal shells. In cylinders the curvature is in one direction, circumferential—the lengthwise curvature is zero. Here we're talking about structures with their surfaces curved in two directions; and, as pointed out in the last chapter, that makes them stronger and stiffer for a given investment of material. The material economy is compounded by the maximization of internal volume for a given surface—for that nothing beats a sphere, and a slight egg-shapedness doesn't make things much worse. We make domes, spherical storage tanks, skylight covers, bowls. Sometimes skin and structure are integral (cast domes); sometimes a thin skin covers a strutted structure (geodesic domes, for example).

Nature puts this shape to use in many places. Our skulls are nearly spherical domes—and the skull bone is surprisingly light and thin, with minimal internal bracing. Similarly, a turtle's shell is a light, strong dome, as are the shells of many bivalve and gastropod mollusks; the thoraces of many insects, spiders, and crustaceans; the eggs of birds; and nut shells. Smashing the inner wall of a coconut takes quite an effort; the pieces, though, don't weigh a lot. Still, small biological domes are probably not quite as spectacularly good as might be expected from the formulas in the last chapter—localized loads are more likely and more troublesome than uniform pressure differences across the dome walls. Perhaps the main limitation of domes lies in their uncompromisingly regular geometry—an organism has a lot to do in life besides resist the slings and arrows of outrageous fortune, and the turtle's shell looks impedimentary when the animal takes arms against a sea of troubles.

Internally pressurized systems

Hydrostats. In a hydrostatic pressurized system a skin is loaded in tension by an internal stuffing having a pressure a bit greater than the outside ambient pressure. As mentioned in Chapter 10, we use this in nonrigid airships (blimps), pneumatic tires, and occasional inflatable buildings. In these cases the internal material is air. Nature makes very extensive use of hydrostatic systems, but her preferred stuffing is water, sometimes together with living tissue. Water, of course, resists compression with less strain than any gas, so these systems can be quite sturdy. Sharks, worms, squid mantle, many plant cells, herbaceous (nonwoody) stems are all examples discussed earlier.

Muscular hydrostats. With muscles involved, any distinction between systems providing support and ones devoted to motility gets thoroughly blurred—as self-supported but fundamentally motile systems, muscular hydrostats will get the attention they're due in the next chapter. These devices are built almost entirely of muscle and perform the two functions with the same machinery; they were relatively recently recognized as a distinct and consistently constructed category of supportive system by Kier and Smith (1985). Naturally enough, there are no known nonliving or botanical examples. In effect, contraction of one group of muscles forces extension of another group, not due to some solid, bony linkage, but on account of the constancy of volume of muscle itself. Squid tentacles and arms, elephant trunks, and reptilian and some other tongues (including our own) work in this manner.

253

The mechanics of motility

> "The centipede was happy quite, until a toad in fun
> Said, 'Pray, which leg goes after which?'
> That worked her mind to such a pitch,
> She lay distracted in a ditch, considering how to run."

Mrs. Edward Craster (1871)

THERE IS more to life than just accommodating imposed forces, than being resigned to some amount of deflection or displacement. There's also the exercise of motion, where deliberate work is done on an organism's surroundings. Such motion we commonly (and inaccurately) take as the achievement of animals that sets them apart from and above mere plants. One can reasonably argue, though, that it was the ability to make rapid motions that elicited the evolution of the quick-acting sense organs and elaborate information-processing equipment of animals—that nervous systems, memory, behavior, thought, even emotion arose because we could move.

It's convenient to separate a brief discussion of motility into two parts, prime movers and linkages. The distinction is essentially that between an engine and a transmission, and it comes with the same understanding that the character of the first constrains the design of the second. We'll start with the prime movers but give somewhat more attention to the linkages, since analysis of the former involves molecular mechanisms largely beyond the scope of the present book.

ENGINES

The most conspicuous natural device that confers motility is, of course, muscle. But it's worth a few words to make the point that nature has other motive machines in her armamentarium—muscle is just the fastest, most massive, and best to eat.

Hydration devices

If you take some dry legume seeds and immerse them in water, they swell up. If they're put in a closed chamber into which water can pass but out of which they can't squeeze or ooze, then their swelling will push outward on the chamber with surprising force. Try it—there's an old device called a "potato ricer" that can be filled with split peas and its handles tied with lots of elastic bands; expansion of the elastic gives an indication of force (you might even calibrate the rubber with weights). Alternatively, a common type of garlic press, one with a pair of handles on one side, will work. The affinity of many polysaccharides and proteins for water is extreme—the resulting pressures may reach thousands of atmospheres (Salisbury and Ross, 1969, give references to experiments and calculations). Wetting of a small quantity of dry seeds accidentally left underneath can lead to the fracture of a concrete pavement.[1] As far as I know, all extracellular mucuses and gels are secreted in concentrated form and then take up water—a conspicuous example is the jelly mass in which the eggs of a frog are suspended. When swollen, the mass is typically larger than the volume of the gravid female.

The phenomenon, officially termed "imbibition," is sometimes used to move material. It is certainly involved in the growth of plants, where substantial amounts of material are lifted upward; the 40-cm upward growth of a bamboo plant I measured during one day in my yard was certainly an immediate result of hydration rather than of new organic synthesis. Imbibition has been put to further use in some plants—the relatively rapid (for a plant) folding of the leaves of the sensitive plant *Mimosa* uses the scheme, as do the closure of the trap of the Venus's flytrap and a few other leaf-folding arrangements among the grasses.

One case of motion driven by hydration is known in animals—the mechanism with which the males of at least one genus of mosquitoes erect the hairs on their antennae (Nijhout and Sheffield 1979). Presumably, an antenna with recumbent hairs has less drag than one with erect hairs, but only with erect hairs can a male detect the hum of a female in flight. (Females, being larger, have a lower wingbeat frequency and thus buzz at a lower pitch; a tuning fork humming with such a siren song will attract males.) Adjacent to the socket of the mosquito's antennal hair is an annular pad of homogeneous protein (Figure 13.1); the angle of erection of the hair is proportional to the angle to which the pad has unfolded. During unfolding and erection the pad increases in volume by 25 to 30% while the cells just beneath it decrease in size.

Osmotic devices

The overall effect of diffusion is to move material away from a location in which the material is concentrated—sugar will slowly diffuse away from a lump and the adjacent region of high concentration. Diffusion of water is no different; it merely takes a slight mental readjustment to view

[1] In one of C. S. Forester's novels, a leak in a ship's hold and consequent swelling of the cargo of rice splits the hull and sinks the ship.

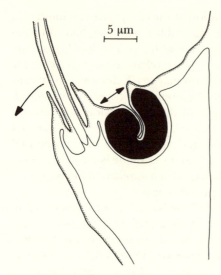

5 μm

FIGURE 13.1. The annular pad (black) near the base of an antennal hair absorbs water, swells, and causes the hair to protrude outward from the axis of the antenna.

a region of high solute (say, sugar) concentration as one of low solvent (water) concentration. For practical purposes the water concentration is lowered in proportion to the number of particles of the solute, whether these are molecules, as of sugar, or ions, as of sodium and chloride dissociated from dissolving table salt.

Consider the apparatus shown in Figure 13.2. A bag with an inelastic wall is filled with water diluted by a solute and is immersed in a large container of pure water. The wall of the bag has a peculiar property—water can diffuse through it but molecules or ions of the solute cannot (or do so very much more slowly). As a result, there is a net inward movement of water that would, if the walls were not inelastic, increase the volume of the bag. In fact, what happens is that the pressure inside increases. Now, however your teacher or textbook of introductory biology may have confused matters, this pressure is a thoroughly ordinary hydrostatic pressure of the type that pervaded Chapter 5. How much pressure is generated? In this somewhat idealized setup, a pressure of 2.26 MPa (over 20 atmospheres) is produced by a unit concentration of solute. The unit, though, is peculiar—it is a concentration equal to 6.0×10^{23} molecules added to a kilogram of water; it's called a one molal solution.

Matters are made easier in practice if a bit more devious in concept by a scheme for counting out the molecules. You look up the so-called gram molecular weight of the solute (for table sugar, sucrose, it is 343) and add that number of grams of the solute to the kilogram of water to get the molal solution—the fact that different molecules have different masses

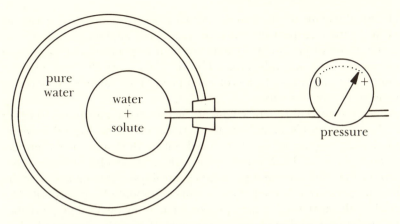

FIGURE 13.2. There will be a net movement of water into a water-permeable bag that contains a solute as well as water. As a result, a positive pressure can be developed within the bag.

has automatically been dealt with. For fully ionizing substances the pressure produced by the molal solution must be multiplied by the number of ions into which each molecule divides in solution—for table salt you multiply by two to account for one ion each of sodium and chloride. (The ions don't act quite independently, so a small but well-known correction must be applied as well.)

A solution capable of producing a hydrostatic pressure of a megapascal with pure water on the other side of a barrier is said to have an "osmotic pressure" of 1 MPa—osmotic pressure thus designates just a capability, not necessarily a manometrically measurable pressure, which is why people get confused by the term. Now these osmotic pressures correspond to very large hydrostatic pressures—about 30 grams of table salt in a kilogram of water (not all that concentrated) can generate 2 MPa—so a closed bag must have a *very* strong wall if it is to avoid explosion. Or the bag might be very small, so the high pressures yield only modest tensions stretching its wall—equation 11.2 still applies.

From pressure to work via a change in volume is a small step; it's the basis of operation of combustion engines, whether the burning of fuel is external (steam engines) or internal (most liquid-fuel engines). Pressure generated osmotically rather than by combustion makes the arrangement no less useful as an engine. Among organisms the scheme is confined (I think) to plant cells, where the relative status of water inside and outside determines the internal pressure (sometimes called "turgor") of individual cells. The interplay of structural constraints, internal pressure, and external forces then determines the shape of a cell at a particular time.

257

The usual scheme for increasing internal pressure involves the release by a cell, or the transport inward from specialized adjoining cells, of ions, mainly those of potassium. (Plants are high in potassium, as people on low-sodium diets are informed.) The cell then passively ("osmotically") absorbs water. Conversely, loss of water lowers internal pressure—wilting is the visible consequence.

The best-studied osmotic engine is the one that regulates the opening of the apertures ("stomata") on the surfaces of leaves. Typical stomates are flanked by pairs of so-called guard cells, with shapes somewhere between kidneys and knackwursts (Figure 13.3). A complex melange of environmental and internal triggers and control systems combines to set the internal pressure (via the potassium concentration) of a guard cell. The width of the aperture is nearly proportional to the internal pressure, but the mechanical details remain controversial. If you inflate an annular balloon (as when an automobile inner tube is grossly overinflated), the increase in pressure *reduces* the area of the central opening. But increasing the pressure in guard cells increases the analogous area, so the tube is not a good model. Either of two schemes seems reasonable. First, the inner walls of the guard cells might be less elastic (higher Young's modulus) than the outer walls, so inflation with water stretches the outer walls more and ends up bending the inner ones more sharply outward. Second, the lengths of the guard cells may be constrained, so increased inflation causes bending; since there's already a little outward bend, further bending of the Euler sort will increase the curvature. Either way,

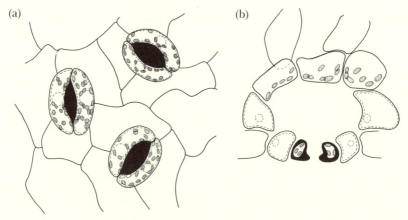

FIGURE 13.3. Stomates and their flanking guard cells on the lower surface of a leaf. In surface view (a) the stomates are dark; in cross section (b) the guard cells are dark and the substomatal chamber conspicuous.

outward expansion pushes on other epidermal cells—the pressure in a guard cell of the leaf of a bean plant is about 3.6 MPa, which is well above the 0.6 MPa of adjacent cells (Raschke 1979).

Other uses of such osmotic devices occur in the process by which the tendrils of climbing plants curl around a support, the reorientation of leaves in species that have different day and night postures (Satter 1979), and in the "gun" of the ballistic fungus *Pilobolus*, about which more will be said in the next chapter.

Protein engines

Muscle is only one of a large group of mechanisms for achieving motility that depend directly on the properties and interactions of proteins. Within cells, microtubules (Figure 13.4) not only serve as supportive elements but are strikingly common wherever cells undergo the chromosome movements associated with division or the shape changes associated with development. Microtubules make up the most conspicuous components of cilia and flagella (except those of bacteria)—most of these organ-

FIGURE 13.4. The arrangement of components in a microtubule. Two types of protein (black and white here) are wound in a long helix about 30 nanometers in diameter.

259

elles do locomotory tasks as they send waves of bending outward from their attachment at the cell surface. At least under some circumstances, microtubules may accomplish motility—pushing things around— through operation as polymerization engines; that is, they may exert their force by adding monomers to their ends while at the same time resisting bending. As structures that resist bending, they are probably the world's smallest functioning beams.

Still smaller organelles, microfilaments, also seem to be associated with movement—especially that curious internal movement of the contents of plant cells (recall the talk about cell size in Chapter 8) called "cytoplasmic streaming" or "cyclosis"—see, for instance, Allen (1974). The speeds of streaming, about 5 μm·s^{-1} (a few much faster cases are known) are far from trivial to a viewer with a microscope (which magnifies space but not time)—we're not talking about something evident only in time-lapse movies. Microfilaments are made of a protein called "actin"; it and another, "myosin," are apparently present in all living eukaryotic (nonbacterial, loosely) cells. The two proteins are, in fact, the basic constituents of muscle. The actual mechanisms by which microtubules and microfilaments generate motion are still unclear, and not for lack of investigators or literature either.

Much more is known about the mechanism of muscle contraction—in this motor, filaments of the two proteins actively shear past each other, ratcheting along through the interactions of a series of cross-bridging structures. Note, though, that the process, inevitably referred to and overtly describable as contraction, is not actually based on a specifically contractile entity—macroscopic contraction simply results from a shearing interdigitation of a large number of filaments of fairly fixed length (Figure 13.5). And even that contraction verges on the illusory—muscle is mainly water and similarly rather incompressible. So contraction in one direction entails expansion in another, as it must in a constant volume system. Try taking enough air in your lungs so you just barely float in a pool; then tense as many of your muscles as you can. Result—nothing; you don't sink. Conclusion—muscle doesn't really contract and become denser. What it does is pull one end toward the other while swelling in the middle.

A worse linguistic abuse, but nonetheless useful term, is "isometric contraction"—the generation of tension by a muscle under conditions where the load is so great it can't be appreciably moved. The isometric condition turns out to be the one in which a muscle generates maximal force. To compare muscles of different sizes it's most useful to consider stress rather than force, and data for maximal stress have been accumulated from a wide variety of animals. The results are satisfyingly monoto-

thick filaments
and cross-bridges

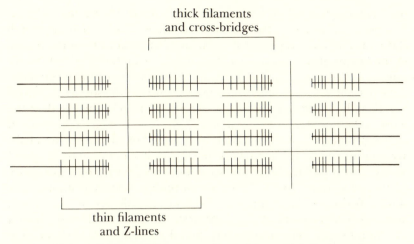

thin filaments
and Z-lines

FIGURE 13.5. A diagram of the contractile machinery of skeletal muscle. Shortening involves the interdigitation of thick and thin filaments (horizontal) brought about by some ratcheting action of the cross-bridges (vertical). The sarcomere is defined as the material between adjacent Z-lines.

nous—so-called striated muscle (which we regard as the ordinary kind) can pull at about 200 kPa, with little variation whether it comes from worms, mollusks, insects, or vertebrates (McMahon 1984). The figure, incidentally, is about 500 times lower than the tensile strength of collagen, which is why fat muscles can be hooked to bones via skinny tendons.

Beyond the matter of maximum isometric stress things get more complicated. Without shortening, a muscle does no actual work against an external load, so enabling a muscle to shorten by reducing the load initially increases the work it will do in a contraction. But at the extreme, with zero load, there's plenty of shortening but again no work is done; maximum work is done at about 40% of maximum load. Furthermore, the less the load, the faster a muscle will contract; maximum power (force exerted times velocity of shortening) is produced at 25 to 35% of the maximum speed of shortening. The gears of bicycles allow riders to keep their muscles shortening in a range of speed that gives a substantial power output even with changes in the slope of the terrain—about one full crank per second is a good rate for the casual cyclist.

The actual speed of shortening depends on the particular muscle—muscles may be equally strong but are otherwise a diverse bunch. Even with identical machinery, speed depends on the length of a muscle; this may seem strange until one realizes that the longer muscle has more contractile units lying end to end. If the filaments of each unit (the unit is

261

the "half-sarcomere," as shown in Figure 13.5) creep over each other at the same rate, whatever the length of the whole muscle, then the overall shortening will be proportional to the number of units in series, that is, to the length of the muscle. To compare the operation of the different units, it's therefore useful to define an "intrinsic speed" independent of the length of the whole muscle. In practice, we use the speed of contraction (length per time, or meters per second) divided by the length of the muscle; the unit is the reciprocal second (s^{-1}) and is referred to as speed in "lengths per second." For a contraction speed of 2.0 $\mu m \cdot s^{-1}$ for a half-sarcomere, and a half-sarcomere length of 2.0 μm, a muscle would shorten by its own length in one second—that's a good, round, ordinary figure for intrinsic speed. Of course, a muscle actually contracts by a shorter fraction of its length—40% is an extreme degree of contraction, and 20% is more usual. With a light load, the latter would take about a fifth of a second. Note, incidentally, that the speed of filament shearing is roughly the same as the speed for cytoplasmic streaming cited earlier—the similarity in speeds ties in agreeably with the notion of a similarity in the underlying macromolecular machines.

Besides a fairly wide variation among the muscles of an individual animal, intrinsic speed varies systematically with differences in animal size, at least for terrestrial mammals, as predicted by Hill (1950). The logic is simple. Speed of running varies approximately with the square root of length (Schmidt-Nielsen 1984)—while the larger go faster, they don't go all that much faster. If body proportions are anywhere near alike, then the muscles of small animals must contract only a little more slowly than those of larger ones. But the muscles of the small animal are a lot shorter (proportional to body length, not its square root) and so have fewer half-sarcomeres in series; therefore their intrinsic speeds must be greater. Specifically, intrinsic speeds of muscles should be inversely proportional to the square root of body length. It seems that the muscles of small mammals, as we found with their cells, are functionally different from those of larger ones. Incidentally, the highest known intrinsic speed is that of the "extensor digitorum longus" (a finger extender) of the mouse—about 60 s^{-1} (Close 1972).

How powerful is muscle relative to its weight? Nicklas (1984) computed a maximum power output relative to mass for striated muscle of 200 watts per kilogram (0.6 horsepower per pound); for comparison, the engine of a passenger car produces about 300 watts per kilogram. But a muscle can't sustain that level of output for reasons related to supply of fuel and oxygen and elimination of heat and wastes, and quite aside from the capabilities of the actin and myosin filaments. Still, muscle seems to be the biological champion (at least above the bacteria); the runner-up, a eukaryotic flagellum, can do only 30 $W \cdot kg^{-1}$.

Other biological motors

Nature has a few more ways of achieving motility up her sleeve. A variety of plants have devices to throw seeds some distance from the parent organism—the most common way of powering the heave is by straining a structure as water evaporates. At some critical degree of dryness, breakage somewhere acts as a trigger, and off flies the seed. A mechanism based on the surface tension of water is used by a fern that forceably discharges spores from its sporangium—evaporation decreases the volume of water and increases the surface curvature in a series of cuplike dead cells until the sustainable cohesion is exceeded; water then vaporizes, and tension is relieved by a movement analogous to that involved in throwing a spear (Steward 1968). The bacterial flagellum is driven by a proton pump driving a reversible rotary motor quite unlike anything elsewhere among organisms. And a variety of microscopic algae (diatoms, desmids, and others) engage in a mysterious sort of gliding motion across surfaces; the motion is usually associated with the secretion of slime, but the source of the shearing force between organism and substratum is unclear and certainly different from any of the mechanisms we've considered so far (Halfen 1979).

TRANSMISSIONS

Here, to me, is the really neat part of motility, the vast diversity of schemes by which organisms contrive to make their motors actually *do* the mundane and profane tasks of daily life, the "cleverlittledetails" that provide the crucial grist to grind exceeding small in all those doctoral dissertations. At this point, a little additional physical background is necessary, together with a few items of terminology.

Levers

Levers appeared briefly in Chapter 2 but have kept a low profile since. First, let us remind ourselves that no conservation law applies directly to force. It's work (finally, energy), not force, that is conserved. A large applied force can, without evident inefficiency, press on something with much diminished force; and a small force can be persuaded, without fear of perpetual motion or divine retribution, to exert a greater force against some load. That persuasion is the accomplishment of various arrangements of pulleys, gear trains, hydraulic couplings, and, simplest of all, levers.

Recall the situation in which a force has a line of action lying to one side of an axis of rotation (Figure 2.4). We call the shortest distance between the force's line of action and the axis the "moment arm" or "lever arm" of the force; we call the product of force and moment arm the "moment" of the force. (The dimensions may be force times distance, but moments do not equate with work—in work, force and distance run along the same line.) In all cases, the rule is that for objects at rest or in steady motion (constant velocity) moments must balance out to zero, where one direction of rotation (clockwise, say) is taken as positive and the other as negative.

Simple levers are composed of a rigid beam, along which three elements are identifiable—(1) a fulcrum, pivot, or hinge point; (2) a point at which a force, the "effort," is applied; and (3) a point at which the force moves something, the "load." The effectiveness of a lever is conventionally expressed in terms of "mechanical advantage," which is simply the ratio of the force exerted against the load to the force of the effort—if by pushing with a force of ten newtons you indirectly apply a hundred newtons, then the mechanical advantage is 10. The mechanical advantage turns out to be equal also to the ratio between the moment arm of the effort (the distance between effort and fulcrum) and the moment arm of the load (the distance between load and fulcrum)—the ratio is a consequence of the rule that moments must add up to zero.

For our purposes, the term "mechanical advantage" can be misleading, since one's inclination is to take "advantage" literally and inadvertently cast muscular machines in a bad light. We'll replace it with the more literally meaningful "force advantage," the factor by which the applied force is multiplied by the lever or transmission. We can contrast the latter with its reciprocal, "distance advantage," the ratio of the distance the load moves to that moved by the effort. ("Distance advantage" proves to be identical to "speed advantage"—if an action takes a given time, then going further is concomitant with going faster.)

Levers are commonly divided into three classes, unimaginatively called "first class," "second class," and "third class." The labels are totally uninformative; and, gallingly, the three don't form a single tripartite division. First class levers are the most "ordinary" sort; as shown in Figure 13.6, their arrangement is effort-fulcrum-load. Common examples abound— the nail-pulling claws of hammers, pliers, toggle switches, drink can piercers, and so forth. The force advantage is most often greater than one, but if the effort arm is shorter (effort closer to the fulcrum) than the load arm, then the advantage will be less than one.

In a second class lever the fulcrum is at one end, and the arrangement is fulcrum-load-effort. Common examples—the puncturing device of hand-held can openers, some nutcrackers, some clamp-and-twist jar top removers, some garlic presses, toilet seats, etc. The force advantage is always greater than one and the distance advantage less than one.

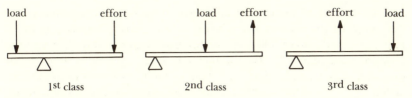

FIGURE 13.6. The three classes of levers.

In a third class lever the fulcrum is again at one end, but the effort arm is now the shorter, with the order being fulcrum-effort-load. The force advantage is always less than one and the distance advantage greater than one. Common examples are not too common—one is a gate latch that is lifted between the hinge and the catchment.

The three classes exhaust the logical possibilities, but second and third class levers are more closely related to each other than either is to the first—both have the fulcrum at one end and divide the world so that second class levers have the better force advantage and third class levers the better distance advantage.

Muscles and bones as levers. A muscle is relatively good at producing force and relatively bad at getting shorter. If it gets only 20% shorter it will have to operate with a substantial distance advantage to move a limb through an angle that may approach 180°. For the good distance advantage it naturally will suffer a poor force advantage—the product of the two must be unity. Agility and strength are functionally antithetical in the design of a skeletomuscular system, whatever the changes we effect through training. A good distance advantage implies use of either third class levers or of first class levers in which the effort arm is much shorter than the load arm—arrangements relatively uncommon in our everyday grab bag of gadgets.

Quite commonly, a first and a third class lever are used together, with one muscle "antagonizing" the other, that is, each muscle acting as the lengthener of the other. (Muscle, be reminded, is a shortening engine, so some other device must intervene to stretch it back out again.) Thus the biceps muscle of the upper arm, which we use to raise the forearm, operates between its bones as the effort in a third class lever (Figure 2.7)—at one end is the pivot; a little way out on the forearm is the attachment of the muscle, defining the effort arm; and at the far end of the forearm is the hand, the site of the load. On the opposite side of the upper arm is the triceps muscle, which attaches to the opposite side of the pivot, just above the elbow. The triceps, then, is the antagonistic muscle operating as part of a first class lever.

265

What sorts of advantages actually result from such arrangements? For the biceps of the human arm, Currey (1984) cited an effort arm of 55 mm and a load arm of 300 mm; the force advantage is thus 0.18 and the distance advantage 5.45. He commented on the contrast between a human and a mole, which has huge flanges on its forearms to get the attachment of the biceps further away from the pivot and thereby to increase the force advantage in a manner appropriate to the mole's fossorial habit. Our triceps, by the way, has a much lower force advantage than our biceps, about 0.05, and a distance advantage of 22—we're not too good at lifting forearm weights while upside down! More realistically, you will find it not at all easy to support your weight on your hands with your hands on supports over a meter apart. It must be borne in mind that the advantages vary from moment to moment depending on the amount by which muscles have shortened.

A few more figures for advantages, these from Tricker and Tricker (1967). For the quadriceps femoris, the large muscle of the front of the thigh, whose tendon runs across the kneecap (patella) and which extends the lower leg, the force advantage is 0.06 and the distance advantage is 17.5. This is, of course, the muscle that mutters nasty remarks when you try to hold a partial knee bend. Further down the leg is the gastrocnemius on the back of the calf; it tapers down to the Achilles tendon, which runs across the heel. You contract this muscle to stand on tiptoe. In the process, half your weight is the load for each leg, with the muscles operating in first class levers. But concomitant, perhaps, with a limited range of motion (a few inches of height gained) and high load, is a relatively great force advantage of 0.43 and a low distance advantage of 2.3. The main muscle for raising the lower jaw is the temporalis, located just in front of the hinge point of the jaw. It's part of a third class lever with its effort arm fixed and load arm determined by which teeth are in action. Thus you bite with incisors at a distance advantage of 3.5 but chew with back molars at a distance advantage of only 1.7, low enough so you can exert forces hazardous to life, limb, and dentures.

These relatively straightforward examples only scratch the surface of the arrangements and applications of living levers. Consider how a small insect might be arranged to beat its wings. The more rapid the beat, the smaller can be the wings, which means less to synthesize, less to injure, and less to attract predators. Muscle contraction is rapid, but not extraordinarily so even at its best. One way to increase the rapidity of motion is to limit contraction to a small fraction of muscle length and then to use a linkage having a very great distance advantage. Another problem—flight is very hard work; and, since the maximum power per unit mass of muscle seems to be unalterable, it inevitably requires a lot of muscle. About

25% of a fly, for instance, is flight muscle, and the breast muscles of birds are impressively large. To add yet another complication, an insect has its muscles on the inside, which restricts the ways in which a large mass of muscle can be hooked to a movable structure with appropriate leverage.

Most insects use an arrangement in which two pairs of muscles run across the inside of the thoracic box (Figure 13.7). Dorsoventral (vertical) muscles drive the upstroke and longitudinal muscles the downstroke, but neither pair actually attaches to the wing base, and both pairs run from wall to wall without tapering to tendons. What they do is distort the thoracic box just a bit, and that distortion is linked through a complex hinge to the wings. Use of the chitinous cuticle as linkage permits a lot of muscle to be fitted into a small container; the particular arrangement of the hinge means that no massive beam has to extend a wing far into the thorax to receive the muscles. The muscles contract only a very short distance, but the wing can swing through a broad arc. For a generic fly,

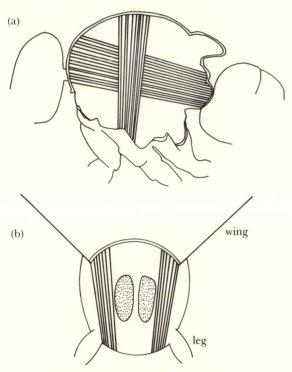

FIGURE 13.7. The flight muscles in the thorax of a fly, in cutaway side view (a) and cross section (b).

267

the dorsoventral muscles might be about 2.5 mm long and contract 4% of that length, while the center of the wing is about 5 mm from the hinge and swings through a third of a circle—about 120°. As you can easily calculate, that comes out to a distance advantage of 100, or a force advantage of 0.01, quite extreme values! Even so, shortening speeds are relatively high for these "indirect" flight muscles. If a fly beats its wings up and down 200 times each second, then the center of the blade is moving 4.2 $m \cdot s^{-1}$; the muscle must be shortening 42 $mm \cdot s^{-1}$, which is an intrinsic speed of 16 lengths per second.

The scheme has some limitations—none of the (up to) four wings can beat at different rates or very far out of phase with any other. Dragonflies and damselflies (Odonata) use a different setup, powering the wings with muscles that attach directly to the wings just inside the hinges; as a consequence they seem to gain unusual maneuverability. It's no trivial matter to be large but still adequately maneuverable to catch small, flying prey. A dragonfly, for instance, can make an unbanked sudden turn when hovering by using its wings in the manner of a rower who turns a boat by pushing on one oar while pulling on the other (D. E. Alexander 1986). Odonata, though, have long wings and relatively low wingbeat frequencies (about 20 per second).

One point about the operation of levers that is easy to overlook is the fact that the lengths of effort and load arms must be measured at right angles to the direction of the forces. A muscle or tendon may attach to a bone or piece of cuticle quite far from a joint, but if the attachment is at an angle other than 90°, then the actual effort arm is less than the distance between attachment and joint, as shown in Figure 13.8a. In the examples picked earlier, the effort arm (at some point in the action) was in fact the distance between attachment and hinge—biceps, triceps, gastrocnemius, and temporalis are nice, tidy cases. But with most muscles in our appendages running along the bones, effort arms are likely to be awkwardly short.

An articulated crane has the same problem. How can a cable from the cab elevate the outer, hinged part of the crane? If the cable just runs over a pulley whose axis is the hinge, then the effort arm will be minute—just the radius of the pulley. What's done is to add an extra strut, a jib nearly perpendicular to the two beams of the crane, which increases the length of the effort arm (Figure 13.8b). Currey (1984) pointed out that the extra bone at the knee joint, the kneecap is an analog of that jib (Figure 13.8c). Its role is to keep the tendon that runs downward from the front of the thigh away from the hinge line of the knee. We're not great at kicking, but we'd be worse without that little bone. For the joint between upper arm and forearm we use an extension of the forearm, the elbow, to accomplish the analogous task for the equivalent extensor muscle.

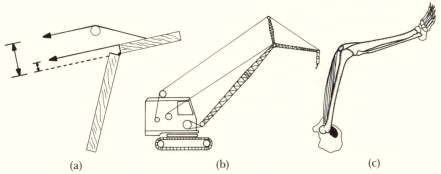

(a)   (b)   (c)

FIGURE 13.8. The uses of a jib. In (a) the moment arm (two-ended arrows) of a force is increased by running a cable over a pulley instead of pulling directly on the hinged member. The force is exerted in the same direction but is much more effective when its line of action is further from the hinge. (b) An articulated crane runs the support cable for its outer, hinged member over a jib for just this reason. (c) In a human leg the tendon of the muscle that extends (straightens) the lower leg runs over an extra bone, the kneecap.

The problem of achieving adequately long effort arms is made more difficult if the muscles have to operate and attach entirely within a tubular skeleton; I haven't specific numbers to cite, but it's clear that the leg muscles of insects, for instance, work with very low force advantages and large distance advantages, as shown in Figure 13.9. Compensation might be achieved with an arrangement similar to that of the flight muscles mentioned earlier—a short, wide muscle produces relatively more force but shortens less than a long, narrow one of the same volume and power. But how to stuff a short, fat muscle into a long and skinny leg? The usual arrangement is something called a "pinnate" (for "featherlike") muscle, whose fibers run obliquely inward from the exoskeleton to a movable apodeme running down the middle and out to the load. The apodemes of crab and lobster claws that you pick out before eating are part of such an arrangement. And insects are not alone in using pinnate muscles— Gans (1974) described how an amphisbaenid (a limbless, burrowing "worm-lizard") uses pinnate muscles back in its trunk to aid its head in tunneling operations. The body muscles of a typical fish seem to be a very much more complicated case of the same underlying scheme—all the muscles are close to the long axis of the body, and very short, and their shortening is conveyed to skin and skeleton by great sheets of tensile tissue.

It ought to be emphasized that the systems so far mentioned (except the fish) are simple mechanisms, having a few solid elements, a hinge or two, and just a few muscles. The reader who wants to see some magnifi-

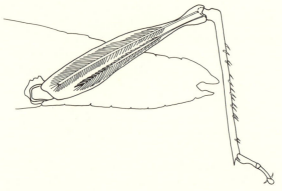

FIGURE 13.9. The main muscles of the hind leg of a
grasshopper. Note that the lever arms must be extremely short because of confinement within a tubular
exoskeleton; that the extensor muscle above is, as
expected from the demands of jumping, larger than
the flexor below; and that the muscle fibers are short
and oblique in this pinnate arrangement.

cently complex mechanisms can do no better than to look at specimens
or descriptions of the skulls and jaws of snakes that swallow prey fatter
than themselves, or at the heads of fish that in one movement open and
protrude both upper and lower jaws for suction feeding.

Muscular hydrostats as antagonists and levers

We saw that biceps and triceps muscles act as antagonists—the contraction of one reextends the other. In the mantle of a squid, short, radial
muscle fibers thin the mantle and thereby reextend the circumferential
muscles. The high internal hydrostatic pressure and slightly stretched cuticle reextend the longitudinal muscles of a nematode as a whole and
permit contraction on one side to effect reextension on the other. Remember that while muscle fibers actively shorten, they do not lengthen
except when pulled upon by some external force. The action is closely
analogous to the piston of an internal combustion engine—burning fuel
pushes on the piston; something else, whether another piston or a fly-
wheel, must alternately push it in the opposite direction. In organisms,
though, the identity of the antagonist isn't always immediately apparent.

Another problem. Where muscular mechanisms work with discrete
hinges and solid beams, it's easy to analyze them in terms of familiar levers. But muscles aren't always used in arrangements with counterparts
in our technological world—we've already mentioned nematodes, sea

anemones, squids, and other nearly or entirely boneless (or strutless) beasts. Still, these systems could be treated as combinations of muscles and fairly simple passive components. What if a system is almost entirely composed of muscle, with only a little passive tension-resisting stuff dispersed here and there? Where is the crucial antagonism, and what meaning now attaches to force and distance advantages? Stick out your tongue—such systems are not exactly obscure.

Human (and lots of other) tongues are examples of these "mostly muscle" systems; other examples, cited in Chapter 12, are the highly extensible tentacles and less extensible arms of squid and the trunks of elephants—Kier and Smith (1985) coined the term "muscular hydrostats" for these arrangements. They (the hydrostats, not Kier and Smith) have an impressive repertoire of motions—they can extend, contract, bend, and twist, and also change their stiffness over a wide range quite independently of any change in shape. Moreover, the motions need not be slow—the typical 70% extension of a squid's tentacle takes place in about a sixtieth to a thirtieth of a second (Kier 1982). Incidentally, this speed is associated with muscles having especially short sarcomeres, around 1 μm (Kier 1985), in proper accord with an earlier generalization.

What these systems can't do is change their volume—muscle, like the water of which it is mainly composed, is for all practical purposes incompressible. And that's the crucial element that enables a muscle to act as both tensile and compressive component of a solid beam or column, and to antagonize another muscle without the aid of either discrete struts or an internal bag of water. Imagine a solid muscular cylinder, with muscle fibers running in various directions within it; the fibers can cause the cylinder to change length and radius, one (of course) at the expense of the other. It's a simple matter to calculate the relationship between radius and length for a fixed volume (the formula is that for the volume of a cylinder, given in Appendix 2); the result is plotted in Figure 13.10. We see that halving the radius (or diameter) quadruples the length. Therein lies a scheme for leverage as well as antagonism.

Consider, first, a short fat cylinder of the proportions of a tuna fish can, with a radius of one unit and a length of one unit. The circumference is then twice pi. If an outer circumferential muscle contracts 16%, a reduction of one unit in girth, what will happen to the length of the cylinder? It will increase by not one unit but only 0.4 units, with a very great force along with this short extension. The force advantage is 1.0/0.4 or 2.5, while the distance advantage is 0.4. Now consider a long skinny cylinder of the same volume as before, with a radius one-twentieth of its length (the radius becomes about 0.37 units, the circumference about 0.74 pi, and the length about 7.4). Again the circumferential muscles

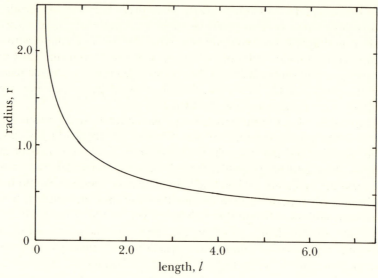

FIGURE 13.10. Radius versus length for a cylinder of fixed volume. In
this example, the volume has been fixed at pi units; thus the radius
equals one over the square root of the length.

shorten by 16%, now only a little over a third of a unit—what happens to
the length? It again increases by about 40%, but this now amounts to a
three-unit extension—great distance but little force. We can calculate a
force advantage of about 0.13 and a distance advantage of 7.9. The situation is quite different from the earlier one. Instead of a short but
forceful push we now have a long, easy, and probably rapid push, a nice
scheme to get some real distance from muscle; notice that it can give a
long axial thrust without a single bone. A squid tentacle, for instance,
need decrease its diameter only 23% to get an extension of 70%.

Squid tentacles and lizard tongues—the latter are more impressive mechanically than the mammalian organ—operate at high speeds and at
great distance advantages. Elephant trunks and squid arms are slower
and extend less far but more forcefully. Despite the fact that tongues,
tentacles, and trunks (the alliteration, for once, is not mine) must represent separate evolutionary origins of muscular hydrostats, both their mechanical capabilities and anatomical arrangements are strikingly similar.
All have muscles oriented (1) perpendicular to the long axis of the appendage, that is, crosswise or circumferentially, (2) parallel to the long
axis and thus longitudinally, and (3) obliquely or helically around the
long axis. Contraction of perpendicular muscles causes overall elonga-
tion; contraction of longitudinal muscles causes overall shortening; con-

traction of one or the other sets of helical muscles causes twisting. In addition, contraction of the helical muscles causes either lengthening or shortening depending on whether, given the shape of the cylinder at a particular moment, their fibers run at angles greater or less than 55° to the long axis, as in the hydrostats of Figure 10.11.

Hydraulic linkages

Apply inward pressure on one wall of a fluid-filled container and the same pressure is exerted outward on all other walls. If the fluid is nicely incompressible and the container is stiff, you don't have to push inward through any appreciable distance to exert the pressure. Moreover, if the fluid moves outward elsewhere, the volume that moves equals the volume you push inward. If the movements aren't too rapid, viscous losses through shearing near walls are minimal, so the work done either to compress or move the fluid can be negligible. You do some work, the increase in pressure times the volume moved, and you get out almost the same work. If the fluid moves outward over a greater area than it is pushed inward, then it advances a shorter distance but exerts a greater force. (Force, remember, is pressure difference times area.) Figure 13.11 illustrates the principle in a model made of two hypodermic syringes, some flexible tubing, and a soap solution. The commonest technological example is, of course, the braking system of an automobile, with a master cylinder replacing one syringe and the slave cylinders of two wheels the other. Notice that the idea of "leverage" applies here also—one just picks the relative cross-sectional areas of the cylinders to get the desired advantages.

Hydraulic linkages occur in a wide variety of biological systems, many of which have already cropped up in these pages. The jetting of a squid is probably the most impressive, but the operation of the tube feet of starfish, the crawling of worms, the erection of a penis, in short, all of the pressurized systems in which motion occurs have, in essence, hydraulic linkages. We even use a hydraulic scheme to make contraction of our left ventricles filter blood through the glomeruli of our kidneys, but I'll leave the details to the physiology textbooks.

FIGURE 13.11. A hydraulic linkage made from two hypodermic syringes. The piston of the smaller syringe moves faster and farther than that of the larger one.

One particular hydraulic device is worth a little more attention here, partly because its existence comes as somewhat of a surprise and partly because it involves just the kind of skeletomuscular system we've been talking about. The eight legs of a spider do not look overtly different from the six of an insect, but they have a curious feature known for almost a century. While properly equipped with flexor muscles (ones that decrease the angle between one segment and another), spider legs lack the antagonistic extensor muscles (ones that increase that angle toward 180°). To the extent that anyone cared, the assumption was that elasticity of the interarticular membranes provided the antagonistic force, not an unreasonable idea on the face of it. But Ellis (1944) remembered that spiders die with legs flexed, not extended as they would if elasticity did the extension. He then found that cutting off the tip of a leg prevented reextension until the tip was resealed, and that mild exsanguination reduced a spider's ability to extend any of its legs. He therefore suggested that extension in spider legs was hydraulic, not muscular or elastic. This idea was confirmed by Parry and Brown (1959), who measured resting pressures of 6.6 kPa and transient pressures of up to 60 kPa (over half an atmosphere) in spider legs. An isolated leg could lift more weight as the pressure inside it was increased, and the spiders turned out to have a special mechanism to seal off a joint when a leg was lost, lest they fatally depressurize.

The system works well enough so one doesn't notice that it's quite different from the scheme used by insects; some spiders (salticids) are even good jumpers. Perhaps the hydraulic arrangement permits larger flexor muscles that can grapple with larger prey—spiders are preeminently predators on creatures of around their own size. But my adaptationist biases show through—present advantage is not necessarily admissible evidence when looking at a question of origins.

Why wheels are rare

One final topic in this fast trip through the appurtenances of motility, a topic better known for its absence than its presence. Elementary biology, when I had to take the course, included the bald assertion that nature had never invented the wheel. I've forgotten what moral followed that particular statement; in any case, it's a fine example of the trap set by inductive reasoning—the statement can't really be proven, only disproven. In fact, it's false. Bacterial flagella are not only smaller but do not wave and wiggle as do the cilia and flagella of other organisms. They truly rotate and are driven, as mentioned earlier, by a rotary engine just inside the cell wall. Berg and Anderson (1973) first argued the case for rotation using decent evidence, of which there is now lots more. For instance, one can make flagellaless mutants stick by their rotary basal hooks to a glass slide—with the slide as an immovable load anchoring the hook, the body rotates instead. And there's the absence of appropriate protein for bending in the shaft together with the presence of what can be little else than an engine at the base.

The fact remains, however, that wheels are uncommon in living systems; if we take a purist's approach and look only for arrangements in which a wheel and axle rotate with respect to the rest of an organism, the bacterial flagellum remains the only clear case. Plenty of structures *translate* in a circular path—a large group of tiny creatures called rotifers, or "wheel animalcules," are typically equipped with a pair of ciliary organs that wave around in circles, and the filtering fans of some aquatic fly larvae (Figure 6.6) are similar but larger. Neither, though, actually rotates.

A few years ago, Gould (1981) devoted one of his engaging essays to the subject of the scarcity of wheels, arguing the difficulty of evolving them *de novo*—moving through nonfunctional intermediates is not nature's style, and it's hard to see the benefit of a partially evolved wheel. He also noted the difficulties of arranging nutrient supplies, axle seals, and so forth—matters of little concern on the scale of bacteria. He asserted that "wheels are not flawed as modes of transport; I'm sure that many animals would be far better with them," and that "animals are debarred from building them by structural constraints inherited as an evolutionary legacy." The goodness of wheels is obvious—modern technology without wheels is unimaginable.[2] Add a bicycle to a person and you reduce threefold the work needed to go a given distance—and that despite the extra dozen or more kilograms of machinery (Tucker 1975). A predator on a bicycle should have a formidable advantage!

Or is there something wrong with this most magnificent device? My fractious friend LaBarbera (1983) wrote an article as a rejoinder to Gould, arguing that "the concept of the general superiority of wheels as a mode of transport is false, and the limitation which constrains the evolution of such systems . . . lies in the limited utility of rotating systems in most natural environments." LaBarbera argued that wheels are wonderful on flat, hard terrain but pretty awful elsewhere, and that appropriate surfaces for effective use of wheels are few and far between—the smaller the organism, the bumpier its world. It takes a lot of power to ascend a curb, and a curb of height greater than the radius of a wheel is an absolute impediment, whatever the power of the engine. The covered wagons of the North American prairies had enormous wheels even for use in flat,

---

[2] How well the Amerindians did without them depends on one's point of view. They apparently knew about wheels and had them on toys but never used them in transport.

dry country, and wheeled cultures are inevitably road-building cultures. Wheeled vehicles can't turn as sharply as legged ones unless they're extremely complex—societies that didn't use wheels could get away with narrow and tortuous streets without functional inconvenience. Or, given such streets, the introduction of wheeled vehicles would hold little appeal. And on even slightly spongy terrain any power advantage is lost. A wheelchair is *very* hard to propel on soft carpet—dense pile without padding is recommended for buildings that must be accessible to the handicapped.

Gould's point is not disproven—in a sense it is not subject to strict disproof. But his point is less self-evident than it has usually been regarded—the dominant constraint on the use of wheels in nature is as likely to be physical as due to some flaw in the adaptive ability of the evolutionary process. Still, the really exciting discovery for the iconoclasts among us is Berg's—the wheel has come full circle, and he noticed it even if it didn't squeak.

CHAPTER 14

Staying put and getting away

"I shot an arrow into the air,
It fell to earth I knew not where."

Henry Wadsworth Longfellow,
confessing his ignorance about
ballistics and trajectories

THE PRESENT protagonists (or, perhaps better, antagonists) are the forces on organisms due to gravity and to the movements of surrounding fluids. Gravity has not been much in evidence here since the discussion in Chapter 4 of the Froude number—it simply isn't a heavy item of concern to most creatures, which, unlike ourselves, are either small, neutrally buoyant, or both. But trees stand erect, people walk, ducks swim on the surface of ponds, fleas jump, and plenty of plants project propagules. Drag was a major concern in Chapters 6 and 7 but has been only a parenthetical presence since. These environmental forces raise, among others, the two general problems that give a title to this chapter—first, how to stay translationally in place and rotationally in orientation; and second, how to get from one spot to another some decent distance away.

STAYING PUT

I find it curious that people interested in the mechanical problems of organisms have been far more concerned with how creatures manage to move than with the converse, how they manage to stay in place. One might even make a devil's advocate argument that the latter is more important in the overall scheme of nature; certainly in a forest of trees or in wave-swept intertidal areas, fixation is the dominant mechanical problem.

In such situations, though, we're looking at a far wider array of problems and arrangements than one might initially guess. There are problems involving gravitational stability—will an object or organism, left to itself, rearrange its position so its center of gravity is a bit lower? How might the downward force of gravity (an organism's weight) be employed to offset the sideways force of drag and perhaps some undesired lift as well? And there are problems of mechanical attachment—how might the parts of an organism be interconnected so they don't become unfastened when pulled, pushed, twisted, or sheared? How might an organism attach itself to a substratum, whether rock, soil, beach, prey, or host?

How to stand stably on the ground

In a world in which gravity is the only external force, the game is simple—its only rule is that the center of gravity, the point at which one's mass behaves as if concentrated (Chapter 2), must be directly above the point of contact with the ground. We disobey at our peril; indeed, a substantial act of will is required (at least for an adult) deliberately to tip over, and the neuromuscular machinery involved in our postural reflexes is enormously complex. At least it's possible, if one's center of gravity is shifted by some lateral force, to move one's legs so the point of contact is again directly beneath. It takes at least three points of contact with the ground to achieve stability; the further requirement is that the center of gravity be above the triangle whose apices are those three points. With two legs or one, the breadth of the foot or feet becomes crucial—on stilts you can walk easily but stand only with great difficulty, and while the great blue heron may extend a skinny leg downward, the leg attaches to a wide foot. With any persistent sideways force the situation gets very much more complicated, but again our neuromuscular machinery permits us to walk in a wind or haul a load.

What about an organism that can't deliberately move its contacts with the ground as its center of gravity wanders around? A tree is the extreme case; the schemes used by trees to avoid falling over ("wind-throw" seems to be the jargon) are varied and, as far as I can tell, surprisingly poorly known. We touched on the matter earlier when talking about drag and about the mechanical properties of wood—let's now try to put some pieces together. First, failure comes in two types—snapping and uprooting; which predominates varies greatly from species to species and forest to forest (Putz et al. 1983). Where I live, failure of an entire tree seems most often to involve uprooting—grabbing the ground is the weak link. No real surprise—the poor tensile strength of soil cannot be countered directly by selection for better-built trees. That means that the strength of the trunk is almost inevitably adequate or else that it may not be the really relevant property.

Consider a hardwood tree such as a large oak. As we weekend woodcutters know well, the main part of the mass of the tree is concentrated in the trunk—the branches are conspicuous but yield much less fuel. And the base is wide, stiff, and shallow—uprooting leaves the lower (now horizontal) trunk well above the ground, and the hole created is not at all deep. I've argued (Vogel 1981b) that what matters for such a tree is not the strength so much as the stiffness of the trunk, and that, as with so

many of our own products, adequate stiffness has the incidental consequence of superadequate strength. Figure 14.1 gives the turning moments on such a tree, a clockwise moment due to its height and drag in a
wind, and a counterclockwise moment due to its weight and the width of
the base. How might natural selection manipulate the variables?

Height in a sense is a given—a consequence, most likely, of both inter-
and intraspecific competition for light. Not that sunlight is appreciably
more intense a hundred feet up, but the trees are in a kind of shade-
avoiding arms race—none is better off for all that height, but there is no
effective evolutionary mechanism for a treaty whereby each tree forgoes,
say, fifty feet of trunk. Drag is a variable subject to manipulation; the
relatively low rate of increase in drag with speed for clusters of leaves
(Chapter 7) suggests a history of intervention by the invisible hand of
natural selection. (Darwin's invisible hand is superior to Adam Smith's
except to fans of Ayn Rand and similar True Believers.)

The highest wind speed at which the moment due to weight remains
greater than the moment due to drag is the maximum speed the tree can
withstand. The alternative to decreasing the moment resulting from drag
is an increase in the countervailing moment of weight times basal width.
Toward this end at least three variables are presumably available to the
selective process. (1) Stiffness—the moment at times of high drag is

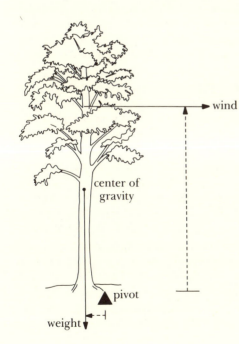

FIGURE 14.1. Moments pro-
moting and resisting wind-
throw of a tree. One moment
is the drag of the wind times
the distance between its line of
action and a pivot point near
the ground about which the
tree might topple. The other is
the tree's weight times the horizontal distance between the
center of gravity and the same
pivot point.

279

greatest if the center of gravity shifts around minimally. If the tree bends, the center of gravity will move laterally and thus reduce the moment opposing drag, clearly a Bad Thing. So the tree, or at least the main trunk, ought to be stiff; if it bends, the bending ought to be restricted to the upper, nonmassive portions. (2) Mass—the resisting moment will be increased by any increase in the weight of the part of the tree that stays above the base in a wind—additional weight is most useful where it has the least freedom to shift laterally. So the tree ought to be heavy, especially in the stiff trunk. Replacing rotten heartwood by concrete would seem a most salubrious therapy—hollow tubes may be stiff and strong, but our tree needs mass per se. (3) Width—the stabilizing moment will be increased by any increase in width, weight, and stiffness of the base. One should perhaps stress the "up" in the word "uprooting"; a stiff, wide base means that the tree will have to rotate upward against its weight before falling downward. These three, of course, are only the main factors. For instance, the upwind roots do grab the soil (with the roots of other plants) to some extent, so there is some tensile resistance to uprooting. And a pile of dirt on the outer part of the base gives extra beneficial weight, so burying the buttressing isn't a bad idea either.

The appearance of an oak tree seems thoroughly rational from this point of view—light and fairly flexible branches on a stiff, solid, heavy trunk above a wide base. There are also, of course, nonmechanical factors—Horn (1971) gives an excellent account of the relationship between geometry and the interception of light. And there are other ways to make a tree. In any wind at all, the staunchly vertical pines in my yard sway disconcertingly; their centers of gravity quite clearly move several feet in all directions. But the pines seem to play a slightly different game—they are equipped with vertical taproots, and any deflection of a trunk must apply a force in the opposite direction (or a moment with the same rotation) to the taproot. The earth may not do well in withstanding a tensile load, but it probably resists a sideways force quite effectively. The one pine that has fallen in the twenty years I've had this property had a rotten taproot.

Yet another kind of tree (a grass to proper botanists) is a bamboo—each stalk holds up less crown than a pine, and the sway is even greater. Bamboo stalks seem very stiff around their bases, and their tubular walls are thickest there. While the bases are not at all wide, bamboos put out a tensilely tough array of roots in all directions and over a range of depths. The stiff region that resists rotation may be small, but the tensile roots go astonishingly far—new shoots appear many feet from any parent stalk, and I assure you that digging out a plant is no fun at all.

Incidentally, the stiffness of most tree trunks may be turned against them in clearing land. In my area the first step in forest fighting is to mow down everything with chain saws. That's easy and satisfyingly destructive. Then a bulldozer has the slow and awkward task of dealing with the stumps. It would seem easier, overall, to attach a cable ten or twenty meters up each tree and then pull with a winch anchored to the base of another tree. In most cases the assaulted tree should uproot with only a modest applied force—its trunk provides a long, stiff lever arm that we're perhaps too quick to amputate.

How to float stably on lake or ocean

To float, a body need only be less dense than the surrounding water. To float with a stable orientation—one that will persist without the aid of an external force and that will restore itself following rotational displacement—requires slightly more complicated conditions. Consider, first, not a floating but a submerged body of the same density as the medium, for instance, a fish with a swimbladder of air (Figure 14.2). It neither rises nor sinks since the upward force of buoyancy equals its weight. In what orientation will it be stable? The upward force is equal to the product of gravitational acceleration and the mass of the water the fish displaces; its line of action runs through the "center of buoyancy," the center of gravity of the displaced water. The weight (downward force) of the fish acts as if concentrated at the actual center of gravity of the fish. In only two orientations do the lines of action of these two forces coincide and thus

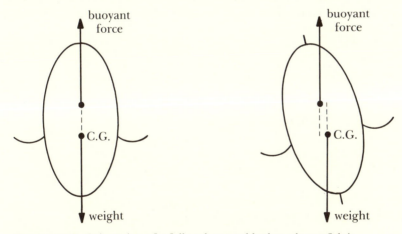

FIGURE 14.2. Orientation of a fully submerged body such as a fish is stable if the center of gravity is below the center of buoyancy—a deflection either way gives rise to a pair of turning moments that act to right the body.

281

generate no turning moments—when the centers of buoyancy and gravity are on the same vertical line. But one of these orientations, in which the center of gravity is above the center of buoyancy, is unstable—any slight roll will create a moment that will increase the roll. Only with the center of buoyancy above is the system truly stable, for only then does a roll generate a moment that opposes the roll. How might a fish keep its center of gravity below the center of buoyancy? If it puts its swim bladder above the center of buoyancy then its weight will be concentrated below, and the condition is met. A dead fish often turns over—that's a matter of gas in the stomach below.

Surface floating, though, is trickier. First, a body (a boat, say) sinks until it displaces a weight of water equal to its own weight. The upward force of the displaced water acts ("action" of displaced water being a convenient linguistic fiction) as if concentrated at a center of buoyancy, defined above. With much of the hull exposed, this center of buoyancy is almost inevitably *below* the center of gravity of the object, which sounds like trouble. But if the object rolls in the water (Figure 14.3), the water line shifts and so (most often) does the location of the center of buoyancy. Because the center of gravity remains fixed, the lines of action of the two forces no longer coincide, and a pair of turning moments is generated. If the turning moments tend to offset the roll ("right" the boat), the body is stable; otherwise, it "turns turtle," an expression certainly offensive to even the thickest-shelled chelonian. So what matters is how the center of buoyancy shifts in a roll, and that depends solely on the shape of the hull. Sides that flare outward as they extend upward make a craft stable; sides

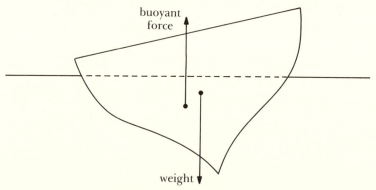

buoyant force

weight

FIGURE 14.3. Cross section of a stable boat hull. A roll shifts the center of buoyancy in the direction of the roll; thus (if the cargo doesn't shift) the centers of buoyancy and gravity generate a pair of turning moments that tend to right the boat.

that curve inward generate instability. It's just the opposite of what our land-based intuition suggests ought to happen, but a few minutes of sketching and counting the squares on graph paper, or a little crude carving of softwood or buoyant soap, ought to be persuasive.

Surface swimming is not especially common among organisms; some of the reasons were explored in Chapter 4. But ducks do it, and they have appropriately flared hulls, at least if they don't swim too low in the water. Sea turtles sometimes swim on the surface; Jeanette Wyneken tells me that the flared, V-shaped bottom is characteristic of buoyant baby sea turtles, which are obligatory surface swimmers. With maturity and the shift to submerged swimming the hull shape changes to one more characteristic of submarines. The change, she says, is most noticeable in greens and least in loggerheads; the differences in hull shapes are evident in the figures published by Davenport et al. (1984).

The problem of adhesion

On many occasions gravity is either unavailable or inadequate. For a tree frog on a pane of glass or a fly on a ceiling, gravity is on the wrong side of the equation describing the balance of forces needed to abide by Newton's first law. Still, organisms stay stuck and do so using a great diversity of schemes for attachment and adhesion.[1] Attachments may be permanent—structural materials are commonly nailed, glued, riveted, or welded; the bones of our skulls never part at their intermeshing sutures and a barnacle ordinarily stays glued to its rock. Or they may be releasable—plugs can be unplugged from sockets, catches and latches disengage, a starfish can walk by alternatively attaching and detaching its hundreds of tiny tube feet. Attachments may be substantially rigid—bolts and welds or the connection between a coral and its substratum—or flexible—buttons, zippers, hinges, adhesive labels, most of our joints, and the connection between the copulatory organs of those species so engagingly equipped.

Wherever nature has a structure, biologists have been painstakingly describing it, but most often paying little attention to mechanical functions. Nachtigall (1974) went the next step, gathering a vast collection of structural schemes for attachment, classifying them by function, and comparing each with its technological analogs. Among interlocking joints he recognizes miters, rabbets, dovetails, and mortises; under releasable attachments he describes plugs and sockets, hooks and eyes, snaps, vises, forceps, anchors, suction cups, and others. The diversity simply defies

[1] A distinction is made between "adhesion," in which two items are held together, and "cohesion," in which a single item (such as water) resists being pulled apart.

summarization. Even Velcro has its biological analog—what Nachtigall calls "probabilistic" attachments, coatings of burrs invented repeatedly as a dispersal device for the seeds and fruits of plants. All burrs require is gentle contact with a sufficiently irregular surface (perhaps fur originally, but cloth is at least as good) and enough of them attach to provide a surprisingly strong connection.

And a lot of biochemistry is involved in attachment. Organisms manufacture good glues, most of them proteins; indeed, until recently most glues used in our technology were mildly modified proteins obtained from blood, hooves (the old horse consigned to the legendary glue factory), bone, skin, fish remains, etc. The similarity of the words "collage" and "collagen" is no mere coincidence—the protein was named for its use as a source of glue. Indeed, the highly hydrated proteinaceous mucuses and the stiffest of biologically produced proteinaceous glues are simply the extremes of a chemical and mechanical continuum. The chemical connections range in specificity and strength from direct covalent bonding through a variety of ionic and hydrogen bonds to the relatively general and weak van der Waals force between molecules; we'll not go further into the mechanisms underlying such complexity.

This van der Waals force may be responsible for a certain amount of macroscopic mischief. Two surfaces brought sufficiently close together most often show some tendency to stick. One rarely notices—few surfaces can be brought into sufficiently intimate contact since their irregularities preclude adequate mating. A pair of scrupulously clean microscope slides behave as if coated with a weak glue, but a few specks of dust from the air are sufficient to minimize the nuisance. Organisms, though, can arrange adequate contact with a rigid surface by using an extremely large number of very small and flexible contact points—the scheme seems to be important to a gecko sticking by its toe pads to a smooth wall, using contacts less than 0.1 μm in diameter (Maderson 1964). Surface irregularities undoubtedly aid attachment by these tiny hooks, but it appears as if the irregularities are not an absolute necessity. Interestingly, the area of the toe pads increases more rapidly with increasing gecko size than does not only body surface but even body volume—the bigger gecko has disproportionately large pads (Maderson 1970). But falling off a wall involves not just mass but a turning moment about a pivot at the bottom of the attachment (consider a picture hung loosely on a wall)—the bigger animal will not only be heavier but project further out from the wall. So scaling more drastically than with volume certainly seems reasonable.

Adhesion—loads and physical mechanisms. A joint or an attachment may be loaded and may fail in any of four different ways—compression, shear, tension, and peel, listed in the order of increasing ease of adhesive failure. Pure compression between two decently mated surfaces requires, of course, no glue at all; it's arguable whether the actual joint is subject to failure. Mortarless walls have stood for thousands of years, and cement is barely adhesive—its tensile strength is an order of magnitude lower than its compressive strength, and it makes a very weak joint even with a rough brick. Cement does, though, distribute stresses and thus minimize the chance of initiating cracks through local tension. Resisting shear takes only a little irregularity in the mated surfaces—a few holes in bricks or blocks, together with a layer of cement, automatically provides sufficient mating. The attachments between our vertebrae and the intervertebral discs do quite well in pure compression and shear; failure seems to be associated mostly with asymmetrical loads, creating bending stresses termed "peeling" in simpler systems.

One particular form of adhesion is at its best against shearing loads. Recall that the viscosity of a fluid was a measure of its resistance to rate of shear (Chapter 6). If two surfaces are separated only by a thin layer of liquid, then even slow motion of one across the other will give a high shear rate; if, in addition, the liquid is very viscous, then movement may take appreciable force applied for respectable lengths of time—the process is called "Stefan adhesion." Its action is of necessity temporary, but so are, say, the wave forces in rocky intertidal areas. I know of no demonstrated case in which Stefan adhesion is the exclusive mode of attachment, but one suspects its involvement when, as Crisp (1960) showed, a steady force will slowly slide barnacles (some at least) across smooth surfaces. But sliding between two surfaces isn't the only way to load liquid in shear—a tensile stress between two objects with liquid between will force inward flow between their confronting surfaces, and the no-slip conditions at the liquid-solid interfaces will ensure the occurrence of shear. Thus Stefan adhesion can play a role in resisting the initial tensile separation of two smooth and well-mated surfaces.

Another physical kind of adhesion is particularly effective for tensile loads in nonsubmerged situations. If a thin layer of water separates two smooth hydrophilic surfaces otherwise surrounded by air, then any attempt to pull the surfaces apart entails the creation of additional air-water boundary in the form of a curved interface (Figure 14.4). Surface tension opposes making more surface and so resists the separation. Thus in such a system, water can act as an adequate glue—the phenomenon is commonly referred to as "capillary adhesion." It ought to be proportional not to the area of contact but to the wetted perimeter as with other phenomena associated with surface tension. It is almost certainly involved when a tree frog placidly rests on a vertical pane of glass, about which more shortly.

Yet another device effective in resisting tension is what is most often

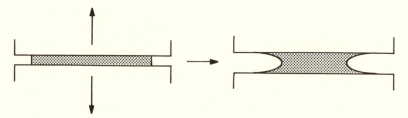

FIGURE 14.4. Adhesion based on surface tension. Pulling apart two hydrophilic surfaces separated by a water film (shaded) increases the area of the
air-water interface, which requires work.

called "suction" but is really a consequence of external pressure. If a flexible surface is pressed against a rigid one, if some force then distorts the
flexible surface so its center is pulled away from the interface, and if a
seal is maintained around the displaced center so fluid cannot move into
the region of reduced pressure, then the external atmosphere will push
the surfaces together. That's the basis of a suction cup. Unless one manages to put an internal liquid in tension, the theoretical maximum stress
holding the surfaces together is the local atmospheric (or, underwater,
atmospheric plus hydrostatic) pressure. Maintaining the seal is crucial,
which limits the use of suction-based devices on rough or porous surfaces. Organisms that use such suction commonly use a bit of mucus as
well, which serves as both sealant and additional adhesive. The maximum
strength of a suction adhesion is, of course, proportional to the area of
contact within the seal. Unlike capillarity, the scheme will work underwater, as every octopus and starfish knows.

The worst sort of load is one that gives rise to the possibility of peel
failure. A piece of tape may adhere very strongly to a surface, but peeling
can remove it with only a small force. What makes such trouble is that
old bugbear, force concentration and the consequent high local stress. In
effect, the small applied force acts over a very small area—the line of
peeling times a short distance in the direction of peel—so the actual stress
in the region of failure is very large. And peeling may be initiated by
loads that are not obviously running across the surface of contact. Consider a tensile load applied in a direction not quite perpendicular to the
surface. It will act almost entirely at one edge of the area of contact (Figure 14.5a), and failure will occur by peeling. And the stiffer the glue, the
worse the situation—the effective area of application of the force will be
smaller and the local stress correspondingly greater since failure will be
occurring along a sharper line.

Loads that might cause peeling are probably quite important in nature.
Neither the holdfast of a large marine alga nor the byssal threads of an

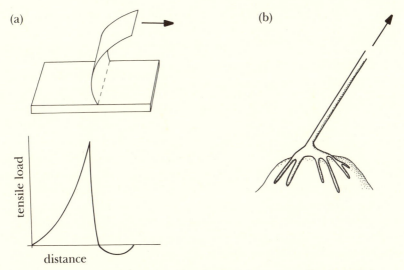

FIGURE 14.5. (a) The concentration of stress when one surface is peeled away from another. (b) The stipe and holdfast of a large alga attached to a rock—the arrangement minimizes the chance of peel failure by making it unlikely that peeling will ever start.

intertidal mussel can be assured of tensile forces perpendicular to their attachment surface. The common response seems to be a tapering disk of attachment that is increasingly flexible toward the periphery (Figure 14.5b)—the whole thing distorts a little to avoid stress concentrations at an edge. Thus a good shape may resemble (and function as) a suction cup even though primarily associated with a different form of adhesion.

Adhesion—unscrambling some mixes. Nothing, naturally, precludes an organism from using some combination of adhesive mechanisms, and determining relative contributions may be tricky. The sucker of an octopus works by suction almost exclusively—breaking force is proportional to sucker area, the forces are properly below (45 to 75% of) the theoretical maximum, and appropriate muscles are present for pulling back the center of a sucker (Parker 1921; Nachtigall 1974). But the tube feet of starfish and other echinoderms are a messier case. Paine (1926) found that about 40% of the adhesion was due to glue rather than suction. Thomas and Hermans (1985) claimed that, at least in one fairly ordinary species, attachment is entirely due to ionic interactions and films of secreted protein—in short, to glue. But demonstration that interference with the protein prevents attachment may just reflect the protein's crucial role as a sealant. Moreover, starfish have the proper retractor muscles for gener-

287

ating a pressure difference—can these possibly be nonfunctional? Still, starfish can attach to wires too fine for their tube feet to act as suction cups, so the best guess is that they use a mix of glue and suction as Paine suggested long ago.

The toe pads of tree frogs provide another mixed and slightly controversial case. Suction seems not to be important—Emerson and Diehl (1980) found that reducing the external pressure had no effect on the attachment. They also found that frogs stick about as well to leaves, glass, and Teflon, but the implied argument against capillarity isn't strong for animals that can make chemically active mucus rather than depend on water. Besides, if a frog on a glass surface is surrounded by water it falls off, which certainly looks like capillarity. But perhaps capillarity holds the frog close enough to the surface for Stefan adhesion, and eliminating the former inevitably disrupts the latter as well. One might expect to resolve the matter with a look at scaling; Green (1981), as well as Emerson and Diehl, determined that the strength of attachment was proportional to the area of toe pads—does this rule out capillarity? Not quite. If (as it appears) toe pads are made of a varying number of flat contacting surfaces of constant size, then the length of edge might be proportional to overall area, just as the total edge length of standard floor tiles is proportional to the area of the floor.

Perhaps the moral of the story is the familiar one that no selective advantage accrues to the use of physical mechanisms in a manner that facilitates analysis by biologists—why shouldn't adhesion be a sticky subject?

GETTING AWAY—ACCELERATION AND TRAJECTORIES

The mechanics of locomotion in animals has received no small amount of attention in recent years. Symposia generate volumes of papers of ever-increasing sophistication, and periodicals such as the *Journal of Experimental Biology* devote a fair fraction of their contents to the subject. I'd like to take a look at one of the less-plumbed aspects of motion. The topic, breaking loose and accelerating against gravity and drag, is the converse of that to which the first half of the chapter was devoted.

Here we'll encounter unbalanced forces—it takes an unbalanced force to produce acceleration according to the familiar $F = ma$ of Newton's second law. Movement will be what's quite descriptively called "unsteady motion." For simplicity, we'll mainly consider constant acceleration, noting the slight artificiality as the price of avoiding some mathematics. It's worth a moment's effort to disenthrall yourself of the common confusion of speed and acceleration in order to give proper attention to the latter. In street usage, a "fast" car is one that can achieve high values of acceleration. (To say "accelerate rapidly" perpetuates this unfortunate tendency to synonymize the two quantities.) We do have contests (drag races and dashes) in which acceleration rather than speed is the chief criterion, but we're only rarely explicit about the distinction.

Nature seems to care about speed in some systems but about acceleration in others. Speed must be more important in any long-distance locomotion, and the notion of most efficient speed has immediate relevance to a migrating bird, a foraging bee, a suspension-feeding whale. Conversely, for the fly escaping the flyswatter, the cat chasing a bird or mouse, the lunging alligator, acceleration must be the crucial variable. The difference is reflected in the appearance of the operative muscles. Pale muscle is the more forceful and, with a greater amount of contractile machinery, characterizes accelerating systems. But its extra fibers replace the devices (mitochondria especially) that provide a sustained supply of fuel. Darker muscle is the engine for sustained motion. Ducks are steady fliers while chickens are at best accelerators, and their flight (breast) muscles show the difference. Many fish swim steadily with muscles adjacent to their vertebral columns and use the rest of their trunk muscle only in short bursts of activity; the distinction between their dark and light muscle characterizes every assay from ultrastructural to gustatory.

Acceleration and size

Speed of movement generally varies with the size of the organism. Thus large fish swim faster than small ones—drag is roughly proportional to surface area, while locomotory machinery increases with body volume. A large creature falls more rapidly than a small one of the same shape and density. What about acceleration? Consider the business of jumping—as mentioned much earlier, in the absence of drag jumpers of any size should achieve about the same maximum height. A rough-and-ready explanation goes as follows: the mass of an animal's muscle is nearly proportional to its body mass (Chapter 3); the work its muscles can do (force times distance) should be proportional to their mass; so the work a jumper can do ought to be proportional to its body mass—put another way, its work per unit mass should be constant. Work is equivalent to mass times acceleration times distance, so work per unit mass should be proportional to the product of acceleration and distance. That product, though, is the square of velocity—the square (and thus velocity itself) should therefore be constant. In short, given muscle of some standard capability, all jumpers should achieve about the same takeoff velocity.

That means that they should achieve the same maximum height. In the absence of drag, initial upward speed determines maximum height—

after leaving the ground, the organism or projectile just decelerates under the influence of gravity or, put the other way, it accelerates steadily downward even as it travels upward. Maximum height is simply given as the square of the upward speed at the end of initial acceleration (termed, militaristically, the "muzzle velocity") divided by twice the acceleration of gravity:

$$h_{max} = v_0^2/2g. \tag{14.1}$$

It must be admitted that even in the absence of appreciable drag, organisms may not quite follow our logic—jumping distance increases somewhat with size, at least in frogs and toads, according to Emerson (1978) and a Useful Fact undoubtedly familiar to the residents of Calaveras County.

So what about acceleration? Bear in mind that once a nonflying creature leaves the ground it can no longer apply force to the ground to further accelerate its mass. If muzzle velocity is constant, then acceleration must be *greater* for the smaller creature—the speed has to be achieved over a shorter period of time because of the effectively shorter muzzle length (l):

$$a = v/t = v^2/2l. \tag{14.2}$$

A shorter leg cannot push a body to as great a height; indeed, acceleration ought to be just *inversely* proportional to body length. What of reality—does it follow this logic? Table 14.1 gathers a few figures for acceleration for systems spanning a wide range of size and diversity of propulsive arrangements. It shows that the values of acceleration in small-scale systems get remarkably high, even when we leave the limited world of aerial jumpers. I should say a few words about the less familiar organisms.

Pilobolus is a tiny fungus that grows on the dung ("cowpies") of grazing mammals; it bears a cluster of spores (a "sporangium") on a cylindrical stalk, or hypha (Figure 14.6). The hypha grows toward a light source and ruptures at maturity, sending the sporangium off atop a cylindrical jet of cell sap. To complete the life cycle, a spore must pass through the intestine of a grazer; since cows, at least, will not graze close to their own excrement (there's a "region of repulsion") the plant contrives to propel its spores away from the endangered feces. Since the hypha grows toward, and thus aims the sporangium at, a light, if a culture is placed in a dark chamber with a hole through which light enters, all the hyphae will shoot their sporangia through the hole.

Hydra is more familiar—this small, freshwater coelenterate is a favorite of biology courses. It bears stinging cells on its surface that are typically triggered by contact with a prey organism—the discharge of a long tu-

TABLE 14.1. MAXIMUM ACCELERATION (M·S⁻²) FOR A VARIETY OF CASES. TO CONVERT THESE TO MULITPLES OF *g*, GRAVITATIONAL ACCELERATION ON EARTH, DIVIDE EACH DATUM BY 9.8.

Car, 0 to 20 m/s (45 mph) in 8 seconds	2.5	(1)
Man, near-record standing jump	15	(2)
Antelope, jump	16	(3)
Lizard, dash	30	(4)
Trout, sudden start	40	(5)
Spider (salticid), jump	51	(6)
Lesser galago, jump	140	(3)
Locust (1st stage), jump	240	(3)
Squid tentacle, extension	330	(7)
Click beetle, jump	400	(3)
Rat flea, jump	2000	(3)
Pilobolus, sporangium expulsion	100,000	(8)
Hydra, nematocyst discharge	400,000	(9)
Rifle bullet, 3000 m/s muzzle speed	500,000	(1)

(1) My calculation; (2) Schmidt-Nielsen 1984; (3) Bennet-Clark 1977; (4) Huey and Hertz 1984; (5) Webb 1976; (6) Parry and Brown 1959; (7) Kier, personal communication; (8) from data of Page 1964; (9) Holstein and Tardent 1984.

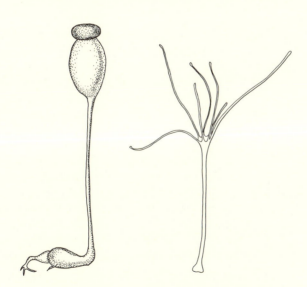

FIGURE 14.6. *Pilobolus* (left) is made up mainly of the sporangium on top, a subsporangial swelling, and a stalk. *Hydra* (right) consists of stalk and tentacles. The scale is the same—each is a few millimeters tall.

bule is powerful enough to punch a hole in the surface of the prey preliminary to giving it a squirt of poison. The mechanism of discharge is
still uncertain but seems to be, at least in part, osmotic.

A human can withstand accelerations of only about 100 m·s^{-2} or so;
thus it seems that these small systems are doing something rather special.
But the appearance is something of an illusion, a consequence of looking
at a quantity once removed from stress, which is all that the material of
the system really experiences. It's much the same sort of scaling illusion
we saw with respect to the high pressures that generated only ordinary
stresses in the cell walls of plants (Chapter 11).

How to do a lot of work in a short time

A more substantial problem in achieving high levels of acceleration is
the necessity for doing a lot of work on a projectile in a very short time—
the power output of the system must be enormous during the period of
acceleration. Muscle, from whatever animal, can produce no more than
a stress of about 200 kPa. It can achieve an intrinsic speed no greater
than about 60 muscle lengths per second, as defined in Chapter 13. But
muscle can't do both at once, so the practical upper limit (assuming that
muscle has the density of water) is a product (power output) not of
12,000 but of less than 500 W·kg^{-1} of muscle (Weis-Fogh and Alexander
1977). If 20% of an animal's mass is jumping muscle, then the figure for
the whole creature becomes only 100 W·kg^{-1}.

According to Bennet-Clark (1977), the observed performance of a flea
requires 2750 W·kg^{-1}—the inescapable conclusion is that a flea must
have some sort of a power amplifier. What it does is contract the muscles
that push its hind legs downward for about 50 milliseconds before it
jumps, deforming small pads of that excellent protein rubber, resilin.
The spring-back of the resilin then powers the actual jump during the
0.7-millisecond period of acceleration. The flea doesn't get something for
nothing—a little energy is actually lost in the process, but work done
slowly, and thus at a lower power level, creates a store of energy that is
released rapidly and thus at higher power. A locust or grasshopper jumping with its hind legs stores up energy in chitinous apodemes and gets a
tenfold power amplification (Bennet-Clark 1975). A click beetle stores up
energy by deforming its external cuticle; its power amplification is fully
a thousandfold (Evans 1973). Each of these creatures has some kind of a
mechanical catch to prevent premature extension while work is being
done to build up stored energy; the specific arrangements, though, are
different for each case.

Such power amplification through the relatively slow storage and sudden release of energy need not be postulated to explain the performance

of larger animals. Strain energy storage still plays a role but in a somewhat different way. A kangaroo jumps repeatedly; it invests the force associated with its deceleration as it hits the ground in stretching tendons—in strain energy storage, for which collagen is notably effective. That stored energy then reappears as work in assisting the acceleration of the subsequent jump; the saving in the cost of locomotion may be as much as a third. Muscles as well as tendons are stretched, but even with their much greater mass they seem to account for little storage of strain energy. The phenomenon is fairly general—kangaroos are merely an extreme case (Alexander 1983). The technological analog is, of course, a pogo stick.

Nothing substantial seems to be known concerning the storage of strain energy in *Hydra* or *Pilobolus*—at least for the latter the only reasonable guess is osmotic generation of pressure, which gradually increases the tension in the walls of hypha and subsporangial swelling. The mechanical properties of these walls would certainly bear investigating.

The ballistic trajectory

Having now explored the ways in which natural missiles are projected, let's inquire about what happens as they travel beyond the gun. We'll limit ourselves to aerial ballistics—a nematocyst is injected directly into prey, so it normally has no free trajectory, and I know of no available data on the path of an underwater intercreature biological missile (ICBM).

Assume that you can send off a projectile at a given speed and that you wish to achieve the greatest possible horizontal distance before it returns to earth at its original elevation. What is the best tactic? Clearly, the launch should be neither vertical—it would go nowhere because it would have no horizontal momentum to carry it—nor horizontal—it would again go nowhere because it would have no vertical momentum to act against the downward acceleration of gravity. It's a standard exercise in elementary physics to calculate that the angle giving the greatest range is just halfway between, 45°, in the absence of any aerodynamic drag. The range, conveniently, depends only on the angle and the initial (muzzle) velocity; it's independent of the mass, size, or shape of the projectile.

What happens in this simple airless ballistic trajectory is that the missile travels horizontally at a constant speed, the horizontal component of the launch velocity—no horizontal forces affect its progress, so Newton's first law applies. At the same time, the unbalanced gravitational force makes it accelerate downward at a constant 9.8 m·s^{-2}. Since it has an initial upward motion, this means that it goes ever more slowly upward until it stops ascending and then goes ever more rapidly downward until impact, which occurs at the same speed and angle as at launch. If you know the launch angle, you can figure the initial vertical and horizontal speeds as components of the overall launch velocity. Since the upward speed at maximum height is zero, the time of ascent (or descent) is the vertical launch speed divided by the acceleration of gravity; the distance (height) of ascent is the product of half that acceleration and the square of the time.

The exact paths of the trajectories are not at all hard to compute using only a little trigonometry. The 45° case is particularly tidy—the maximum height is a fourth of the horizontal range and half the height that would have been achieved had the projectile been projected (at the same muzzle speed) directly upward. If you simultaneously drop a weight and shoot a projectile horizontally from the same elevation, they will strike the earth at the same time—the downward acceleration of gravity is quite unaffected by simultaneous horizontal motion.

Size, range, and real trajectories

But we'll not do such computations here. Not to avoid the trigonometry, but because of the assumption that drag is of no consequence. That's tolerable to some extent for big, dense cannon balls; for smaller, less dense, athletic and biological items, the assumption is distinctly unsafe. Reckoning with drag, though, makes the analysis most untidy, although not much more difficult conceptually. Recall Figure 7.8—the relationship between drag, size, and speed does not easily encapsulate in a simple equation. Brancazio (1984) calculated some trajectories for real baseballs; they at least operate at Reynolds numbers high enough so a constant drag coefficient isn't a bad assumption. I've adopted his procedure, trying the patience of a home computer that steps incrementally through a trajectory, but I make it calculate the Reynolds number at each step and then find the drag coefficient from the equivalent of Figure 7.8. The program is given in Appendix 2.

What one does in this iterative procedure is specify a size and density for an assumedly spherical projectile; these data permit calculation of mass (for gravitational force) and cross-sectional area (for drag). One specifies, as well, an initial speed, an initial angle of projection, and a time interval. The time is a period short enough to divide the total time of flight into fifty to a hundred parts, determined by trial and error. The computer then figures the drag and from that the acceleration due to drag at the initial speed, separating this initial drag acceleration into horizontal and vertical components. To the vertical component is added the acceleration of gravity. Then, using the chosen time interval and the accelerations, the computer figures a new horizontal and vertical velocity at the end of the interval; from these it figures the horizontal and vertical coordinates (as on a graph) of the new position. Using the new velocity and position, it repeats the procedure and continues until the projectile returns to earth, that is, until the computer calculates a vertical position equal to or less than zero height. The accumulated horizontal distance is then displayed, together with any other figures the programmer wants.

Table 14.2 gives the results of such simulations—the maximum horizontal travel, the initial angle that produces that maximum, the greatest height achieved, and the height that would be achieved if the sphere had been initially hurled directly upward at the same speed. It's clearly a different world for small projectiles! Drag shifts from being a small tax to forming the preeminent physical phenomenon—for the tiniest projectile, the range with drag is less than a ten-thousandth of the range without it. Direction of firing has little effect on overall distance—the smallest projectile can go as far upward as it can go horizontally. Take that, gravity! Finally, the angle for maximum range gets lower for smaller projectiles; due to drag they decelerate drastically as they travel, so the best tactic is to put some horizontal distance behind you while you still have some speed.

The unusual shapes of the trajectories of small projectiles are most evident if the particulars of several are plotted, as in Figure 14.7, with an abscissa scaled in fractions of maximum range. The symmetry of the no-drag case is gradually lost and the optimal angle drops as size decreases. What doesn't show well in either table or graph are the speeds during travel. The large projectile appears to move at a nearly constant speed; for the smallest ones, the gentle ascent is accompanied by a drastic de-

TABLE 14.2. TRAJECTORIES OF SPHERES WITH AND WITHOUT DRAG. INITIAL SPEED IS 20 M·S^{-1}; PROJECTILE DENSITY IS 1000 KG·M^{-3}.

		Diameter, mm				
	No Drag	100	10	1	0.1	0.01
Max. range, m	39.6	37.4	17.4	3.11	0.158	0.00295
Angle for max. range, degrees	45	39	37	25	10	1
Max. height en route, m	9.82	7.64	4.56	0.767	0.0218	0.000048
Max. height vertical shot, m	20.2	19.1	11.3	2.53	0.163	0.00294
Range loss to drag, %		5.6	56.0	92.1	99.6	99.99

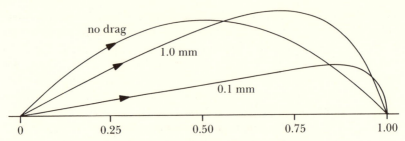

FIGURE 14.7. Trajectories for spheres shot at angles picked to give the greatest horizontal range. The sphere with no drag may have any size, mass, and initial speed; the others have an initial speed of 20 m·s⁻¹, have the density of water, and move through air. The axes are scaled in fractions of the maximum horizontal range of each projectile.

celeration reminiscent of the flight of a roman candle, while the final descent is steep and steady but painfully slow.

After that lengthy preamble we're equipped to examine some real projectiles. First, baseballs and golf balls, whose sizes and masses are regulated with fanatic seriousness. A baseball hit at 40 m·s⁻¹ would go 160 meters in a vacuum; it goes only 130 meters in air, a 19% penalty. The best initial angle is about 43°, assuming (unnaturally) that the ball can leave the bat with the same speed in any direction. A golf ball struck at 60 m·s⁻¹ would go 360 meters in a vacuum and 230 meters in air, a 36% decrease. The best angle is lower, about 37°. (Brancazio's figures are slightly different, for reasons perhaps connected with the choice of drag coefficients.)

Jumping locusts are smaller but slower and stay in the air only briefly, so drag plays a minor role. I've assumed a spherical animal 5 millimeters in diameter (an underestimate to compensate for the normally streamlined shape) with an effective density half that of water and an initial speed of 3 m·s⁻¹ (from Bennet-Clark 1975). The range comes out to 0.9 meters without drag and 0.75 meters with drag, a mere 16% decrease. As with the baseball, the best angle is 43°.

A flea is still smaller; from the data of Bennet-Clark and Lucey (1967) I've assumed a size of 0.5 millimeters, again half the density of water, and a takeoff speed of 4 m·s⁻¹. The small size makes real trouble—the 1.6 meters without drag is now degraded to 0.27 meters with, a reduction of 83%; the best angle is down to 30°. All that acceleration and muzzle speed, and the flea goes less than a foot. Still, the range is an impressive 500 times the length of the flea itself.

But things get yet worse before nature gives up on ballistics. According to Buller (1934), the *Pilobolus* sporangium has an impressive muzzle speed of about 20 m·s^{-1}. Assuming a diameter of 0.3 millimeters and the density of water, I calculate that it should go only 0.75 meters in air. In fact, it does somewhat better, about 2 meters (a mere 95% instead of 98% drag tax)—the explanation seems to be that the projectile carries a substantial droplet of cell sap with it, so it's really bigger and heavier than assumed. The cell sap has the additional virtue of sticking the sporangium to the surface it hits—falling down between blades of grass would keep it from being satisfactorily eaten. The best angle for shooting is now down to 18°. A sporangium may contain as many as 90,000 spores—small individual spores would do better for wind dispersal, but they'd be hopeless as projectiles.

More fungal artillery. Ingold and Hadland (1959) looked at a small ascomycete called *Sordaria*, which shoots spores singly or in clusters of up to eight. Since all combinations seem to start at the same speed, the largest cluster naturally goes the farthest. The largest cluster is about 0.04 millimeters in diameter and typically goes about 6 centimeters, from which I calculate the mighty muzzle speed of 30 m·s^{-1}. The best pitch is now only 6° above horizontal, and drag has reduced the range by more than 99.9%. So why should *Sordaria* bother, which is to say, how might we rationalize the evolution of such an inefficient scheme? Probably the arrangement is not ineffective—recall the business about boundary layers (Chapter 6). Dispersing windborne spores puts a premium on getting them out through most or all of that velocity gradient and into a decent breeze. Some plants grow tall; some, like mosses, bear spores atop stalks; *Sordaria* chooses to shoot. It's far from unique—lots of species play the game, according to Ingold (1961). But they've received little attention; *Pilobolus* is the fashionably physical fungus.

Energy and afterthoughts

> "The energies of our system will decay, the glory of
> the sun will be dimmed, and the earth, tideless and
> inert, will no longer tolerate the race which has for
> a moment disturbed its solitude."
>
> Arthur James Balfour, inadvertent
> thermodynamicist

IT WAS force that gave coherence to the Newtonian world. Gravitational forces maintained the orderly motions of the cosmos, and mechanical forces linked carriages and horses. Stresses and pressures provided the application of forces to immediate practical problems. Momentum, the quantity of motion, contributed a conservation law for explosions, collisions, and the like. It was (and is) a remarkably consistent world—at least internally consistent. With only a few emendations it has been the world of the past fourteen chapters, a world quite adequate for our mainly mechanical bits of biology.

Where the mechanical system of Newton, his contemporaries, and his successors proves inadequate, the inadequacy is a result of the system's isolation from nonmechanical aspects of reality. In the nineteenth century the disability became glaring and then was eliminated in a grand synthesis, impelled by the imperatives of the industrial revolution. The shift of power sources from horses and falling water to deliberately contrived combustion engines made the interrelationships among temperature, pressure, fuel economy, and mechanical work into profoundly practical matters. But the mechanics with which the century began had no place for dealing with such problems. Thomas Young (1773–1829), whom we mentioned earlier in connection with his elastic modulus, first borrowed a vernacular word, "energy," for the product of mass and the square of velocity (to which product we now apply a factor of 1/2). This quantity eventually proved central to the unification of mechanics and thermodynamics in the elaborate structure we associate with people such as Carnot, Joule, Kelvin, Boltzmann, and others.

The subtle and intellectually difficult concept of energy, with its associated mathematical splendor, has thus permitted the integration and use of classical mechanics in areas concerned with heat, wave phenomena, electricity, and even quantum mechanics and relativity. It has also provided a link between mechanics and the biological sciences, especially metabolic physiology and trophic ecology. This book is concerned with a part of biology that best fits between the conventional areas of physiology and ecology. We may have no especially compelling need for the notion of energy within our limited domain, but the notion is useful to tie it to established biological fields in precisely the same way that the engine builders of the nineteenth century tied mechanical devices to heat and fuel.

HEAT, TEMPERATURE, AND SPECIFIC HEAT CAPACITY

To be any more specific in talking about energy we need some additional variables; again we have to unlearn some household half-truths and be a little more precise in saying what we mean. Of special importance is the distinction between heat and temperature, which becomes somewhat clearer if a third variable, specific heat capacity, is introduced to specify the relationship between the first two.

Heat

Bluntly, if somewhat uninformatively, heat is a nonmaterial quantity which, when added to a body, ordinarily raises its temperature. Of course, we beg the question of just what a "nonmaterial quantity" might be; we'll simply admit ignorance and move on. The definition presumes we know what temperature is—let's quietly pass over that matter, too, for the moment. And let's also defer the "ordinarily." Heat is a form of energy; as such it has the same dimensions and units as mechanical energy or work. Unfortunately, the proper SI unit, the joule, is all too rarely used for heat—instead one encounters calories, kilocalories (or Calories, or nutritional calories, confusingly), and British thermal units (Btus). Appendix 1 gives a few conversion factors.

Temperature

Temperature is a measure of the "intensity" or the "potential" of heat. Under most circumstances it determines the direction of heat flow—heat ordinarily moves from a body or a part of a body at a higher temperature to one at a lower temperature. Of two otherwise identical bodies (same size, mass, and material), the hotter one contains the more heat; the statement is analogous to the assertion that of two identical buckets, the heavier contains the more water. Another way to view the matter is to analogize heat with the amount of something and temperature with its concentration—pour a cup of cold water into a bowl of hot; the hot is

299

diluted by the cold (less hot, really), and so the temperature in the bowl is lowered.

So Sydney Smith should have used "high temperature" instead of "heat" when he wrote, "Heat, ma'am . . . it was so dreadful here that I found there was nothing left for it but to take off my flesh and sit in my bones."

With temperature we encounter a fourth basic dimension, as promised when mass, length, and time were first encountered. The symbol "Θ" (Greek capital theta) is often used for the dimension; we'll use "T" for the specific quantity and "K," "C," or "F" for the units. Again, Appendix 1 gives conversion factors.

symbol:	T
dimension:	Θ
SI unit:	Kelvin degree (K)

Kelvin degrees are units above absolute heatlessness, the point where molecular motion is absent. Since temperature is ultimately a measure of the intensity of molecular motion, there can be no negative temperatures on the Kelvin scale. The Kelvin scale is unfamiliar to most—it's awkward for everyday use on account of its low zero point. Thus water freezes at 273K, room temperature is about 293K, and our body temperatures are around 310K. But it's easy to use since it borrows the magnitude of its degree from the Celsius (formerly centigrade) scale; thus a Kelvin degree is exactly the same size as a Celsius degree, and in a formula with a degree increment in the denominator ("per degree") they are interchangeable. Only the base is different—to get Kelvin temperature from Celsius, just add 273 (actually 273.15) degrees. Also, Kelvin temperatures drop the degree sign (°) still used for Celsius. The common Celsius scale (C) sets its base and the size of the degree by proclaiming that water freezes at 0°C and boils at 100°C. The Fahrenheit scale, now just an American anachronism, sets the freezing point of water at 32°F and the boiling point at 212°F.

Specific heat capacity

Specific heat capacity (or, sometimes, just heat capacity or thermal capacity) is the amount of heat needed to produce a unit rise in the temperature of a unit mass of material. Put another way, specific heat capacity is the change in heat content per unit change in temperature—a high heat capacity means that a lot of heat must be put in to raise the temperature of a material only a little bit. The SI unit for specific heat capacity is the heat needed to raise the temperature of one kilogram of a material by one Kelvin (or Celsius) degree.

symbol: c_p (the subscript specifies that pressure, not volume, is kept unchanged by any addition of heat)

dimensions: $L^2 T^{-2} \Theta^{-1}$

SI unit: $J{\cdot}kg^{-1}{\cdot}K^{-1}$

We will use °C instead of K.

Fortunately, the specific heat capacity of most materials changes only minimally with the temperature at which it is measured, at least where the material stays in the same phase. Of common materials, liquid water has the highest capacity. As expected, flesh has nearly as high a value as water. Air has a somewhat lower capacity than water, although the comparison can be rather misleading. Specific heat capacity is ordinarily defined, as above, relative to the *mass* of material; a given mass of gas naturally occupies far more volume than the same mass of liquid. So for some problems it's much more useful to consider, not this normal mass-specific heat capacity, but instead a special volume-specific capacity, the heat needed to raise a cubic meter of material a Celsius degree. For instance, you might be interested in the relative volumes of air and liquid coolant that must flow through an automobile radiator, or the amount of blood and air that must pass under and over hot skin. Table 15.1 compares these capacities for air and water at ordinary temperature and pressure; a few other data for less precisely defined materials—metals, generally less than 400 $J{\cdot}kg^{-1}{\cdot}°C^{-1}$; body fat, about 2100; a mammal as a whole, 3400; and wood, 1700.

The high specific heat capacities of water and the bodies of organisms are of considerable biological consequence. A store of water or flesh, whatever its other uses, confers protection against excessively rapid change in body temperature. A camel in a hot desert is subjected to a substantial radiative heat input during the day; it might, as would a human, maintain a nearly constant body temperature by evaporating water, at least until it reached its tolerable limit of desiccation. The camel's relative tolerance for water loss is in fact far greater than our own, but it

TABLE 15.1. HEAT CAPACITIES OF AIR AND WATER—THE CONVENTIONAL MASS-SPECIFIC CAPACITY COMPARED WITH CAPACITY RELATIVE TO THE VOLUME OF MATERIAL

	Mass Specific ($J{\cdot}kg^{-1}{\cdot}°C^{-1}$)	Volume Specific ($J{\cdot}m^{-3}{\cdot}°C^{-1}$)
air	1003	1204
water	4169	4,169,000

also takes steps to minimize the loss itself. A dehydrated camel permits its body temperature to drop at night to 34°C or a little lower; it permits elevation during the day to 40°C or a little higher. Schmidt-Nielsen (1964), who did the classic work on heat and water balance in desert animals, calculated that by tolerating a 6°C temperature rise a 500-kg camel could store as much heat as would have been dissipated by evaporating 5 liters (0.005 m³) of water. (You might check the calculation, using a heat of vaporization for water of 2.4 MJ·kg^{-1}.) The scheme can be depended upon as surely as the fact that night follows day.

But a camel is a very big animal, with lots of volume for storage of heat relative to its surface for absorption. The opposite extreme is the leaf of a sunlit tree. When the sun emerges from behind a cloud in windless air, or when there is a lull in even a slight breeze, leaf temperature may rise by a degree every few seconds. Some plants, especially those characteristic of fairly dry habitats, have very thick, so-called succulent leaves. A few measurements done by Dwight Kincaid and me in a pulsing wind tunnel suggest that succulence is, in part, a scheme analogous to the camel's but works on a very much shorter time scale. With more water per unit surface area and with the high specific heat capacity of water, the succulent leaf should be tolerant of longer or more extreme lulls in the wind and should require less evaporation of precious water to avoid lethal temperatures.

What these tales boil down to is that for object or organism the practical relationship between heat and temperature depends on how much there is of the organism and of what it is made. And nothing beats water as a constituent to damp the temperature fluctuations caused by variations in input of thermal energy.

THE FIRST LAW—CONSERVATION OF ENERGY

It's hard to start with anything better than Feynman's (Feynman et al. 1963) statement. "There is a fact, or, if you wish, a *law*, governing all natural phenomena that are known to date. There is no known exception to this law—it is exact so far as we know. The law is called the *conservation of energy*. It states that there is a certain quantity, which we call energy, that does not change in the manifold changes which nature undergoes. That is a most abstract idea, because it is a mathematical principle; it says that there is a numerical quantity which does not change when something happens. It is not a description of a mechanism, or anything concrete; it is just a strange fact that we can calculate some number and when we finish watching nature go through her tricks and calculate the number again, it is the same."

A more formal statement of the first law of thermodynamics is that "in an isolated system, the total energy content is fixed." An isolated system is defined as one in which neither material nor energy is exchanged with the surroundings—as opposed to a "closed system," which can exchange only energy, or an "open system," which can exchange both material and energy. The first law permits energy to be used as the currency in a useful sort of account keeping, which certain ecologists find attractive. Organisms are, on any reasonable time scales, open systems. But the law is still applicable—dealing with open systems merely requires that one include terms for import and export of energy and possibly material.

The origin of the law traces to odd observations such as the heat inevitably generated in the boring of cannon barrels and the increase in the temperature of water when stirred rapidly. These led to the determination (by Joule) of a precise equivalence of mechanical work with what we now call thermal energy; later, electrical work and energy were brought into the picture. If you agitate water rapidly, say in a blender at top speed, it gradually warms up; with care it can be shown that the heat has for the most part not merely been conducted or convected up from the motor. In wind tunnels and flow tanks with return circuits for the air or water, the effect is a substantial nuisance (say I with the passion of the afflicted)—most such devices are unusable for all but brief periods without some provision for cooling.

Earlier (Chapter 2), the utility of conservation laws was proclaimed. The present conservation law is clearly the most general we've yet encountered; indeed, that's probably what makes us put up with that queer stuff, energy. A few examples will illustrate the point.

Deforming fluids

Viscosity is a measure of a fluid's resistance to the rate of deformation. So it takes work to deform a fluid at any appreciable rate; the work must go somewhere, and that somewhere is ordinarily the transformation to heat. The amount of heat can be precisely linked to the viscosity of the fluid and the rate at which it is being sheared. The shock absorbers of an automobile use viscosity to resist rapid movement—compression or extension of an absorber requires that a liquid pass through a narrow aperture. Shock absorbers are noticeably warmer after one drives along a bumpy road.

Chilling air by rapid expansion through an aperture

Exhale—the air leaves your mouth at body temperature. Purse the lips and blow—the air is warm close to the lips but rapidly cools as it moves away. How come? Work must be done on a gas to compress it; the gas

303

gets warmer in the process (feel the bottom of a tire pump after using it). Conversely, releasing the pressure on a compressed gas allows it to expand; in expanding, work is done by the gas on the surroundings, and the gas gets cooler. Work, after all, is a manifestation of a transfer of energy.

A bouncing ball

The ball goes up and down—at the top of its trajectory its distance from the center of the earth is greatest and its speed is least (zero). There, its "gravitational potential energy," the product of its weight (not mass) and the distance above some reference height, is at a maximum. At the bottom its speed is greatest—its "kinetic energy," half the product of its mass times the square of its speed, is at a maximum. So the sum of these two somewhat less-than-obvious quantities is constant—knowing one and their sum (or one, where the other is zero, as at the top of the trajectory) we can calculate the other for any point in the cycle. Of course, the average distance between ball and floor slowly decreases. But conservation of energy meets that challenge—it sends us looking for some other sinks for energy, and we discover that the ball, the floor, and the air all warm up slightly in the process. Collisions can be analyzed similarly. Both momentum and energy are conserved, but some of the energy appears as heat unless the collision is (as none is *truly*) perfectly elastic.

Heating an auditorium

A sedentary human needs to consume about 2400 kilocalories per day according to the books on nutrition. That's 100 kcal/hr, or 25 cal/s, or 100 J/s, or 100 watts. Since the human isn't slowly warming or cooling, he or she must be losing heat at the same rate—you are heating the room in which you read this by just as much as a 100-watt light bulb; a class of ten is just as effective in heating the room as a kilowatt heater. An audience of a few hundred puts a heavy load of heat on an auditorium—except in very cold climates and just prior to use, heaters are far less useful than air conditioners in places where people congregate. Which, incidentally, seems to be why the traditional theater and concert seasons begin after the summer.

We've now introduced several forms of energy—kinetic, gravitational potential, and thermal. We alluded to the energy involved in compressing a gas—often called "pressure-volume energy." There are various others, including the energy released (or consumed) in chemical reactions, that made available when some device is connected between the poles of a battery (electrical energy), and so forth. The first law of thermodynamics implies that all of these can be interconverted *and* that they can all be expressed in the same units—joules, in SI.

It's interesting to tabulate the energy content of various substances not ordinarily viewed in this context (Table 15.2). "Energy content," it must be admitted, is really a bit misleading—whether the substances actually *contain* the energy is a moot point. What we mean is the energy released in a certain kind of chemical reaction, here oxidation or (approximately) combustion, done under a standard set of conditions.

In all but one of these cases the combustion involves some external source of oxygen; the exception is gunpowder, which contains its own oxidant and can blow up independently of the local atmosphere. Hydrogen is clearly the best fuel relative to mass, but it does need eight times its own mass of oxygen to do the job. Biological fats and oils are slightly oxidized to begin with, so they release less energy when fully oxidized to carbon dioxide and water than, say, heating oil. Proteins and carbohydrates are yet more oxidized compounds and so release a little less energy still.

TABLE 15.2. ENERGY CONTENT, IN JOULES PER KILOGRAM, OF A VARIETY OF SUBSTANCES

Hydrogen	122,000,000	
Natural gas	55,000,000	
Heating oil	42,000,000	(gasoline the same)
Coal	30,000,000	(lots of ashes)
Protein, pure dry	24,000,000	
Wood, air dried	17,000,000	
Starch, pure dry	17,000,000	(pure sugar the same)
Gunpowder	3,000,000	(contains own oxidant)
Salad oil	37,000,000	(or any fat)
Walnuts	27,000,000	(lots of oil)
Wheat flour	15,000,000	(mostly starch)
Lima beans, raw	5,000,000	(67% water)
Potatoes, raw	3,000,000	(80% water)
Spinach, raw	1,000,000	(over 90% water)
Beef, T-bone steak	17,000,000	(37% fat)
Beef, flank steak	6,000,000	(6% fat)
Chicken, skinless	4,000,000	(2% fat)

To convert a figure to kilocalories per 100 grams, the units common in nutrition books, divide by 42,000. To convert to British thermal units per pound (Btu/lb), the units used by American engineers, divide by 2300.

THE SECOND LAW—IRREVERSIBILITY OF REAL PROCESSES

The principle of conservation of energy was termed the "first law of thermodynamics." "First" turns out to imply neither historical precedence nor intellectual primacy—a little older and a lot more interesting is what we usually call the "second law." As we'll see, it lends itself rather poorly to purely verbal statement; perhaps as a consequence, it is not widely appreciated outside the scientific community. And where it is invoked, it too often merely forms the basis of one or another inappropriate analogy—the presence of the word "entropy" is the inevitable finger pointing to use or abuse of the second law. As C. P. Snow, the novelist (and physicist) once complained

> A good many times I have been present at gatherings of people who, by the standards of the traditional culture, are thought highly educated and who have with considerable gusto been expressing their incredulity at the illiteracy of scientists. Once or twice I have been provoked and have asked the company how many of them could describe the Second Law of Thermodynamics. The response was cold: it was also negative. Yet I was asking something which is about the scientific equivalent of: *Have you read a work of Shakespeare's?*

The second law is perhaps best approached stepwise. First, a few words about what might be called "reversibility." The Newtonian laws are essentially symmetrical with respect to time—deceleration is simply negative acceleration, motion or momentum or force in one direction is just a negative version of motion, momentum, or force in the opposite. If time were to run backwards, the resulting world would not look much different. But all our experience proclaims the fact that the world is overwhelmingly irreversible—whether with respect to our aging and personal decay, or to the impracticality of coasting down a hill in a car and thereby refilling the gas tank, or to a futile effort to unscramble an egg. *Real processes are always to some extent irreversible*—in the natural or spontaneous direction of things, the bouncing ball eventually settles onto the floor.

And then a distinction between two sorts of motion. If you throw a ball you put it in motion. If you heat a cup of tea you increase the average motion of its molecules. Both processes require energy, and neither changes the total energy of its whole system, ball plus surroundings or tea plus surroundings. But the two sorts of motion prove to be fundamentally different—the ball is in *coherent* motion, with all of its constituent parts moving together. In the tea, what has increased is merely the average rate of random, *incoherent* motion of the parts. Motion, in short, can be either disordered or ordered. It turns out that this difference is the essence of the distinction between heat and work, indeed between heat and all other forms of energy.

With easily envisioned equipment, you can throw a ball upward and have it land on a platform that moves earthward under the weight of the ball, and in doing so turns a generator that makes electricity that heats the cup of tea. Conversely, you can heat the tea hot enough so that it produces steam, which pushes a piston that hoists the ball as high as you would have thrown it. The ball comes earthward again onto the platform harnessed to the device that imparts heat to the tea. Reversible? Not really. All the energy imparted to the moving ball may be converted to heat (although it won't all heat the tea!). But only *some* of the heat put into the tea can be harnessed to do mechanical work—some will merely reemerge as heat in the components of the system.

We have thus illustrated the basic irreversibility of real processes. Any other form of energy can be converted completely to heat, but heat cannot be completely converted to any one or a combination of other forms of energy. If you throw a ball, it heats the air through which it passes. By contrast, applying the same amount of heat to the air cannot be made to move the ball at the speed of the throw. We have here a rule, not about *states*, to which the first law applies, but about *processes*. It is one of several verbal statements of the second law of thermodynamics:

> No process is possible that results in the complete conversion of heat into work.

This proscription, together with the implied lack of an equivalent proscription for the process that moves in the opposite direction, is one of the very few (perhaps the only) physical law that is asymmetrical with respect to time—it's the law that ultimately rationalizes every example of irreversibility. The second law involves still another proscription:

> No process is possible that results solely in the transfer of energy from a cooler to a hotter body.

The two statements rule out all sorts of neat perpetual motion machines! What they come down to is the notion that heat is an inevitable byproduct of real processes, that heat is fundamentally a special and least generally usable form of energy, and that the higher the temperature, the higher the "quality" of the heat. The link between heat quality and temperature can be put into a quantitative statement of the limit on the efficiency of a heat engine. Efficiency, recall, is the ratio of work output to work input (multiplied by 100); it obviously cannot be greater than 100%. The very best one can do with heat is no better than

$$E = \frac{100(T_1 - T_2)}{T_1}.\tag{15.1}$$

T_1 is the higher temperature (on the absolute, Kelvin scale), that of the heat source, and T_2 is the lower temperature, that of the heat sink. Among other things, the equation underlies the use of steam at the highest possible temperature in electrical generating plants. So energy has a kind of quality as well as its obvious quantitative meaning—heat is the lowest quality stuff, and the lower the temperature, the lower the quality. Which is why we can't do much except heat living space with the so-called waste heat of many industrial processes—the temperature of the heat (really, of the hot output material) is impractically low.

As far as we know, nature has never evolved a heat engine. The "why nots" of evolution may be truly unknowables, but one might venture a guess here. Organisms carry out their activities in a temperature range extending from the freezing point of liquid water up to where their proteins become chemically or structurally unstable. On top of any other inefficiencies, a biological heat engine would have to suffer a thermodynamic "tax" of around 90% as a consequence of the application of equation 15.1. And a primitive version using a lower temperature difference would have even greater inefficiency.

The simple metaphorical statement of the second law is that, left alone, everything just goes to hell. Note the appropriate thermal allusion! Better, perhaps, is an encapsulation of the laws, in which the first states that you can't win, and the second states that you can't even break even. There is a third law to the effect that you can't quit the game either. A good, nonmathematical introduction to the mysteries of the second law is a book entitled, unambiguously, *The Second Law*, by Atkins (1984).

THE BIOLOGICAL SIDE OF ENERGY DEGRADATION

By the first law, energy is conserved—so the frictional losses of blood flowing through vessels or of a bird flying through air must turn up as some other entry in the accounting of energy. By the second law we understand that, in the absence of specific transducing machinery, the energy will turn up as an increase in the heat content and (ordinarily) the temperature of the system.

Life must make heat

From time to time the suggestion recurs that organisms manage to evade the second law, but persuasive evidence never seems to be forthcoming, and the notion again goes dormant. So as far as we know or expect, this phenomenon—the escape of substantial amounts of energy into thermal form in real processes—is universal among organisms. One can measure the metabolic rate of an organism by following its heat production; it's trickier than measuring, say, oxygen consumption, but it entails fewer assumptions and will work for anaerobic situations. As a further result, organisms are usually at least a little hotter than their surroundings—except for a few oddities such as a Finn in a sauna (losing heat by evaporating water) and a leaf on a tree on a clear night (losing heat by radiation to the cold sky). Retention, and regulation of the retention, of such "waste" heat might reasonably constitute steps toward "warmbloodedness"—the recurrent evolution of superambient, constant internal temperature is in part rationalized.

The energy recovered in the relaxation of a strained collagenous tendon or resilin wing hinge is less than the work done in straining it—resilience is always less than unity. It's just another case of the same degradation of energy; again part of the work done goes into heat; again the heat may be put to some secondary use such as space heating in a building or thermoregulation above ambient temperature in an organism. All our muscle-driven activities are grossly inefficient. Every step in the sequence from ingestion of food and inhalation of air inevitably involves losses to heat—by the second law the processes can be driven at appreciable rates only if such losses occur. If one defines efficiency as mechanical work done divided by chemical energy consumed, then the best a muscular machine can do is about 25% (Margaria 1976).

The cost of locomotion

But specifying values for efficiency is rarely simple and only occasionally useful—metabolic rate may be a reasonably operational specification of input, but determining output is distinctly peculiar in at least two ways. First, as mentioned back in Chapter 2, it takes energy for a muscle to produce force—the harder a muscle pulls, the more oxygen and fuel it uses, *even if the load is so great that the muscle cannot shorten at all*. Still, unless a muscle changes length, it cannot be said to have done work on the load, and its energetic efficiency must be zero. In fact, a muscle does require a greater input if it contracts than if it simply produces the same tension against an immovable load—doing work isn't just incidental to generating force. More peculiar yet, if work is being done on a muscle—that is, if the muscle is opposing some external force that is stretching it—still less input is needed to develop force than if the muscle is merely developing tension. Thus this "negative work" is more efficient (although it's perhaps a misleading use of efficiency) than ordinary, "positive work."

The second peculiarity concerns the nature of the external work done

309

by a muscular machine. If you climb a hill, you do work against gravitational force in moving your weight away from the center of the earth, so a physical measure of output can be simply specified. By contrast, if you run on a level track, the same definition of output implies that no external work is being done and that your efficiency must therefore be zero! In fact, running takes considerable effort (masochism, say some) and generates quite a noticeable quantity of heat. Your muscles are mostly busy accelerating and decelerating various components of body mass. Perhaps as much as half this decelerative work is conserved in a nonthermal form, as we've noted earlier, either by using resilient materials to store the strain energy for use in the subsequent acceleration or through simple gravitational storage as you go up and down, step by step. But however highly evolved runners we might be, running is still an expensive proposition—the energetic advantage of a bicycle has already been mentioned. Whether running is "inefficient," though, is mainly a matter of definition.

In recent years, a biologically reasonable kind of output has been recognized, one that circumvents the argument that no real work is done in steady, horizontal motion. This quantity is the product of body mass and distance. Thus the cost of locomotion as measured by fuel consumption can be corrected for scale by dividing by that product—the resulting "cost of transport" is the energy input required to move a unit mass a unit distance, in SI units, joules per kilogram-meter. If weight is used instead of mass, one gets a dimensionless cost of transport—Tucker (1975) prefers that version—but the conclusions are no different. In practice, oxygen consumption is the measure of input; the approximate factor for converting volume of oxygen to energy, according to Schmidt-Nielsen (1979), is 4.8 kcal·liter^{-1} or 20 MJ·m^{-3}.

What, then, does locomotion cost? Figure 15.1 summarizes a large amount of information. First, for a given size of animal, running or walking is about three times as costly as flying to cover a given distance, and flying is about three times as costly as swimming. The numbers may seem odd until one remembers that flyers go much faster than runners, so the very high rate of energy expenditure in flying is offset by a greater speed, and that we're talking about real swimmers, not clumsy and inefficient humans or surface craft such as muskrat or duck. Running is a bad business, whether bipedal or quadrupedal—too much up and down, too much acceleration and deceleration. A lot of money has been spent on experimental walking vehicles, with little to sell even in the most profligate times of military procurement—one suspects that after the problems of mechanical complexity are dealt with there must remain a disablingly high fuel consumption. Nature has been building walkers for a

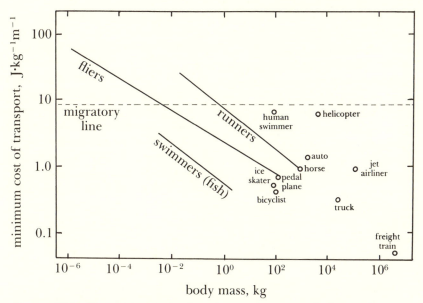

FIGURE 15.1. The cost in energy of moving a unit mass a unit distance in relation to size and mode of locomotion. Migration appears to be practical only below the dashed line.

long time, and even hers are poor relative to swimmers, flyers, or wheeled vehicles—a cyclist is three times as good as a runner, and an ice skater almost as good as the cyclist.

Second, this relative cost of transport drops with increasing body size—it's roughly proportional to one over body length. For swimmers and fliers, the relationship makes immediate sense—the main opposing force is drag, and drag (an area function) relative to body mass (a volume function) should be about inversely proportional to length. For walkers and runners, the explanation is less obvious, but it must be connected with frequency of acceleration and deceleration of the limbs and torso.

Third, the distance a creature can go without refueling depends directly on this cost of locomotion. And that distance, in turn, determines the practicality of migration. Tucker (1975) identified what he called a migratory line on the graph of cost of transport versus mass or length—migration is practical only below the line. Eligible are all swimming fish, all flying birds but only the largest flying insects (and only a few insects such as monarch butterflies and desert locusts do really long, uninterrupted flights), and runners of the size of cats but not of rats. Among machines, the range in cost is about 130-fold, from freight trains and steamers, at the low end, to helicopters, which are fully as bad as a swim-

311

ming human. As a very coarse estimate, we might expect a helicopter and a swimmer to have about the same range on a load of fuel.

There's another interesting implication of these data. The relative cost of ascending, doing work against gravity, ought to be independent of an animal's size. But the cost of terrestrial transport per se decreases with size. It looks as if the cost of ascending relative to the cost of level transport must be greater for the larger animal. And the look is no illusion—either in life or on a treadmill, up and down is almost all the same for a mouse or squirrel and only a minor matter for a cat or dog but a running human certainly notices a difference! And the situation for a horse is even worse. I live in a city full of small hills in which the older radial roads weave tortuously. I've been told that the two conditions are related—for horse-drawn vehicles it paid to avoid grades by taking longer routes.

Chaos and the second law

Thermal motion is the ultimate in disorder—it's completely incoherent, and we can only specify some average rate of movement and cannot predict what any given molecule is likely to do. In short, future states of individual components of a system are quite unpredictable from past states. This notion of randomness, disorder, or (what amounts to the same) unpredictability takes on great significance in applications of the second law. It turns out that disorder can be usefully quantified—it varies with the number of ways a system might arrange itself. With a large number of components unconstrained by much large-scale structure, the disorder is great—the system has a large element of unpredictability.

Besides its implications for heat transfer and thermal processes, the second law applies to processes in which the degree of randomness, disorder, or unpredictability changes. The second law requires that the spontaneous direction of real processes—the direction they'd take if left to themselves—be toward increasing disorder. The spontaneous degradation of the various forms of energy to heat is merely a particular case of a more general dictum that things, left to themselves, deteriorate.

Living systems are the most complexly structured, and hence least disorderly, entities of which we have any knowledge. The spontaneous direction of real processes is away from the creation and toward the destruction of such systems—thermodynamics implies that, while we're not exactly illegal, we are indeed very special. How then might life have evolved and how might it perpetuate itself in a universe fundamentally inimical to high levels of organization?

The general answer is not unknown or even very controversial. In the form we've discussed, the second law of thermodynamics applies to isolated systems—as mentioned, these exchange neither material nor energy with their surroundings. It certainly doesn't prohibit the local accumulation of order within a larger system. It does, though, tacitly ask for some special rationalization of the development of such orderly patches. In our case the general explanation is based on the presence of a hot sun radiating to an earth that in turn radiates to a cold void—heat is being continually transferred from higher to lower temperature. Just as with a heat engine (the formula for whose efficiency was given earlier), this spontaneous transfer can be harnessed to do work—evolving and maintaining life is as legitimate as turning a turbine. We absorb high quality (hot) solar energy, transform it, and reradiate lower quality (colder) energy off into space, enough so that the overall direction of the larger system of sun plus earth plus space deteriorates in the proper manner.

But this general answer is not sufficient. We are asking something rather unlikely, and the general rationalization really says only that life isn't ruled out by thermodynamics—we are still pretty special! Specific mechanisms have to exist for the diversion of the solar energy into biological systems and for its transfer within these systems. It is with these specific mechanisms that almost every area of inquiry within biology is concerned, our present one certainly not excepted.

*   *   *

AND AFTERTHOUGHTS

We began with the contrasting themes of constraints and opportunities—and with them the notion that, to an extent not commonly appreciated, the functional characteristics of organisms must depend on the nature of an inescapable physical world. I hope I have convinced the reader that a fair chunk of biology finds a decent context when juxtaposed with even a very superficial view of a few physical phenomena or a few geometrical imperatives. I'll not reiterate the cases here, nor will I erect some grand and abstract edifice—abstraction and philosophizing must be done by writers less self-conscious than I.

Besides, the themes were, in part, just a way to get going and to make two other points, one personal and the other more general. The personal point is that I really *enjoy* this particular view of biology—it provides explanations for many of the questions that I happen to ask about living systems. Perhaps the coincidence is no coincidence—I've simply learned over the past thirty years which questions to ask. Or perhaps not.

The general point is that I think I have been writing about an area of biology with an internal coherence not greatly different from that of more traditional areas—an area with major ideas, a detectable history,

313

particular heroes, contemporary activity (quite a renaissance, in fact), and which asks interesting, operational questions. The information the field uncovers has immediate relevance to quite a number of other areas of inquiry. And it's an area with enough contemporary activity to provide stimulation yet not so much that investigators bump heads too often, with the inevitable sociological unpleasantness.

While the field is literally "biomechanics," this word has been preempted by an incorrigibly anthropocentric branch of medical or athletic science, as I've argued elsewhere (Vogel 1986). This book has been about "biological" biomechanics, but one can't use that redundant expression without making people suspicious of concealed insults and antagonisms. I prefer to call the present subject "comparative biomechanics," the "comparative" to be taken as a euphemism for "biological" by analogy with "comparative anatomy" or "comparative physiology."

The reader who is not involved in some area of functional biology may be a bit overwhelmed by all the phenomena and organisms that have strutted and fretted on the present stage. The writer, at this point, feels more than a little frustrated by the opposite. Within the general topics of the book there are so many other interesting phenomena and beguiling biological cases, and yet even the items included here received only the briefest treatment. A cursory look at the original literature or at one of a number of more advanced books will emphasize both the brevity and superficiality of the present discussions. At best you've been introduced to a certain kind of question that can be asked about the operation and arrangements of organisms—despite having read this book, you are barely a neonatal biomechanic.

And I must emphasize the limit of breadth as well as that of depth. Organisms bump into physical reality at far more interfaces than those mentioned here. Thus almost no space was given to thermal problems, to the implications of variables such as thermal conductivity and thermal expansion coefficient. Mention was made of locomotion, but a rich literature on just the immediately mechanical aspects of swimming, running, flying, crawling, burrowing, and so forth has been almost totally ignored. Support was left lying with talk of little more than properties of materials, since adequate space could not be given to the anatomical detail needed to take the subject much further. And of sound, of light, of waves, and of electrical phenomena in general, just silence and darkness. I've talked mainly about subjects with which I felt comfortable; it would be most pleasant to discover someone else's book on topics with which I am not.

Mark Twain begins *Huckleberry Finn* with a "Notice"—that "Persons attempting to find a motive in this narrative will be prosecuted; persons attempting to find a moral in it will be banished; persons attempting to find a plot in it will be shot." All of which seems rather harsh. I would, though, offer a related admonition. We have repeatedly either analogized or contrasted living with nonliving technology, and it is a great temptation to look toward nature's mechanics for schemes that we might copy. Popular publications and granting agencies like any hint of practicality, and possible technology transfer is part of the Faustian bargain we've made to obtain funding. In fact, the transfer has rarely happened as envisioned—usually a device, once invented, has stimulated recognition of its preexistence in nature. Copying nature has rarely had good results[1]—for very practical reasons ships don't wiggle their sterns, cars aren't the least bit equine, and airplanes don't flap their wings. Otto Lilienthal (1848–1896) wrote a book entitled (in English) *Bird Flight as the Basis of Aviation*; he designed very birdlike hang gliders and was killed in one of his thoroughly unstable craft. A bird, as Maynard Smith (1952) pointed out, turns the instability to good advantage through increased maneuverability. But a person lacks the bird's highly evolved neuromuscular control.

There are lessons in nature that might be applied to our designs—how to achieve material economy through the use of adequately strong but unusually unstiff structures, for instance. Or the reminder that one might take advantage of flows in the environment to reduce the cost of operation of filtration or ventilation equipment. Or an occasional hint that rectilinearity isn't always necessary. But these lessons are general ones, not specific recipes to be copied.

The real attraction of comparative biomechanics is, I'd argue, purely intellectual, no different from the attraction of literature or history. Not that it lacks a distinctly human perspective. This age, which can send hardware beyond the earth, has ended up making extraterrestrial civilizations seem more rather than less remote—their presumed accessibility was mostly wishful thinking and romanticism. And our particular industrial technology is, with trivial differentiation, globally distributed. So how can we judge which aspects of human technology are historical accidents and which are inevitable consequences of, say, the composition of the earth's crust or the locally operative physical constants such as its gravitational acceleration? Where is there a technology with which to compare our own? But that's just the comparison made available by the present subject—as mentioned at the start, looking at life's devices provides a comparison between our human technology and the only other of which we have any knowledge. What more fascinating prospect could there be?

[1] "Velcro" is apparently one of those rare instances; it was modeled on plant burrs.

Notes on numbers

O NE SOURCE of the great explanatory and predictive power of contemporary science is its quantitative character. A very large number of our most interesting questions begin with "how much" or "how often." Less obviously, even outwardly qualitative questions often yield answers only in response to investigations that are quantitative at every stage. We're all taught to use numbers, although with some curious practical omissions, but the skills involved even in simple mathematics corrode swiftly. The present book may not be highly mathematical, but it is thoroughly quantitative; if you don't commonly work with numbers or if you are, say, an accountant, and use them in quite a different way, the following quick brushup and commentary might prove useful.

ARITHMETIC

Numbers can be divided into "positive," "negative," and "zero." If positive numbers are envisioned as a scale running from left to right, then negative numbers mark a continuation of the same scale from right to left beginning just to the left of zero. Numbers are assumed to be positive unless preceded by a minus sign.

Equipped with negative numbers, it is easy to see that the operation called "subtraction" is a special case of a more general operation called "addition"—subtraction is merely the addition of a negative number.

$5 - 3 = 2$ will henceforth mean "add -3 to 5"

$-3 + 5 = 2$ is the same as $5 - 3 = 2$

So subtraction needn't concern us further.

Similarly, it is easy to see that the operation called "division" is a special case of a more general operation called "multiplication"—division is merely multiplication by a fraction.

6 divided by $3 = 2$ is the same as

$6 \times 1/3 = 2$ which is very close to

$6 \times 0.333 = 2$ (really 1.998, but close enough to 2 for government work)

So we don't really need division either. My word processor lacks the division symbol, but until now I never even noticed! A horizontal bar is all we need to indicate division—in the example above (second line), one can multiply the 6 by the 1 and then use the bar for the subsequent division.

$6 \times 1/3 = 6/3 = 2$

The appropriate signs to use in multiplication (and division) aren't instantly obvious. Just remember that the product of two negative or two positive numbers is positive, while the product of a positive and a negative number is negative.[1]

$-4 \times -2.5 = 10$ and $-4 \times 2.5 = -10$

Two more operations need mention at this point, and again the second is just a special case of the first—"exponentiation" and "extraction of roots." (The former sounds hortative, the latter dentistical.) Linguistical conventions vary—"three to the second power," "three squared," or "three exponent two" all mean "three multiplied by three," or the product of two threes, and they correspond to

$3^2 = 9$

The same notation works equally well for roots; thus we can write the "cube root of 8" as "8 to the power 1/3" and, if desired, put the latter in decimal notation (the actual operation is not readily done without a calculator).

$8^{1/3} = 2$ or $8^{0.333} = 2$

Dividing by a number that has an exponent is equivalent to multiplying by that number with the exponent reversed in sign (times -1).

$1/2^3 = 2^{-3} = 1/8$

So these exponents can be integers, fractions, or combinations; they can be positive or negative; and they can even take a value of zero. The latter is a rather peculiar case—anything with an exponent of zero is equal to one.

$17^0 = 1$

Note that a number to the power 1 is just the number itself—to every number with no explicit exponent there attaches an implicit exponent of 1.

$17^1 = 17$

---

[1] Certain of the statements here quite shamelessly lack even the merest pretense of honest explanations. While they aren't really just conventions, for present purposes they're most easily treated as Revealed Truth.

Multiplying a set of identical numbers with exponents is the same as adding their exponents. It's obvious in the first example, less so in the subsequent ones.

$3^1 \times 3^1 \times 3^1 = 3^3 = 27$

$3^1 \times 3^2 = 3^3 = 27$

$3^{4.2} \times 3^{-1.2} = 3^3 = 27$

Raising a number, say 3, to a nonintegral power such as 4.2 ($3^{4.2}$) is easily done with a hand calculator that has a key labeled "y^x." Otherwise it's awkward.

To avoid ambiguity, it is necessary to establish rules for precedence of operations. The conventional sequence, proceeding from initial to later operations, is

exponentiation and extraction of roots
multiplication and division
addition and subtraction

As an example, the following expressions yield 8, not 12.

$2 + 2 \times 3$ or $3 \times 2 + 2$

The way to circumvent these conventions is with another convention involving parentheses (in pairs, always, which may be nested symmetrically like Fabergé eggs when there are more than one pair). The transcendent rule is always to perform the operation in the innermost parentheses first, proceeding to outer parentheses. Beyond that, the previous precedential rules apply. Parentheses also replace the sign "x" for multiplication.

$3(2 + 2) = 12$ $3 \times 2 + 2 = 8$

$(3 \times 2)^2 = 36$ $3 \times 2^2 = 12$

$(3(2 + 2))^2 = 144$

Note that with enough parentheses the "×" need never be used at all. In fact, it becomes handy to avoid the "×" designation for multiplication in algebra since "×" is commonly used to represent an unknown quantity, and using the same symbol for two different items is a Bad Thing.

SCIENTIFIC NOTATION

Scientific notation gets around the inconvenience of writing very large and very small numbers. Its essential feature consists of expressing a number in two parts linked by an "×"—in one, the significant figures led by one digit to the left of the decimal point, and in the other, the number of places by which the decimal point has been moved in the shift from ordinary notation. Each place, of course, represents a power of ten in our counting system, and the ten is given explicitly in the second part.

126,000,000 can be written as (1.26×10^8)

0.000,000,126 can be written as (1.26×10^{-7})

Note that the exponent is obtained just by counting the number of places by which the decimal point is shifted, shifts to the left being positive and shifts to the right negative.

The notation also simplifies calculations—for instance, in multiplying two numbers, the significant figures (on the left) are multiplied first and then the exponents (on the right) are added.

$(3.4 \times 10^6)\,(2.3 \times 10^{-3}) = 7.82 \times 10^3 = 7820$

$(5.2 \times 10^6)\,(7.3 \times 10^{-3}) = 37.96 \times 10^3 = 3.796 \times 10^4 = 37{,}960$

Division is as easy—one just switches the sign of the divisor's exponent (what something is being "divided by") before adding the exponents.

MEASUREMENT

Numbers are used in "scales of measurement." It's useful to distinguish among four basic sorts of scales and to ask which arithmetic operations are permissible with each (Stevens, 1946).

(a) *nominal scale*—numbers are used only as labels. Examples are the numbers on football jerseys and serial or model numbers. The only permissible operation is the determination of equality (is this model a 610 or not?).

(b) *ordinal scale*—numbers are used to rank order things. Examples are acquisition numbers in a library, lumber grades, and Mohs' scale of hardness of minerals. In the latter, rank is based on the pecking order of who scratches whom. Additional permissible operations are determinations of inequalities: "greater (or harder) than" ($>$) and "smaller (or less hard) than" ($<$). One can't assert that something numbered 10 is twice as hard or twice as recent as something numbered 5, or that a 7 is two standard units of hardness or time less than a 9.

(c) *interval scale*—equal differences between pairs of numbers imply equal intervals between them, so the scale is truly quantitative. But

319

the zero point is a matter of convention or convenience. The commonest cases are our temperature scales, Fahrenheit and Celsius, and calender dates. Addition and subtraction work nicely—40 degrees is as much hotter than 30 as 25 degrees is hotter than 15. But 40 is not twice as hot as 20 degrees in any sense at all—multiplication and division are precluded.

(d) *ratio scale*—numbers imply both equal intervals and an absolute zero. So multiplication and division are possible, along with the other operations. Most physical scales are of this type, as is the scale of numbers with which we count things. Five apples are indeed half as many as ten. One set of values (say in feet) can be converted to another (in meters) with a simple multiplication by a constant conversion factor. The absolute temperature scale (Kelvin degrees) is of this type—zero is the temperature at which molecular motion ceases, and the scale is a measure of the intensity of such motion.

Strictly speaking, no measurement is of any value unless information is available about its accuracy. No measurement is absolutely accurate, and, indeed, the accuracy required depends on the use to which the measurement will be put. It's no trick to obtain a balance that will weigh five pounds to the nearest thousandth of a pound, but would you insist that your bag of apples be so accurately determined? It would require either an expensive selection process or the use of fractional apples, certainly a rotten idea.

Accuracy, though, reflects the effects (or lack of effects) of two quite distinct components. First, "imprecision," a term used here for the scatter among a series of separate determinations. Put a bunch of thermometers in boiling water—their readings may vary from 97 to 101 degrees Celsius. The commonest measure of this scatter is something called the "standard deviation"—one figures the average of the measurements, then the average of the squares of the deviations of the individual measurements from the first average, and then takes the square root (1/2 power, remember) of the latter average. Often the standard deviation is given following a datum and preceded by a plus/minus: 5.67 ± 0.03 kg for items made to a very close tolerance. Obviously it would be a joke to report the mass above as 5.668734 kg even if the balance could be read this closely!

The other thing limiting accuracy is "systematic error." It amounts to a measure of the accuracy of the calibration of the measuring device; its estimation, though, requires some knowledge of an absolute (or much more accurate) standard. Weigh yourself on a bathroom scale ten times—you can, with care, read it to the nearest half-pound, and the standard deviation ought to be somewhere around such a value. But there may be a concealed error of a pound or two, observable only if you have access to a standard weight whose value is trustworthy to perhaps a few ounces. No measurement is better than the calibration of the measuring device, no matter how impressive the flashing digital readout!

In the primary scientific literature, it is important that sufficient indication be given concerning the imprecision and systematic error of every item of data, either with the data or in the section presenting the methods used. In a general discussion such as this book one can get away with looser practices; still, it's a good idea to provide, in context, sufficient distinction among numbers meaningful to an order of magnitude (a tenfold factor), to about 50% either way, to about 10%, and to about 1%. Thus, saying something is "of the order of 100,000," something else is "very roughly 100,000," another number is "about 100,000," and still another is "101,000" makes (in this book at least) those distinctions.

ALGEBRA—EQUATIONS

Algebraic statements, which we've been tacitly using, are essentially simple sentences, with the ordinary sequence, subject-verb-object. Subject and object are mathematical expressions. The commonest verb is "equals"; a statement using "equals" is called an "equation."

Two plus four equals six.
$2 + 4 = 6$

With ordinary algebra, an equation is equally valid read either from subject end or from object end. In an algebraic statement, one or more numerals are replaced by letters. A letter (Roman, Greek, or other) stands for a number which is, at the moment of writing, unspecified (almost always—π, the ratio of the circumference to the diameter of a circle, with a value of 3.1416, is an exception). Queerly enough, as we'll see, it proves very useful to symbolize something unknown! The choice of letter is fairly arbitrary—it may be a de facto abbreviation (r for radius, V for volume) or a convention (S for area). Beyond that, letters early in the alphabet (a, b, c) most commonly refer to constants; letters late in the alphabet (x, y, z) most commonly refer to variables.

$2 + a = 6$ (a is obviously 4)

Two (or more) letters run together are (with very few exceptions) treated as if separated by a multiplication sign; thus the expression $F = ma$ implies that m is multiplied by a. An equation might describe how to figure the surface area of a rectangular block

321

$$S = 2lw + 2lh + 2wh,$$

where S is area, l is length, w is width, and h is height. Notice how much more economical of space and easily perused is the equation compared to a verbal statement such as the one earlier for standard deviation.

Letters or numerals below the level of the others ("subscripts") are just labels put there for convenience. Thus one way of stating that the net pressure across a barrier is the difference between outside and inside pressures is

$$\Delta p = p_\text{o} - p_\text{i}.$$

(The Greek "delta" is thus a shorthand for "difference between values of.")

By far the most important rule for working with algebraic statements is the following. *You can always do the same thing to both sides of an equation without changing the essential meaning or truth of the statement.* Put in terms of equations, it says that you can do just about anything as long as you do it simultaneously to the expressions on both sides of the equal sign.

The commonest thing one has to do with an equation is to "solve" it—to find the value or values of a particular unknown compatible with the algebraic statement. It's usually done by combining things on each side of the equation wherever possible and by applying the rule above.

$2a + 4 = 5 + 5$	Combine 5's and you get
$2a + 4 = 10$	then subtract 4 from both sides to get
$2a = 6$	then divide both sides by 2 and you have
$a = 3$	and the equation is solved!

One can get quite accustomed to "reading" equations for some idea of how one quantity changes in response to a change in one or more others. It may be useful to rearrange an equation (again applying the rule above) to assist the process. We can consider a few cases as examples.

(a) A hypothetical relationship between the number of organisms that can feed in a given area and the mass of each organism is

$$C = n\,(M_\text{b}),$$

where C is a constant, n is the number of creatures, and M_b is the body mass of each. Clearly, a doubling of numbers would necessitate a halving of mass; a tripling of mass would require reducing numbers by a third, etc.

(b) A fast fox catches a slower rabbit. How is the time it takes related to the distance between them when pursuit starts and to the differ-ence in their speeds? For steady speeds, distance is the product of speed and time. Pursuit takes a time t, the rabbit runs at speed U_r, the fox at U_f, and they're initially a distance d_0 apart. A little thought should convince you that

$$d_0 = U_f t - U_r t.$$

A little additional messing around gives

$$d_0 = t(U_f - U_r).$$

And one might then divide both sides by $(U_f - U_r)$ to get

$$t = \frac{d_0}{U_f - U_r}.$$

All of which says that the time it takes is equal to the initial distance divided by the difference in their speeds. Or that the initial dis-tance can be no greater than the time available for pursuit times the difference in their speeds. And all of which is relevant to when stalking can stop and pursuit begin, and to how far away must the prey detect the predator for a certain distance between prey and refuge (rabbit and hole, here).

(c) The surface area S is being compared to the volume V for a set of cube-shaped objects of different sizes. The following equation is obtained:

$$S = 6V^{2/3}.$$

What the equation says is that for any change (increase or decrease) in volume, surface area changes in the same way (the exponent is positive) but by a smaller factor (the exponent is less than one). (It's true, in fact, for any set of similarly shaped solids—only the 6 is specific for cubes.) If you triple the volume you (roughly) double the area. If you double the length of each side, you increase the volume by a factor of 8 but increase the surface by only a factor of 4. The biological implications are truly profound, believe it or not.

ALGEBRA—PROPORTIONALITIES

Consider an equation relating two variables and containing a constant of the form

$$y = kx$$

What this says is that any multiplicative operation done to x implies the same operation done to y—triple x and you must thereby triple y, for instance. To avoid the distraction of that mild red herring, k, we can state the relationship between x and y as a so-called proportionality: "y is proportional to x," and write it using a symbol for "is proportional to":

$$y \propto x.$$

For this particular proportionality we can make the more specific statement that "y is *directly* proportional to x," distinguishing it from

$$y \propto 1/x,$$

for which we say that "y is *inversely* proportional to x," or, for that matter, to distinguish it from any other proportionality such as

$$y \propto x^3,$$

which states that "y is proportional to the cube of x" and indicates that doubling x implies an eightfold increase in y.

Note that you can multiply or divide both sides of a proportionality by the same number or you can raise both sides to the same power, but you violate the relationship if you add the same thing to both sides.

Graphs

A great aid in envisioning the implications of equations is a graph. In the most ordinary sort, one considers two variables. As one (the "independent variable") changes, the other (the "dependent variable") changes in consequence according to the rules embodied by the underlying equation. This is so even if the equation can't easily be written down or exists only in theory or is just a summary of some experimental data. It ought to be noted that the distinction between independent and dependent variables can be somewhat arbitrary at times. But usually the distinction is meaningful if less than crucial—for the fox mentioned earlier, the distance covered increases as time advances. Time is the independent variable, distance dependent, since it seems less sensible to say that elapsed time increases as the distance increases, although it does.

On the ordinary sort of plot, the independent variable is expressed as horizontal distance and marked on the so-called x-axis, or "abscissa." The dependent variable is given as vertical distance and marked on the y-axis, or "ordinate." A line is then drawn through all the points that satisfy the relationship given by the equation or through a set of experimentally determined points from which an equation might be concocted.

Frequently the "slope" of a line on a graph is of great interest. This is a measure of how far upward (or downward) the line goes for a given distance from left to right. In a graph of the equation

$y = 6x + 4,$

the slope is 6. And the line intersects the vertical (y) axis at a value of 4 (when $x = 0$, $6x = 0$, and thus $y = 4$). In short, the equation can be derived from the straight line of the graph. It is common practice to use y for the dependent variable and for it to be the sole item on the left of an equation, and to use x as the independent variable, with other items, on the right (as above).

FORMULAS

A few specific formulas will be useful.

(a) The "Pythagorean theorem." Consider a triangle in which one of the angles is a right angle (90°). If one calls the two sides adjacent to that angle a and b and the side opposite c, the relationship among the lengths of the three sides is

$c^2 = a^2 + b^2.$

(b) For a circle of radius r, diameter d, circumference C, and enclosed area S,

$C = 2\pi r = \pi d,$

$S = \pi r^2.$

(c) For a sphere of radius r, surface area S, and volume V,

$S = 4\pi r^2,$

$V = (4/3)\pi r^3.$

(d) For an open-ended cylinder of radius r, height h, surface area S, and volume V,

$S = 2\pi r h,$

$V = \pi r^2 h.$

CONVERTING UNITS

As explained in Chapter 1, this book uses SI units, some of which are everyday items and others of which are uncommon outside the scientific literature. For definitions of the SI units, see Table 2.1. A few conversion factors are given below—to convert *to* the SI unit, divide by the indicated factor; to convert *from* the SI unit, multiply by the factor.

1 meter	=	39.37 inches
		3.28 feet
		0.000621 miles
1 kilogram	=	35.3 ounces (avoirdupois)
		2.20 pounds
1 cubic meter	=	1,000,000 cubic centimeters
		1000 liters
		264.2 U.S. gallons
1 meter/second	=	3.28 feet/second
		2.24 miles/hour
		3.60 kilometers/hour
1 kg/cubic meter	=	0.0624 pounds/cubic foot
		0.001 grams/cubic centimeter
1 newton	=	100,000 dynes
		0.225 pounds weight (on earth)
		0.102 kilograms weight (on earth)
1 pascal	=	0.000145 pounds/square inch (on earth)
		0.00750 millimeters of mercury (on earth)
		0.00000987 atmospheres (on earth)
1 joule	=	10,000,000 ergs
		0.738 foot-pounds (on earth)
		0.238 calories
		0.000238 kilocalories (Calories)
		0.000948 Btu
1 watt	=	0.00134 horsepower
		3.412 Btu/hour

To convert temperature in degrees Fahrenheit to degrees Celsius, subtract 32 and *then* multiply by 5/9.

To convert temperature in degrees Celsius to degrees Fahrenheit, multiply by 9/5 and *then* add 32.

To convert degrees Celsius to degrees Kelvin, add 273.

The most useful source of conversion factors I've ever seen for the types of problems raised here is a booklet by Pennycuick (1988). I use its earlier incarnation and can't imagine doing much at all at a desk that lacks a copy.

APPENDIX 2

Problems and demonstrations

THE FOLLOWING is a miscellany of items supplementary to the various bits of open-endedness in the text. Equipment and supplies for the demonstrations (calling them "experiments" does violence to real science) are mainly household items or things easily obtained from grocery, stationery, toy, or hardware store. The computer programs assume a minimal machine that accepts programs in some fairly generic BASIC; since machines vary widely in how they do graphic work on their screens, I've just given general suggestions and left the rest for the reader to work out. Answers to quantitative problems have not been provided; I'll supply them upon request, limiting their distribution within a specific institution or department if requested to do so.

CHAPTER 1

(1) It is occasionally suggested that the emergence of evolutionary novelty is constrained by the incrementalism implicit in the process of natural selection. It is further suggested that this constraint is one of the main differences between natural design and human technological progress. Argue the converse, citing specific instances if possible, that human technology (whether primitive or advanced) ends up in practice with a very similar constraint.

(2) Metallic materials have excellent mechanical properties, certainly not inferior to those of most of the substances used by organisms. And all organisms do use metal atoms in their metabolic machinery—we, for instance, require iron, copper, zinc, vanadium, chromium, tin, and other metals. Suggest one or more reasons why the use of metals in bulk for mechanical applications has apparently never arisen in the living world.

CHAPTER 2

(1) Of the two sorts of devices, beam balances and spring scales, one will not work properly on the moon. Tell which and explain why.

(2) Consider the problem of knowing whether you are moving. How

might translational and rotational motion differ in their requirement for some external frame of reference? Why does rotational motion of necessity involve some sort of acceleration?

(3) a. Two cars suffer the same degree of damage when run into a rigid wall. Why is the heavier one still safer for passengers in a two-car collision?
b. A "biological" version. A 600-kg bull moose ("Theodore") moving east at 10 m/s collides with a 300-kg moose moving south at the same speed. What's the final speed and direction of the impacted and combined moose-mass (at least approximately).

(4) Joe Dumbwaiter likes to show off by pouring coffee into cups in a fast, narrow stream from a pot held several feet above the table. Use the principles of conservation of mass and conservation of momentum to design a spout that will promote success in the stunt. Test your household spouts using this criterion.

(5) Practice in vector arithmetic. Two people equipped with hand winches are trying to pull a car out of a muddy ditch. The hand winches have to be attached to distant trees, and the only available trees force the winches to pull at a right (90-degree) angle to each other. The first winch exerts a maximum force of 10,000 newtons, the other 5000 newtons. How much more force can be exerted by the two than by the first one alone?

(6) Elevation of your pulse rate is a decent, if crude, indication of the rate at which you are working, at least if the work is sustained at a steady level for a minute or more. Devise a demonstration to show that a human must do work to exert force against an unyielding obstacle, but must do even more work actually to move something against a resisting force.

CHAPTER 3

(1) Consider a spherical orange. What fraction of the overall radius will be made up of rind if the rind has a third the volume of the whole orange?

(2) A practical matter. A given house is found to require a furnace capable of producing 30,000 watts (100,000 Btu/hr). How big a furnace would be required for a larger building, 10 times as wide, 10 times as long, and 10 times as high, of the same sort of construction?

(3) There exists a kitchen frill used to estimate how much raw spaghetti

to cook for various numbers of portions; it consists of a plate with four round holes through which bundles of spaghetti can be stuffed. If the smallest hole corresponds to one portion and is 22 mm in diameter, how wide should the holes be for two, three, and four portions?

(4) Obtain, from fishmonger or ichthyologist, a batch of fresh or preserved fish whose body lengths span at least a twofold range. You can use a mixed bag or a single species, although the results may differ between the two. For each fish (a dozen or so) measure body length and mass. Figure the exponent that relates length (x-axis) to mass (y-axis)—the "a" if mass is proportional to length to the power "a." So-called log-log paper is handy for graphing; use the slope of the best straight line through the data points and equation 3.7, or any other appropriate procedure.

Chapter 4

(1) Figure out the speed at which you might be expected to shift from a walk to a jog or vice versa, based on the Froude number (equation 4.6) and your hip height. Convert to minutes per mile. Does the figure seem to correspond to practical reality?

(2) Figure out your ponderal index, assuming a density of 1000 kilograms per cubic meter (equation 4.5); in addition, get figures for your nearest and dearest or any other cooperative and honest victims. Again, do the figures seem to correspond to reality—do they arrange people along a perceptually persuasive scale from skinny to obese?

(3) Consider an object speeding up from rest with a constant value of acceleration. Plot graphs of, first, acceleration versus time, second, velocity versus time, and then distance versus time. Ignoring units, use the graphs to explain the origin of the two formulas

$$v = at,$$

$$x = at^2/2.$$

(4) Ecologists distinguish between the concepts of "standing crop," or how much biological material is present in a given area at a given time, and "productivity," or how much biological material is produced by a given area over a given period of time. Introductory textbooks in biology very often confuse the two. How might the basic notion of the calculus clarify the relationship between the two

concepts? And what might be the crucial agricultural distinction between them? Finally, think of habitats that represent extremes—lots of standing crop but low productivity, and low standing crop but high productivity.

(5) A bit of backwards music loudly proclaims the preference of our auditory system for highly positive temporal gradients of sound. But getting a tape of music played backwards is a bit tricky. The easiest way is to find an old "AR" turntable—it can be persuaded to turn counterclockwise by giving an initial push in that direction—and taping a record. Alternatively, an external elastic band can be used to connect a more ordinary turntable to the shaft of an external motor, but one then has to worry about motor speed and so forth. Switching a few wires on an auto-reverse cassette deck might work, but I've not tried it. I find a tape done backwards on one side and frontwards on the other convenient for class use; there is, though, the odd illusion that one is doing the reversal by turning over the tape cassette! As for music, solo piano seems best—everyone is familiar with how it ought to sound, and people don't even notice the sharp termination of the backwards notes.

(6) Make a pattern of parallel gray stripes of increasing darkness. Producing such a set of uniformly gray stripes is most easily done by exposing strips of a sheet of photographic paper of average contrast grade to a constant light for successively longer intervals—I've used an enlarger as a light source and intervals increasing by a factor of the square root of two. Thus, if the exposure of the first strip is 1.0, then the others are 1.4, 2.0, 2.8, 4.0, 5.6, 8, 11.3, 16, and 22.6. If one starts with a covered piece of paper and removes the covering in increments, the exposure factors for the steps are 6.6, 4.7, 3.3, 2.4, 1.6, 1.2, 0.8, 0.6, 0.4, and for the fully uncovered sheet, 1.0. Alternatively, photographic film can be used to make a projection slide.

(7) Illustrate (if somewhat crudely), by extending the telescoping antenna of a portable radio, flow in which annuli of successively smaller radii travel faster and in which the one with the largest radius is fixed.

CHAPTER 5

(1) Devise a scheme whereby a hand pump of the sort commonly used to inflate tires is used as the centerpiece of a simple device to weigh

things. Indicate the sort of scale of numbers that might be marked on the device from which the weight of an object might be read.

(2) In the early days of electron microscopy of biological material, sections of tissue sufficiently thin for an electron beam to penetrate were cut with knives made from broken pieces of plate glass. Not only did these knives get dull through use, they got unusably dull just sitting on a shelf for a few months. Old window panes are a little thicker near their lower edges than further up. What is happening in these cases to which the material in Chapter 5 might be relevant?

(3) The device for illustrating the force of surface tension (Figure 5.7) can be made by bending a few wires—it can be surprisingly crude and still work. For the downward force arrow substitute a small "V" of wire in which the ends are formed into loose loops around the slider. You can pull on this wire and note that the force needed to move the slider is independent of the slider's position; alternatively you can tip the whole rig slowly up from a horizontal position. Any elevation sufficient to move the slider ought to be enough to move it any distance. The solutions sold in toy stores for blowing bubbles seem to be a little better than diluted liquid dishwashing detergent.

(4) There seems to be no readily available technological example of a countercurrent exchanger; I have one designed for renal dialysis, but it's not cheap. If one is wanted badly enough, instructions for constructing a demonstration heat exchanger are given in the appendix of Vogel and Wainwright (1969).

CHAPTER 6

(1) A very soft flatworm is aware of the no-slip condition. It finds a shallow depression on the top surface of a very smooth rock, a depression just large enough for the worm to fit with its upper surface coplanar with the rock. The worm figures that it thereby ought to avoid entirely the force of water currents. But it finds itself still pulled downstream. What is the origin of this drag on the worm?

(2) Imagine that you are a "particle of fluid" midway down in the velocity gradient over a flat surface oriented parallel to flow. Flow above you goes one way and flow below you goes the other way with respect to your own position as you move along the surface. As a result, you rotate as if you were rolling along the surface, even though you don't ever touch it. How might this situation explain why snow fails to accumulate on the upwind side of a tree trunk exposed to a

331

windy snowstorm? To what other phenomena in the real world might this have relevance?

(3) Golf balls with dimples go further than smooth ones became they have less drag under some circumstances. Assume the dimples on a golf ball are effective in reducing its drag at Reynolds numbers between 75,000 and 200,000. Golf balls are 4.3 cm in diameter. In what range of speeds are the dimples effective?

(4) A biological version of the last question. Assume that a tree is exposed to a maximum wind of 25 $m \cdot s^{-1}$. For what range of trunk diameters might the roughness of the bark reduce the drag of the tree? Comment on the hypothesis that rough bark has adaptive significance in reducing the chance that a tree will be blown over.

(5) A spigot has an orifice 6 millimeters in diameter. At what rate of discharge of water (volume per unit time) will flow through the spigot change from laminar to turbulent? To get some feel for such a rate, convert your result to something familiar, such as quarts per minute.

(6) An everyday device based on the viscosity of a fluid is a shock absorber, as used in parallel with the springs of a car, on some lifting tailgates of hatchbacks and station wagons, and as the resistive elements of rowing exercisers. In current designs, a liquid is forced through a small aperture—the greater the force, the *faster* the liquid passes. Obtain a shock absorber and examine its behavior, noting the fundamental difference between it and any kind of spring.

(7) Reversible flow at very low Reynolds numbers can be demonstrated, as mentioned, with corn syrup and molasses. Commercial corn syrup at room temperature is about a thousand times as viscous as water, so the Reynolds numbers get low indeed. Fill a clear glass container (such as a pint measuring cup) with corn syrup; inject a blob of molasses near but not touching the outer wall with the aid of a soda straw. Then insert a cylindrical stirrer such as the handle of a wooden spoon into the center of the container and move it radially about half-way to the outer wall. Slowly and steadily stir clockwise about three full turns, watching the distortion of the blob of molasses. Then reverse things, moving the stirrer counterclockwise in the same pattern. The blob of molasses should reform, showing that it was distorted but not disorganized by the shear in the fluid.

CHAPTER 7

(1) A sailplane with a lift-to-drag ratio of 40:1 is towed 1000 m up in still air; it glides at 25 m·s⁻¹. How long can it stay aloft? A monarch butterfly, by contrast, glides at 2 m·s⁻¹; what must its lift-to-drag ratio be for it to stay up as long, starting at the same height?

(2) Some idea of the location of the separation point in flow around a nonstreamlined object (as in Figure 7.7) can be obtained by exposing a cylinder of chalk to a steady flow and observing the pattern of erosion of the chalk by the flow. The easiest scheme is to stick the end of a piece of blackboard chalk into the substratum beneath a moving stream so the rest of the piece protrudes vertically into the flow. A good exposure time will range from 3 to 24 hours, depending on the speed of flow and the particular chalk. Separation is associated with a low local speed, so erosion is low. Note also the greater erosion on the *downstream* side of the cylinder and the differences in erosion near the substratum. For speeds at which chalks stay erect, separation may occur upstream of the maximum width. Eroded chalks can be dried and saved. Soap or candy cylinders will work also.

(3) Low Reynolds number flow around an object, where vortices, turbulence, and even separation are quite unknown, may be demonstrated as follows. Mix 20 drops (1 ml) of food coloring in about a tablespoon, 13 ml, of corn syrup. Fill a tall, transparent glass with uncolored corn syrup, then dip a glass marble in the colored syrup and allow it to fall down the middle of the glass.

(4) At high Reynolds numbers, flow around a nonstreamlined object leaves a turbulent wake, with irregular shedding of vortices. The result is a high and unsteady drag. For a streamlined object the drag is both lower and steadier. A demonstration and practical application consists of making two antenna ornaments to assist location of a car in a parking lot. Compare the behavior of a sphere with some approximately streamlined object (a water rocket or a wood carving) when the car is traveling rapidly.

CHAPTER 8

(1) The quantitative crux of diffusion is the peculiar relationship between time and net distance covered. The following program, in BASIC, is a model that illustrates how one can obtain that relation-

ship from a random walk. It has three parts—first, a random walk (which might be run repeatedly); second, a display of the end points of a hundred such walks (and the emergence of a semblance of statistical order); and third, a graph of the results of running the simulation for fifty walkers, taking their average distances from the start after 50, 100, 150, etc. steps. This final graph represents several hours of looping on my computer; the reader might check it by applying the Pythagorean theorem to the end points in a modification of the second part of the program. That the walkers are constrained to move in the four compass points makes little difference. Graphics will have to be added—I've managed to do so for a Sinclair ZX81, an Apple 2+, and a Kaypro 2X; the program has a few notes.

```
10  REM - TURN OFF CURSOR AND ERASE THE SCREEN
20  REM - PRINT AN "X" IN MIDDLE OF SCREEN
30  REM - DESIGNATE ITS COORDINATES H AND V
40  FOR A = 0 TO 200
50  Z = INT(4*RND)
60  N = 0
70  IF Z = 0 THEN V = V + 1
80  IF Z = 1 THEN V = V − 1
90  IF Z = 2 THEN H = H + 1
100 IF Z = 3 THEN H = H − 1
110 REM − TURN OFF PIXEL AT H,V (TO BLINK DISPLAY)
120 REM − TURN ON PIXEL AT H,V
130 N = N + 1
140 IF N < 4 THEN GOTO 70
150 REM − MAKES 3 PIXEL PATH; CHANGE 4 ABOVE TO ADJUST
    SIZE
160 NEXT A
170 REM − MAKE SYSTEM PAUSE ABOUT HALF A SECOND
180 GOTO 40
190 STOP
200 REM − ERASE THE SCREEN
210 REM − PRINT "ENDS OF 100 WALKS, 50 STEPS EACH"
220 REM − AGAIN PRINT "X" IN MIDDLE OF SCREEN
230 REM − CALL ITS COORDINATES C AND D
240 N = 0
250 A = 0
260 B = INT(RND*8)
270 IF B = 0 THEN C = C + 3
280 IF B = 1 THEN C = C − 3
290 IF B = 2 THEN D = D + 3
300 IF B = 3 THEN D = D − 3
310 IF B = 4 THEN C = C + 2: D = D + 2
```

```
320 IF B = 5 THEN C = C + 2: D = D − 2
330 IF B = 5 THEN C = C − 2: D = D + 2
340 IF B = 7 THEN C = C − 2: D = D − 2
350 A = A + 1
360 IF A < 50 THEN GOTO 260
370 REM − TURN ON PIXEL AT NEW C,D
380 N = N + 1
390 REM − RESET C,D TO MID-SCREEN VALUES
400 IF N < 100 THEN GOTO 250
410 STOP
420 REM − ERASE THE SCREEN
430 REM − DRAW GRAPH'S AXES; LABEL "TIME" (X); "DISTANCE"
    (Y)
440 REM − PRINT "AVERAGE CROW-FLIGHT PATH LENGTHS, 50
    WALKERS"
450 REM − PRINT "EACH TIME INCREMENT ADDS 50 STEPS"
460 REM − TURN ON PIXELS FOR GRAPH POINTS, ONE BY ONE
470 REM − (PAUSE FOR EFFECT BETWEEN TURNING EACH ON)
480 REM − SUCCESSIVE COORDINATES FOR GRAPH POINTS ARE
    BELOW
490 END
```

Coordinates (time, distance): 50, 18.7; 100, 28.0; 150, 35.6; 200, 36.7; 250, 41.0; 300, 49.1; 350, 50.8; 400, 60.0; 450, 62.7; 500, 59.0; 550, 71.5; 600, 64.0; 650, 64.3; 700, 65.1; 750, 67.2; 800, 74.4.

(2) Using the procedure described in Chapter 3 and the coordinates above, draw a graph on geometrically scaled axes and figure the exponent that relates distance to time. Is it around the expected value?

(3) Demonstration of the leisurely pace of diffusive transport over macroscopic distances is easiest if convection is eliminated by using a solid—adding a trace (about 1.5%) of gelatin gives functionally non-convecting solid water. The kitchen scheme consists of dissolving a tablespoon (7 g, or one packet) of gelatin in 2 cups (450 ml) of water, half filling several small clear glasses, and chilling to set the gel. Then pour on top of the gel a tablespoon or two of diluted food coloring (any color except yellow)—about 20 drops (1 ml) of colorant in 1/4 cup (55 ml) of water. Do this at various times before a final comparison—50 hours, 5 hours, and 1/2 hour are appropriate. Further comparisons might involve different concentrations of colorant, different colorants, and transfer of colorant from gels made with added food color into gels made without. There's a latent art form here.

CHAPTER 9

(1)   The bones of young humans are relatively tougher, while those of older and larger humans are relatively stiffer. Several questions: (a) What might be the material basis for the change with aging? (b) What might be the practical advantages and disadvantages for the young? (c) What might be the contrasting advantages and disadvantages for adult folk?

(2)   Why can even a short stack of bricks joined with mortar withstand a greater compressive load than a stack without mortar? But why, by contrast, is the resistance of a stack of wooden blocks to compression quite indifferent to whether there's glue between blocks?

(3)   It's of interest to compare two tension-resisting structures of about the same size used by the same animal. One is the sturdy collagenous tendon that can be dissected from a lamb shank (obtainable as the lower part of an untrimmed leg); the other is the nuchal ligament which, as mentioned, runs along the upper part of the neck and can be freed without much trouble. Lying on a cutting board they look about the same—the nuchal ligament is a bit yellow, so they won't get mixed up. Pull on them, and a substantial difference ought to be apparent.

The remainder of the neck makes a fine lamb stew or the basis of a gigantic scotch broth. The leg has other biomechanics to teach before being subjected to thermal degradation—see below (Chapter 13).

(4)   Even a very crude composite material will have mechanical properties quite different from those of its components; a crude one has the advantage of requiring only modest loads to cause some form of failure. Such a composite can be made by combining fiberglass as a stiff component with gelatin as a compliant component. Form two troughs of aluminum foil—about 10 × 2 × 1 inch (25 × 5 × 2.5 cm) is an appropriate size—and spray their insides with no-stick cooking spray ("Pam"). In one, lay a strip or two of fiberglass batting—the material sold in rolls for wrapping pipes is a good source—and place the troughs in a baking pan. Then dissolve one ounce (28 g) of gelatin in 3 cups of boiling water and nearly fill the troughs. Adding water to the baking pan will offset the flimsiness of the foil. Chill until set, drain the pan, and fold back the foil molds. Explore the properties of the resulting composite, particularly not-

ing differences between composite and pure gel attributable to alterations in the way cracks propagate.

CHAPTER 10

(1) Gibbons both walk on their feet and "brachiate," hand over hand—the weight supported is the same in either case. Suggest why it is functionally reasonable for their arm bones to be more slender than their leg bones.

(2) Consider the slightly V-shaped cross section of a long, thin leaf such as that of a lily or willow. How might this shape, with thin edges a bit above and a thicker midrib below, be a reasonable design for such a beam?

(3) You wish to build a deck on one side of your house. The surface (assumed to contribute nothing to the deck's ability to resist sagging) will rest on either 2″ × 6″ or 2″ × 8″ wooden beams. The former are 13.5 cm top to bottom, the latter 18.5 cm. Testing a 2 × 6, you find that the intended span bends 1 cm downward in the middle under the load of a person. How much will a 2 × 8 bend under the same circumstances?

(4) You're interested in the relative hazards of Euler and local buckling in hollow, circular cylindrical columns. From the relevant formulas for I and the forces required to initiate the two forms of buckling, concoct a dimensionless ratio that provides an index to the relative chance of either mode of failure. Consider how this index changes with shape, size, and material stiffness.

(5) Spaghetti is supplied as brittle, solid cylinders in a range of thicknesses; it is handy for small-scale mechanical testing even if the results are hardly noteworthy. Obtain spaghetti of several different diameters (capellini, vermicelli, spaghettini, spaghetti, etc.) and determine their diameters by measuring the width of ten or twenty apposed strands. Then extend a strand between two supports and hang a weight (a half- or one-ounce fishing sinker) in the middle, arranging matters to get a deflection very roughly a tenth of the distance between the supports. From equations 10.1 and 10.3, figure out the flexural stiffness EI and stiffness E of strands of different thickness; compare the results with the data cited in Table 9.1.

(6) Demonstration of the importance of the orientation of fibers in a

fiber-wound cylinder requires only that sleeves be sewn from a piece of fabric and cylindrical balloons inflated within them. Loose fabric of ordinary weave is best—burlap or the flimsiest cotton, for instance. If the fabric has a pattern that indicates the direction of the threads, so much the better; otherwise the direction might be emphasized with an indelible marker. Sleeves about 6 or 7 cm (2½″) in diameter, 20 cm (8″) in circumference, and about 35 cm (14″) long work well. Both ends should be left open. One trick—stretch the cylindrical balloon through the sleeve before inflating it and hold it stretched while inflating.

CHAPTER 11

(1) If rubber were a perfectly "Hookean" material, that is, if the line of its stress vs. strain graph extended to the breaking point without curvature, what limitations would that place on the possible shapes of practical balloons?

(2) Image and describe two worlds in which all liquid water had enough detergent in it so that the surface tension at air-water interfaces was only 2% of the value for pure water. In the first world the detergent pollution is sudden and contemporary; in the second it has been a feature throughout the evolution of living forms.

(3) Attempting to inflate two balloons on a Y-tube is simple and dramatic; the only obscure item is the Y-tube. It may be obtained as a garden hose accessory; some of these even have individual shut-off valves in each leg.

CHAPTER 12

(1) You wish to build a lightweight, freestanding tent, one that might need staking around its perimeter but not ropes running to trees or remote points on the ground. Your notion is to sew pockets into the fabric into each of which a single fiberglass rod or piece of fresh wood may be inserted to erect the tent. To simplify erection, you decide to have no joints connecting the rods; to minimize stress concentrations, you decide that the rods should not even abut each other directly. The number of rods isn't important so long as the individual rods are all of the same size and not inconveniently long for backpacking. See how far you can get in working out a practical design.

CHAPTER 13

(1) Some kitchen appliance (take your pick) is manufactured in two versions, one motorized and one hand-operated. Either version requires about a tenth of a horsepower (75 watts) to operate. Besides the interchange of motor and crank, what additional differences in design might you expect to see?

(2) You are a biologist interested in the way birds put on weight before their fall migrations. Design a bird feeder with a perch that indicates the mass of a bird—it should be something that can be made from simple materials such as string, plywood, screws, hinges, etc. (but no springs) in an ordinary workshop. Design as part of the feeder an indicating device that can be read by the intrepid scientist through a small telescope from a comfortable blind.

(3) The lamb leg that made its contribution to Problem 3 of Chapter 9 can be further violated to any desired degree—it illustrates the linkages of muscles to bones, the angles and mechanical advantages, two sorts of joints, and a kneecap. It's a handier size than a cow's leg and, being longer, it makes the connections easier to see than in the leg of a pig.

(4) Some idea of the reality of osmotically generated pressures can be obtained by looking at systems that have (as described in Chapter 10) a central core of water-filled cells resisting compression from peripheral elements. Several ordinary vegetables fill the bill—carrots and celery are handiest—and the role of concentration differences is best seen by deliberately altering them. Fill four containers with a graded series of salt solutions—0, 1, 2, and 4 teaspoons (0, 7, 14, and 28 g) of table salt in 2 cups (475 ml) of water; in each put as long a piece of carrot and of celery as will fit. Soak for two hours and note the changes (or lack of changes) in the texture of the vegetables. Make lengthwise slits halfway down the stalks and observe whether the cut ends flare out.

CHAPTER 14

(1) The following program calculates several variables for the trajectory of a sphere projected in air; with a general scheme for calculating drag (steps 220–290) it will work for spheres of any size or initial speed. Pick time increments (step 90) short enough to get at least 30

iterations (steps 410, 480). To get data for graphs of the trajectories, merely add

405 PRINT "X = ";X,"Z = ";Z

To draw the trajectories on the screen it is necessary to specify (after step 90, perhaps) some appropriate scale for axes on the screen—trial and error is the simplest way. Then turn on a pixel at the co-ordinates for X and Z after each iteration (step 405 again) and delete steps 440–490.

```
10   PRINT "DIAMETER OF PROJECTILE (METERS)?"
20   INPUT DP: RP = DP/2: REM - PROJECTILE RADIUS
30   PRINT "DENSITY OF PROJECTILE (KG/CU M)?"
40   INPUT PP: M = PP*4.189*RP^3: REM - PROJECTILE MASS
50   PRINT "INITIAL SPEED (M/S)?"
60   INPUT VP
70   PRINT "PITCH ANGLE WITH HORIZ (DEGREES)?"
80   INPUT PA: PR = PA*3.1416/180: REM - ANGLE IN RADIANS
90   PRINT "TIME INCREMENT (SECONDS)?"
100 INPUT TI
110 VV = VP*SIN(PR): REM - INITIAL VERTICAL SPEED
120 VH = VP*COS(PR): REM - INITIAL HORIZONTAL SPEED
130 PRINT "INCLUDE AIR RESISTANCE? Y OR N"
140 INPUT A$
150 PM = 1.2: REM - AIR DENSITY
160 MM = .0000181: REM - AIR VISCOSITY
170 G = 9.8: REM - GRAVITY
180 ET = 0: REM - ELAPSED TIME
190 IT = 0: REM - NUMBER OF ITERATIONS
200 X = 0: REM - INITIAL RANGE
210 Z = 0: REM - INITIAL ALTITUDE
220 S = 3.1416*RP^2: REM - FRONTAL AREA
230 RE = PM*DP*VP/MM: REM - REYNOLDS NUMBER
240 REM - "CD" IS DRAG COEFFICIENT; "D" IS DRAG
250 IF RE>1E5 THEN CD = .1: GOTO 280
260 IF RE>1E3 THEN CD = .5: GOTO 280
270 CD = .4 + 24/RE + 6/(1 + SQR(RE))
280 IF A$ = "N" THEN CD = 0
290 D = .5*CD*PM*S*VP*VP
300 DA = D/M: REM - ACCEL DUE TO DRAG
310 DAV = DA*SIN(PR)
320 DAH = DA*COS(PR)
330 VH = VH - DAH*TI
340 IF VV>0 THEN VV = VV - TI*(G+DAV)
350 IF VV>=0 THEN ZM=Z
```

```
360 IF VV<=0 THEN VV = VV − TI*(G-DAV)
370 VP = SQR(VV*VV + VH*VH)
380 PR = ATN(VV/VH)
390 X = X + VH*TI
400 Z = Z + VV*TI
410 IT = IT + 1
420 ET = ET + TI
430 IF Z>0 THEN GOTO 230
440 PRINT "HORIZONTAL RANGE = ";X;" METERS"
450 PRINT "MAXIMUM HEIGHT = ";ZM;" METERS"
460 PRINT "ELAPSED TIME = ";ET;" SECONDS"
470 PRINT "# OF ITERATIONS = ";IT
480 PRINT "FINAL SPEED = ";VP;" M/S"
490 END
```

CHAPTER 15

(1)   It takes work to shear a fluid at a nonzero rate; that's what viscosity
       is all about. And the work has to appear in some form of energy
       afterwards. A quick and qualitative demonstration: Put a cup of mo-
       lasses or corn syrup in a blender and agitate it at top speed for one
       or two minutes. (Enough air will be introduced so the syrup will
       have to be returned to a larger container.)

(2)   The only appliances in which all the energy consumed is put to the
       desired use are electrical heaters—they alone have an efficiency of
       100%. Electricity can be made with gas, coal, oil, etc.—the same fuels
       used by other home heating arrangements. But electrical heating is
       typically more expensive than the use of combustile fuels. Suggest
       how the economics and thermodynamics might be reconciled.

(3)   There's a nursery rhyme that speaks directly to the subject of the
       second law of thermodynamics. What is the rhyme and what is the
       message?

List of symbols

(To FOLLOW common practice in the different fields from which material has been drawn, some symbols have been used for several different quantities; in practice, no serious ambiguity should arise.)

a	acceleration	PI	ponderal index
a	radius, overall radius	Q	total flow (volume/time)
C	concentration (mass/volume)	r	radius, position along radius
c	celerity (wave propagation speed)	Re	Reynolds number
C_D	drag coefficient	S	area (surface)
C_L	lift coefficient	T	the dimension *time*
c_p	specific heat capacity, constant pressure	T	tension
C_p	pressure coefficient	T	temperature
D	distance arm of a moment	t	time
D	drag	V	volume
D	diffusion coefficient	v	velocity (speed)
E	efficiency	W	work
E	stiffness (Young's modulus of elasticity)	w	weight
EI	flexural stiffness	w	width
F	force	x	independent variable, general
FI	flatness index	x	distance, in direction of material movement
Fr	Froude number	y	dependent variable, general
g	gravitational acceleration	y	deflection of a beam
h	height, depth	γ	surface tension
I	second moment of area	Δ	difference between two values of
k	constant, unspecified value	δ	boundary layer thickness
L	the dimension *length*	ϵ	strain
L	lift	Θ	the dimension *temperature*
l	length, characteristic length	Θ	magnitude of an angle
M	the dimension *mass*	λ	wavelength
m	mass	μ	viscosity
MA	mechanical advantage	π	3.1416
P	power	ρ	density (mass/volume)
p	pressure	σ	stress (force/area)

342

References and index of citations

(Asterisks indicate useful general sources; italic numbers in parentheses refer to citations in the text.)

Alexander, D.E. (1986) Wind tunnel studies of turns by flying dragonflies. *J. Exp. Biol. 122*: 81–98. (*268*)

Alexander, R.M. (1981) Factors of safety in the structure of animals. *Sci. Prog., Oxf. 67*: 109–30. (*88, 243, 245*)

———. (1983) *Animal Mechanics*. Oxford, UK: Blackwell Scientific Publications. 301 pp. (*80, 184, 191, 293*)

———. (1984) Walking and running. *Amer. Sci. 72*: 348–54. (*69*)

———. (1986) The ideal and the feasible: Physical constraints on evolution. *J. Linnean Soc. Lond. 26*: 345–58. (*3*)

Alexander, R.M. and A.S. Jayes (1983) A dynamic similarity hypothesis for the gaits of quadrupedal mammals. *J. Zool., Lond. 201*: 135–52. (*68*)

Allen, N.S. (1974) Endoplasmic filaments generate the motive force for rotational streaming in *Nitella. J. Cell Biol. 63*: 270–87. (*260*)

Andersen, N.M. and J.T. Polhemus (1976) Water striders (Hemiptera: Gerridae, Veliidae, etc.) Pages 187–224 in L. Cheng, ed., *Marine Insects*. New York: American Elsevier. (*100*)

Arnold, G.P. and D. Weihs (1978) The hydrodynamics of rheotaxis in the plaice (*Pleuronectes platessa* L.). *J. Exp. Biol. 75*: 147–70. (*156*)

*Atkins, P.W. (1984) *The Second Law*. New York: Scientific American Library. 230 pp. (*308*)

*Bascom, W. (1980) *Waves and Beaches*, 2nd ed. Garden City, NY: Anchor/Doubleday. 366 pp. (*101*)

Bennet-Clark, H.C. (1975) The energetics of the jump of the locust *Schistocerca gregaria. J. Exp. Biol. 63*: 53–83. (*292, 296*)

———. (1977) Scale effects in jumping animals. Page 185–201 in T.J. Pedley, ed., *Scale Effects in Animal Locomotion*. London: Academic Press. (*291, 292*)

Bennet-Clark, H.C. and E.C.A. Lucey (1967) The jump of the flea: A study of the energetics and a model of the mechanism. *J. Exp. Biol. 47*: 59–76. (*296*)

*Berg, H.C. (1983) *Random Walks in Biology*. Princeton, NJ: Princeton University Press. 142 pp. (*163*)

Berg, H.C. and R.A. Anderson (1973) Bacteria swim by rotating their flagellar filaments. *Nature 245*: 380–82. (*274*)

Billington, D.P. (1983) *The Tower and the Bridge*. New York: Basic Books. 306 pp. (*209*)

*Boys, C.V. (1902) *Soap Bubbles: Their Colours and the Forces Which Mold Them*. New York: Dover Publications (reprint, 1959). 192 pp. (*236*)

*Brancazio, P.J. (1984) *Sport Science*. New York: Simon & Schuster. 400 pp. (*294*)

Bridgman, P.W. (1961) Dimensional analysis. *Encyclopedia Britannica* 7: 387–387J. (*61*)

Briggs, L.J. (1950) Limiting negative pressure of water. *J. Appl. Phys. 21*: 721–22. (*93*)

Buller, A.H.R. (1934) *Researches on Fungi*, Vol. VI. London: Longmans, Green. 513 pp. (*296*)

Burton, A.C. (1954) The relation of structure to function of the tissues of the wall of blood vessels. *Physiol. Rev. 34*: 619–42. (*232*)

*Calder, W.A. (1984) *Size, Function, and Life History*. Cambridge, MA: Harvard University Press. 431 pp. (*49*)

Carey, D.A. (1983) Particle resuspension in the boundary layer induced by flow around polychaete tubes. *Can. J. Fish. Aquat. Sci. 40* (Suppl. 1): 301–308. (*113*)

Carpita, N.C. (1985) Tensile strength of cell walls of living cells. *Plant Physiol. 79*: 485–88. (*235*)

Chance, M.M. and D.A. Craig (1986) Hydrodynamics and behaviour of Simuliidae larvae (Diptera). *Can. J. Zool. 64*: 1295–1309. (*113*)

Chapman, R.F. (1982) *The Insects: Structure and Function*, 2nd ed. Cambridge, MA: Harvard University Press. 919 pp. (*100*)

Charters, A.C., M. Neushul, and C. Barilotti (1969) Functional morphology of *Eisenia arborea*. *Sixth Int. Seaweed Symp.*, Madrid, pp. 89–105. (*247*)

Clark, R.B. and J.B. Cowey (1958) Factors controlling the change of shape of certain nemertean and turbellarian worms. *J. Exp. Biol. 35*: 731–48. (*222*)

Clements, J.A. (1962) Surface tension in the lungs. *Sci. Amer. 207 (6)*: 120–30. (*235*)

*Close, R.I. (1972) Dynamic properties of mammalian skeletal muscle. *Physiol. Rev. 52*: 129–97. (*262*)

Cowey, J.B. (1952) The structure and function of the basement membrane muscle system in *Amphiporus lactifloreus* (Nemertea). *Quart. J. Micro. Sci. 93*: 1–15. (*222*)

Craig, C.L. (1987) The significance of spider size to the diversification of spider-web architectures and spider reproductive modes. *Amer. Nat. 129*: 47–68. (*185*)

Crisp, D.J. (1960) Mobility of barnacles. *Nature 188*: 1208–1209. (*285*)

*Currey, J.D. (1984) *The Mechanical Adaptations of Bones*. Princeton, NJ: Princeton University Press. 294 pp. (*185, 186, 188, 193, 199, 213, 243, 244, 245, 266, 268*)

Davenport, J., S. A. Munks, and P.J. Oxford (1984) A comparison of the swimming of marine and freshwater turtles. *Proc. Roy. Soc. Lond. B220*: 447–75. (*283*)

Denny, M. (1980) Silks—their properties and functions. Pages 247–72 in J.F.V. Vincent and J.D. Currey, eds., *The Mechanical Properties of Biological Materials—Symp. Soc. Exp. Biol. No. 34*. Cambridge, UK: Cambridge University Press. (*190, 195*)

———. (1988) *Biology and the Mechanics of the Wave-Swept Environment*. Princeton, NJ: Princeton University Press. 400 pp. (*101*)

Denny, M. and J.M. Gosline (1980) The physical properties of the pedal mucus of the terrestrial slug, *Ariolimax columbianus*. *J. Exp. Biol. 88*: 375–93. (*199*)

Dimery, N.J., R. M. Alexander, and K.A. Deyst (1985) Mechanics of the ligamentum nuchae of some artiodactyls. *J. Zool. Lond. A206*: 341–51. (*184*)

Dixon, H.H. and J. Joly (1894) On the ascent of sap. *Ann. Bot. 8*: 468–70. (*94*)

DuBois, A.B., G.A. Cavagna, and R. Fox (1974) Pressure distribution on the body surface of swimming fish. *Biol. Bull. 60*: 581–91. (*138*)

Eckman, J.E., A.R.M. Nowell, and P.A. Jumars (1981) Sediment destabilization by animal tubes. *J. Mar. Res. 39*: 361–74. (*113*)

Ellis, C.H. (1944) The mechanism of extension of the legs of spiders. *Biol. Bull. 86*: 41–50. (*274*)

Emerson, S. (1978) Allometry and jumping in frogs: Helping the twain to meet. *Evolution 32*: 551–64. (*290*)

Emerson, S. and D. Diehl (1980) Toe-pad morphology and mechanisms of sticking in frogs. *Biol. J. Linn. Soc. 13*: 199–216. (*288*)

Emlet, R.B. (1982) Echinoderm calcite: A mechanical analysis from larval spicules. *Biol. Bull. 163*: 264–75. (*210*)

*Eshbach, O.E., ed. (1952) *Handbook of Engineering Fundamentals*, 2nd ed. New York: Wiley. (*186*)

Evans, M.E.G. (1973) The jump of the click beetle (Coleoptera, Elateridae)—Energetics and mechanics. *J. Zool., Lond. 169*: 181–94. (*292*)

Eylers, J.P. (1976) Aspects of skeletal mechanics of the starfish *Asterias forbesi*. *J. Morphol. 149*: 353–67. (*217*)

Fairchild, L. (1981) Mate selection and behavioral thermoregulation in Fowler's toads. *Science 212*: 950–51. (*46*)

*Feynman, R.P., R.B. Leighton, and M. Sands (1963) *The Feynman Lectures on Physics*, Vol. 1. Reading, MA: Addison-Wesley. (*8, 302*)

Fish, F.E. (1982) Aerobic energetics of surface swimming in the muskrat, *Ondatra zibethicus*. *Physiol. Zool. 55*: 180–89. (*68*)

Foelix, R.F. (1982) *Biology of Spiders*. Cambridge, MA: Harvard University Press. 306 pp. (*92*)

Fuller, R.B. (1962) *Tensile-integrity structures*. Washington, DC: U.S. Patent Office, No. 3,063,521. (*216*)

*Gans, C. (1974) *Biomechanics: An Approach to Vertebrate Biology*. Philadelphia, PA: Lippincott. 261 pp. (*269*)

Gibo, D.L. and M.J. Pallett (1979) Soaring flight of monarch butterflies, *Danaus plexippus* (Lepidoptera: Danaidae), during the late summer migration in southern Ontario. *Can. J. Zool. 57*: 1393–1401. (*156*)

*Goldstein, S. (1938) *Modern Developments in Fluid Dynamics*. Oxford, UK: Clarendon Press. 702 pp. (Reprinted, New York: Dover Publications, 1965). (*95*)

*Gordon, J.E. (1976) *The New Science of Strong Materials*. Harmondsworth, UK: Penguin Books. 287 pp. (Reprinted, Princeton, NJ: Princeton University Press, 1984). (*178, 189*)

———. (1978) *Structures; or, Why Things Don't Fall Down*. New York: Plenum Press. 395 pp. (*178, 185, 188, 191, 193, 206, 212, 233, 242*)

*Gould, S.J. (1980) *The Panda's Thumb*. New York: W.W. Norton. 343 pp. (*11*)

———. (1981) Kingdoms without wheels. *Nat. Hist. 90 (4)*: 42–48. (*275*)

Green, D.M. (1981) Adhesion and the toe-pads of tree frogs. *Copeia 1981*: 790–96. *(288)*

*Haldane, J.B.S. (1928) On being the right size. Pages 20–28 in *Possible Worlds*. New York: Harper. *(7)*

Halfen, L.N. (1979) Gliding movements. Pages 250–67 in W. Haupt and M.E. Feinleib, eds., *Physiology of Movements*. New York: Springer-Verlag. *(263)*

*Happel, J. and H. Brenner (1965) *Low Reynolds Number Hydrodynamics*. Englewood Cliffs, NJ: Prentice-Hall. 553 pp. *(124)*

Harris, J.E. and H.D. Crofton (1957) Structure and function in the nematodes: Internal pressure and cuticular structure in *Ascaris*. *J. Exp. Biol. 34*: 116–30. *(223)*

Harvey, E.N., D.K. Barnes, W.D. McElroy, A.H. Whiteley, D.C. Pease, and K.W. Cooper (1944) Bubble formation in animals. I. Physical factors. *J. Cell. Comp. Physiol. 24*: 1–22. *(237)*

*Hayward, A.T.J. (1971) Negative pressure in liquids: Can it be harnessed to serve man? *Amer. Sci. 59*: 434–43. *(93)*

Hill, A.V. (1950) The dimensions of animals and their muscular dynamics. *Sci. Prog., Oxford 150*: 209–30. *(262)*

*Hinton, H.E. (1976) Plastron respiration in bugs and beetles. *J. Insect Physiol. 22*: 1529–50. *(101)*

*Hoerner, S.F. (1965) *Fluid-Dynamic Drag*. S.F. Hoerner, 2 King Lane, Greenbriar, Brick Town, NJ 08723. *(145, 147)*

Holstein, T. and P. Tardent (1984) An ultrahigh-speed analysis of exocytosis: Nematocyst discharge. *Science 223*: 830–32. *(291)*

Horn, H.S. (1971) *The Adaptive Geometry of Trees*. Princeton, NJ: Princeton University Press. 144 pp. *(280)*

*Horowitz, N.H. (1986) *To Utopia and Back: The Search for Life in the Solar System*. New York: W.H. Freeman. 168 pp. *(12)*

Huey, R.B. and P.E. Hertz (1984) Effects of body size and slope on acceleration of a lizard (*Stellio stellio*). *J. Exp. Biol. 110*: 113–23. *(291)*

Hunter, T. and S. Vogel (1986) Spinning embryos enhance diffusion through gelatinous egg masses. *J. Exp. Mar. Biol. Ecol. 96*: 303–308. *(164)*

Ingold, C.T. (1961) Ballistics in certain ascomycetes. *New Phytol. 60*: 143–49. *(297)*

Ingold, C.T. and S.A. Hadland (1959) The ballistics of *Sordaria*. *New Phytol. 58*: 46–57. *(297)*

Jensen, A. (1986) Why the best technology for escaping from a submarine is no technology. *Amer. Heritage of Invention and Technology 2 (1)*: 44–49. *(87)*

Jeronimidis, G. (1980) Wood, one of nature's challenging composites. Pages 169–82 in J.F.V. Vincent and J.D. Currey, eds., *The Mechanical Properties of Biological Materials—Symp. Soc. Exp. Biol. No. 34*. Cambridge, UK: Cambridge University Press. *(193)*

Jones, D.R. and D.J. Randall (1978) The respiratory and circulatory systems during exercise. Pages 425–501 in W.S. Hoar and D.J. Randall, eds., *Fish Physiology*, Vol. 7. New York: Academic Press. *(140)*

Kier, W.M. (1982) The functional morphology of the musculature of squid (Loliginidae) arms and tentacles. *J. Morph. 172*: 179–92. *(271)*

———. (1985) The musculature of squid arms and tentacles: Ultrastructural evidence for functional differences. *J. Morph. 185*: 223–39. (*271*)

*Kier, W.M. and K.K. Smith (1985) Tongues, tentacles and trunks: The biomechanics of movement in muscular-hydrostats. *Zool. J. Linn. Soc. Lond. 83*: 307–24. (*219, 253, 271*)

*Kleiber, M. (1961) *The Fire of Life. An Introduction to Animal Energetics*. New York: Wiley. 454 pp. (*38*)

Koehl, M.A.R. (1977) Mechanical organization of cantilever-like sessile organisms: Sea anemones. *J. Exp. Biol. 69*: 127–42. (*199, 212, 246*)

———. (1982) Mechanical design of spicule-reinforced connective tissue: Stiffness. *J. Exp. Biol. 98*: 239–67. (*218*)

Koehl, M.A.R. and J.R. Strickler (1981) Copepod feeding currents: Food capture at low Reynolds number. *Limnol. Oceanogr. 26*: 1062–73. (*122*)

Koehl, M.A.R. and S.A. Wainwright (1977) Mechanical adaptations of a giant kelp. *Limnol. Oceanogr. 22*: 1067–71. (*195*)

*Krogh, A. (1941) *The Comparative Physiology of Respiratory Mechanisms*. Philadelphia, PA: University of Pennsylvania Press. 172 pp. (*174*)

*LaBarbera, M. (1983) Why the wheels won't go. *Amer. Nat. 121*: 395–408. (*275*)

*LaBarbera, M. and S. Vogel (1982) The design of fluid transport systems in organisms. *Amer. Sci. 70*: 54–60. (*171*)

*Langhaar, H.L. (1951) *Dimensional Analysis and Theory of Models*. New York: Wiley. 166 pp. (*61*)

Lapennas, G.N. and K. Schmidt-Nielsen (1977) Swimbladder permeability to oxygen. *J. Exp. Biol. 67*: 175–96. (*91*)

Lilienthal, O. (1910) *Der Vogelflug als Grundlage der Fliegekunst*. Munich: Verlag R. Oldenbourg. 186 pp. (*315*)

Lucas, J.R. (1982) The biophysics of pit construction by antlion larvae (*Myrmeleon*, Neuroptera). *Anim. Behav. 30*: 651–64. (*199*)

McCutchen, C.W. (1977) The spinning rotation of ash and tulip tree samaras. *Science 197*: 691–92. (*150*)

Macdougal, D.T. (1925) Reversible variations in volume, pressure, and movement of sap in trees. *Carnegie Inst. Wash. Publ. 365*, pp. 1–90. (*95*)

McMahon, T.A. (1973) Size and shape in biology. *Science 179*: 1201–1204. (*52*)

———. (1984) *Muscles, Reflexes, and Locomotion*. Princeton, NJ: Princeton University Press. 331 pp. (*261*)

*McMahon, T.A. and J.T. Bonner (1983) *On Size and Life*. New York: Scientific American Books. 255 pp. (*49*)

Maderson, P.F.A. (1964) Keratinized epidermal derivatives as an aid to climbing in gekkonid lizards. *Nature 203*: 780–81. (*284*)

———. (1970) Lizard glands and lizard hands: Models for evolutionary study. *Forma et Functio 3*: 179–204. (*284*)

Margaria, R. (1976) *Biomechanics and Energetics of Muscular Exercise*. Oxford, UK: Clarendon Press. 146 pp. (*309*)

Marquis, D. (1927) *archy and mehitabel*. Garden City, NY: Doubleday. 264 pp. (*207*)

*Marsden, E.W. (1969) *Greek and Roman Artillery: Historical Development*. Oxford, UK: Clarendon Press. 218 pp. (*194*)

REFERENCES AND CITATIONS

*Marsden, E.W. (1971) *Greek and Roman Artillery: Technical Treatises.* Oxford, UK: Clarendon Press. 277 pp. (*194*)

Martin, R.D. (1981) Relative brain size and basal metabolic rate in terrestrial vertebrates. *Nature 293*: 57–60. (*65*)

Mastro, A.M. and A.D. Keith (1984) Diffusion in the aqueous compartment. *J. Cell Biol. 99*: 180s–87s. (*164*)

Maynard Smith, J. (1952) The importance of the nervous system in the evolution of animal flight. *Evolution 6*: 127–29. (*315*)

Morowitz, H.J. and M.E. Tourtellotte (1962) The smallest living cells. *Sci. Amer. 206(3)*: 117–26. (*170*)

*Morrison, P. and P. Morrison (1982) *Powers of Ten.* New York: Scientific American Books. 150 pp. (*5*)

*Motokawa, T. (1984) Connective tissue catch in echinoderms. *Biol. Rev. 59*: 255–70. (*251*)

*Nachtigall, W. (1974) *Biological Mechanisms of Attachment.* New York: Springer-Verlag. 194 pp. (*283, 287*)

Nicklas, R.B. (1984) A quantitative comparison of cellular motile systems. *Cell Motility 4*: 1–5. (*262*)

Nijhout, H.F. and H.G. Sheffield (1979) Antennal hair erection in male mosquitoes: A new mechanical effector in insects. *Science 206*: 595–96. (*255*)

*Nobel, P. (1974) *Introduction to Biophysical Plant Physiology.* San Francisco: W.H. Freeman. 488 pp. (*240*)

Page, R.M. (1964) Sporangium discharge in *Pilobolus*: A photographic study. *Science 146*: 925–27. (*291*)

Paine, V. (1926) Adhesion of the tube feet in starfish. *J. Exp. Zool. 45*: 361–66. (*287*)

*Pais, A. (1982) *"Subtle is the Lord. . . ." The Science and the Life of Albert Einstein.* Oxford, UK: Oxford University Press. 552 pp. (*159*)

Parker, G.H. (1921) The power of adhesion in the suckers of *Octopus bimaculatus* Verrill. *J. Exp. Zool. 33*: 391–94. (*287*)

*Parkes, E.W. (1965) *Braced Frameworks, An Introduction to the Theory of Structures.* Oxford, UK: Pergamon Press. 198 pp. (*213*)

Parry, D.A. and R.H.J. Brown (1959) The hydraulic mechanism of the spider leg. *J. Exp. Biol. 36*: 423–33. (*274, 291*)

Pennycuick, C.J. (1988) *Conversion Factors: SI Units and Many Others.* Chicago: University of Chicago Press. 47 pp. (*326*)

*Peters, R.H. (1983) *The Ecological Implications of Body Size.* Cambridge, UK: Cambridge University Press. 329 pp. (*49*)

Peterson, J.A., J.A. Benson, M. Ngai, J. Morin, and C. Ow (1982) Scaling in tensile "skeletons": Structures with scale-independent length dimensions. *Science 217*: 1267–70. (*249*)

*Petroski, H. (1985) *To Engineer is Human.* New York: St. Martin's Press. 247 pp. (*243*)

Pickard, W.F. (1974) Transition regime diffusion and the structure of the insect tracheolar system. *J. Insect Physiol. 20*: 947–56. (*158, 166*)

*Prandtl, L. and O.G. Tietjens (1934) *Applied Hydro- and Aeromechanics.* 311 pp. (Reprint, New York: Dover Publications, 1957). (*109*)

348

Prange, H.D. and K. Schmidt-Nielsen (1970) The metabolic cost of swimming in ducks. *J. Exp. Biol. 53*: 763–77. (*68*)

*Purcell, E.M. (1977) Life at low Reynolds number. *Amer. J. Physics 45*: 3–11. (*121, 165*)

Putz, F.E., P.D. Coley, K. Lu, A. Montalvo, and A. Aiello (1983) Uprooting and snapping of trees: Structural determinants and ecological consequences. *Can. J. For. Res. 13*: 1011–20. (*278*)

Randall, D.J. and C. Daxboeck (1984) Oxygen and carbon dioxide transfer across fish gills. pp. A263–314 in W.S. Hoar and D.J. Randall, eds., *Fish Physiology*, Vol. 10. New York: Academic Press. (*140*)

Raschke, K. (1979) Movements of stomata. pp. 383–441 in W. Haupt and M.E. Feinleib, eds., *Physiology of Movements*. New York: Springer-Verlag. (*259*)

Raup, D.M. (1966) Geometric analysis of shell coiling: General problems. *J. Paleont. 40*: 1178–90. (*56*)

Reynolds, O. (1883) An experimental investigation of the circumstances which determine whether the motion of water shall be direct or sinuous, and the law of resistance in parallel channels. *Trans. Roy. Soc. Lond. 174*: 935–82. (*114*)

Rohrer, F. (1921) Der Index der Korperfulle als Mass des Ernahrungszustandes. *Muenchner Medizinische Wochenschrift 68*: 580–82. (*66*)

*Rose, C.W. (1966) *Agricultural Physics*. Oxford, UK: Pergamon Press. 230 pp. (*240*)

*Rubenstein, D.I. and M.A.R. Koehl (1977) The mechanisms of filter feeding: Some theoretical considerations. *Amer. Natur. 111*: 981–84. (*112*)

*Salisbury, F.B. and C. Ross (1969) *Plant Physiology*. Belmont, CA: Wadsworth. 747 pp. (*255*)

Satter, R.L. (1979) Leaf movements and tendril curling. Pages 442–84 in W. Haupt and M.E. Feinleib, eds., *Physiology of Movements*. New York: Springer-Verlag. (*259*)

*Schmidt-Nielsen, K. (1964) *Desert Animals: Physiological Problems of Heat and Water*. Oxford, UK: Clarendon Press. 277 pp. (*302*)

———. (1972) *How Animals Work*. Cambridge, UK: Cambridge University Press. 114 pp. (*129*)

———. (1979) *Animal Physiology: Adaptation and Environment*. 2nd ed. Cambridge, UK: Cambridge University Press. 560 pp. (*88, 164, 310*)

———. (1984) *Scaling*. Cambridge, UK: Cambridge University Press. 241 pp. (*49, 87, 262, 291*)

Scholander, P.F. (1954) Secretion of gases against high pressures in the swimbladder of deep sea fishes. 2. The rete mirabile. *Biol. Bull. 107*: 260–77. (*91*)

Scholander, P.F., H.T. Hammel, E.D. Bradstreet, and E.A. Hemmingsen (1965) Sap pressure in vascular plants. *Science 148*: 339–46. (*95*)

*Schrödinger, E. (1944) *What is Life?* Cambridge, UK: Cambridge University Press. 91 pp. (*5, 160, 170*)

Sheehan, S. (1986) *A Missing Plane*. New York: Putnam. 224 pp. (*66*)

*Sprackling, M.T. (1985) *Liquids and Solids*. London: Routledge and Kegan Paul. 237 pp. (*93*)

Stevens, S.S. (1946) On the theory of scales of measurement. *Science 103*: 677–80. *(319)*

*Steward, F.C. (1968) *Growth and Organization in Plants*. Reading, MA: Addison-Wesley. 564 pp. *(263)*

Strickler, J.R. (1977) Observation of swimming performances of planktonic copepods. *Limnol. Oceanogr. 22*: 165–70. *(123)*

Strickler, J.R. and S. Twombly (1975) Reynolds number, diapause, and predatory copepods. *Verh. Internat. Verein. Limnol. 19*: 2943–50. *(122)*

Stride, G.O. (1955) On the respiration of an aquatic African beetle, *Potamodytes tuberosus* Hinton. *Ann. Ent. Soc. Amer. 48*: 344–51. *(91, 141)*

*Tabor, D. (1979) *Gases, Liquids and Solids*, 2nd ed. Cambridge, UK: Cambridge University Press. 301 pp. *(84, 166)*

Taghon, G.L., A.R.M. Nowell, and P.A. Jumars (1980) Induction of suspension feeding in spionid polychaetes by high particulate fluxes. *Science 210*: 262–64. *(113)*

Tarkov, J. (1986) Engineering the Erie Canal. *Amer. Heritage of Invention and Technology 2 (1)*: 50–57. *(33)*

Taylor, C.R. and E.R. Weibel (1981) Design of the mammalian respiratory system. I. Problem and strategy. *Resp. Physiol. 44*: 1–10. *(168)*

*Teale, E.W. (1949) *The Insect World of J. Henri Fabre*. New York: Dodd, Mead. 333 pp. *(111)*

Telewski, F.W. and M.J. Jaffe (1986) Thigmomorphogenesis: Field and laboratory studies of *Abies fraseri* in response to wind or mechanical perturbation. *Physiol. Plant. 66*: 211–18. *(146)*

Thomas, L.A. and C.O. Hermans (1985) Adhesive interactions between the tube feet of a starfish, *Lepasterias hexactis*, and substrata. *Biol. Bull. 169*: 675–88. *(287)*

*Thompson, D'Arcy W. (1942) *On Growth and Form*. Cambridge, UK: Cambridge University Press. 1116 pp. *(11, 48, 65, 158, 210, 213, 222)*

————. (1961) *On Growth and Form*. Abridged, J.T. Bonner, ed., Cambridge, UK: Cambridge University Press. 345 pp. *(48)*

Thorp, W.H. and D.J. Crisp (1947) Studies on plastron respiration. I. The biology of *Aphelocheirus* [Hemiptera, Aphelocheiridae (Naucoridae)] and the mechanism of plastron retention. *J. Exp. Biol. 24*: 227–69. *(240)*

*Trefil, J.S. (1984) *A Scientist at the Seashore*. New York: Scribner's. 208 pp. *(101)*

*Tricker, R.A.R. and B.J.K. Tricker (1967) *The Science of Movement*. New York: American Elsevier. 284 pp. *(266)*

Tucker, V.A. (1969) Wave-making by whirligig beetles (Gyrinidae). *Science 166*: 897–99. *(103)*

————. (1975) The energetic cost of moving about. *Amer. Sci. 63*: 413–19. *(275, 310, 311)*

*Van Dyke, M. (1982) *An Album of Fluid Motion*. Stanford, CA: Parabolic Press. 176 pp. *(106)*

Vincent, J.F.V. (1975) Locust oviposition: Stress softening of the extensible intersegmental membranes. *Proc. Roy. Soc. B188*: 189–201. *(187)*

————. (1982a) *Structural Biomaterials*. New York: Wiley. 206 pp. *(182, 186, 189, 190, 191, 197)*

————. (1982b) The mechanical design of grass. *J. Materials Sci.* 17: 856–60. (*198*)

Vincent, J.F.V. and P. Owers (1986) Mechanical design of hedgehog spines and porcupine quills. *J. Zool., Lond.* A210: 55–75. (*6, 75*)

Vogel, S. (1978) Organisms that capture currents. *Sci. Amer.* 239 (2): 128–39. (*134*)

————. (1981a) Behavior and the physical world of an animal. Pages 179–98 in P.P.G. Bateson and P.H. Klopfer, eds., *Perspectives in Ethology*, Vol. 4. New York: Plenum Press. (*18*)

————. (1981b) *Life in Moving Fluids*. Boston: Willard Grant Press; Princeton, NJ: Princeton University Press. 352 pp. (*106, 153, 278*)

————. (1983) How much air flows through a silkmoth's antenna? *J. Insect Physiol.* 29: 597–602. (*112*)

————. (1984) Drag and flexibility in sessile organisms. *Amer. Zool.* 24: 37–44. (*146*)

————. (1986) Subtlety and suppleness. *Mechanical Engineering* 108 (11): 60–68. (*314*)

————. (1987) Flow-induced mantle cavity refilling in jetting squid. *Biol. Bull.* 172: 61–68. (*117*)

Vogel, S., C.P. Ellington, Jr., and D.C. Kilgore, Jr. (1973) Wind-induced ventilation of the burrow of the prairie dog, *Cynomys ludovicianus*. *J. Comp. Physiol.* 85: 1–14. (*117, 134*)

Vogel, S. and C. Loudon (1985) Fluid mechanics of the thallus of an intertidal red alga, *Halosaccion glandiforme*. *Biol. Bull.* 168: 161–74. (*146*)

Vogel, S. and S.A. Wainwright (1969) *A Functional Bestiary*. Reading, MA: Addison-Wesley. 106 pp. (*172, 235, 331*)

*Von Karman, T. (1954) *Aerodynamics*. Ithaca, NY: Cornell University Press. 203 pp. (*109, 145*)

*Wainwright, S.A. (1988) *Axis and Circumference*. Cambridge, MA: Harvard University Press. 132 pp. (*12, 252*)

*Wainwright, S.A., W.D. Biggs, J.D. Currey, and J.M. Gosline (1976) *Mechanical Design in Organisms*. London: Edward Arnold (reprinted, Princeton University Press). 423 pp. (*178, 182, 184, 188, 190, 212, 213, 222*)

Wainwright, S.A., F. Vosburgh, and J.H. Hebrank (1978) Shark skin: Function in locomotion. *Science* 202: 747–49. (*224*)

*Wallace, J.B. and R.W. Merritt (1980) Filter-feeding ecology of aquatic insects. *Ann. Rev. Entomol.* 25: 103–32. (*135*)

Walsby, A.E. (1980) A square bacterium. *Nature* 283: 69–71. (*57*)

Ward, D.V. and S.A. Wainwright (1972) Locomotory aspects of squid mantle structure. *J. Zool., Lond.* 167: 437–49. (*224*)

Waters, N.E. (1980) Some mechanical and physical properties of teeth. Pages 99–135 in J.F.V. Vincent and J.D. Currey, eds., *The Mechanical Properties of Biological Materials—Symp. Soc. Exp. Biol. No. 34*. Cambridge, UK: Cambridge University Press. (*184, 192*)

Webb, P.W. (1976) The effect of size on the fast-start performance of rainbow trout *Salmo gairdneri* and a consideration of piscivorous predator-prey interactions. *J. Exp. Biol.* 65: 157–77. (*291*)

Weis-Fogh, T. (1960) A rubber-like protein in insect cuticle. *J. Exp. Biol.* 37: 889–907. (*80*)

REFERENCES AND CITATIONS

Weis-Fogh, T. and R.M. Alexander (1977) The sustained power output from striated muscle. Pages 511–45 in T.J. Pedley, ed., *Scale Effects in Animal Locomotion*. London: Academic Press. *(292)*

Weis-Fogh, T. and W.B. Amos (1972) Evidence for a new mechanism of cell motility. *Nature 236*: 301–304. *(123)*

*Went, F.W. (1968) The size of man. *Amer. Sci. 56*: 400–413. *(4, 44)*

West, L.J. et al. (1962) Lysergic acid diethylamide: Its effects on a male Asiatic elephant. *Science 138*: 1100–1103. *(38)*

Wetmore, K.L. (1987) Correlations between test strength, morphology and habitat in some benthic foraminifera from the coast of Washington. *J. Foram. Res. 17*: 1–13. *(243)*

White, C.M. (1946) The drag of cylinders in fluids at low speeds. *Proc. Roy. Soc. Lond. A 186*: 472–79. *(122)*

Wilcox, R.S. (1979) Sex discrimination in *Gerris remigis*: Role of a surface wave signal. *Science 206*: 1325–27. *(104)*

Wu, T.Y.-T. (1977) Hydrodynamics of swimming at low Reynolds numbers. *Fortschr. Zool. 24 (2–3)*: 149–69. *(122)*

Ycas, M., M. Sugita, and A. Bensam (1965) A model of cell size regulation. *J. Theoret. Biol. 9*: 444–70. *(169)*

Young, C.M. and L.F. Braithwaite (1980) Orientation and current-induced flow in the stalked ascidian, *Styela montereyensis. Biol. Bull. 159*: 428–40. *(135)*

Subject index

abductin, occurrence, use, 180; resilience, 190

acceleration, definition, 19–20; detection, 58; dimensions, 19–20; and force, 21, 30; gravitational, 20; size and, 290; speed and, 28, 70, 289; table, 291; unit, 19

accuracy, 14–15, 320–21

actin, 260–61

adaptationism, 10–11, 274

adhesion, 283–88; liquid-solid, 97–98

aggression, 42–43

air: density, 108; diffusion in, 164, 165; heat capacity, 300; viscosity, 108. *See also* atmosphere, gases

airfoils, 150–53; gliding, 155–56; lift, 150–51; lift-to-drag ratio, 152–53

albumin; diffusion coefficient, 164

algae: drag and flexibility, 146; floats, 218; gliding movement, 263; habitat, 246, 247; *Halosaccion*, 146; holdfast, 286–87; shape, 171; size range, 39; tensile systems, 248

allometry, 40–41, 45–54; exponents, 47–48; graphical display, 46–47; within cells, 42

aluminum: crack propagation, 146; density, 27

alveoli: Laplace's law, 235; number, 41; oxygen movement, 74; size, 167–68; stress-strain relation, 235; surface area, 41; wetting agent in, 235

ammonia, solubility, 90

anemometer, Pitot tube, 135

aneurisms, 185

angle of attack, 151–52

angle of repose, 199

annelids, hydroskeleton, 220

ant: use of fire, 4; weight lifting, 4

antelope, acceleration, 291

antennae, insect: air passage, 111–12; erecting hairs, 255

Antheraea (silkmoth), 178

antler, density, 27

ant lion, larval pit, 199

aorta: continuity, 36; cross-sectional area, 174; diameter, 173; non-turbulent flow, 127; toughness, 191

apodemes: energy storage, 292; lobster, 179; pinnate muscles, 269; safety factor, 245; Young's modulus, 184

area, 19

area, second moment. *See* second moment of area

Argyroneta, diving and web, 92

arterial wall: collagen of, 178, 232; deposits, 108; elastin of, 232; fibers in, 232; pliancy, 181; strength, 185; stress-strain relation, 80, 232

arteries, internal pressure, 4

arthropods. *See* crustacea, insects, spiders, *etc.*

Ascaris (roundworm), internal pressure, 223–24

ascidians: cellulose in, 179; Pitot tube use, 135; spicules in, 218

atmospheric pressure, 24, 85

atoms. *See* molecules and atoms

attachments, 283–84

autotomy, 244

axonal transport, 171

backbone, as truss, 210–11

bacteria: feeding, 165; flagella, 74, 276; internal pressure, 235; propulsion, 121; size, 5, 38, 41, 170; square, 57

balance, beam, 22, 34

ballistae: collagen in, 179; performance, 193–94

bamboo: growth, 255; stability, 280

barnacles: adhesion, 283, 285; spicules, 218

baseball, trajectory, 296

beams: biological, 206; box, 209; flexural stiffness, 203–206; neutral plane, 202–204; second moment of area, 203–205; shapes, 207–209; trusses as, 209

bear, overwintering, 52

INDEX

fracture. *See* bones, crack propagation, *etc.*
frameworks: articulated struts, 250–51; braced, 213–16; trusses, 209–211
free stream speed, 75, 110
freezing: bubbles and, 239; water column and, 95
frog: adhesion, 285, 288; interdigital webbing, 36; toe pads, 288
Froude number, 66–69, 103
fruit-fly: as glider, 155; wing mass, 252; wing performance, 153

gait: Froude number and, 68–69; on moon, 22
galago, acceleration, 291
gases, 84–92; Boyle's law, 85; compressibility, 88, 105; Henry's law, 88; incompressibility in flow, 36; mean free path, 159; partial pressure, 86; pressure and density, 85; pressure and volume, 85; in pressurized structures, 13, 218; resisting stresses, 84; solubilities, 88; as structural materials, 13, 218; supersaturation, 90. *See also* air, oxygen, *etc.*
gasoline, density, 27
gecko, adhesion, 284
geodesic domes, 214, 215
gills: flow between, 128; pressure and, 140
glass: bubbles formation on, 237–38; crack propagation, 196–97
gliding, algal, 263
gliding in air, 148, 154–56
glue, 284, 287
glycine, diffusion coefficient, 164
gradients. *See* concentration gradients, rates and gradients, temperature gradients, velocity gradients
graphs, 324–25; areas and summations, 77–78; geometric axes, 46–47; showing rates and gradients, 71–72; slopes, 47
grass: resistance to tearing, 198; surface area, 41
grasshoppers. *See* locusts
gravitational forces: in Froude number, 67; in Jesus number, 70
gravitational stability, 35, 58, 277
gravity: acceleration of, 3, 7, 20; force

of, 21–22; pendulum and, 61–62; on structures, 206, 208; trajectories and, 293–97
gravity, center of. *See* center of gravity
gravity waves, 101–102
gray stripe illusion, 76
guard cells, motility, 258

Hagen-Poiseuille equation, 126–27, 130, 167, 172
hair, 181, 185
Halobates (bug), water walking, 100
Halosaccion (alga), drag, 146
heart: and continuity, 36; fish, 140
heartbeat: frequency, 49, 53; per lifetime, 54
heat, 299–302; as energy, 299, 307; from organisms, 307–308; and second law of thermodynamics, 307, 308
heat capacity, specific, 300–301
heat conduction, 13; analogy with diffusion, 164; through fat, 73
hedgehogs, shock absorbing spines, 6, 76
helices of fibers, 219–21
helicopter, 154
helium, solubility, 90
Henry's law, 89; and bubbles, 236–39
hollowness: advantages, 205; box beams, 209; columns, 212–13
holly, leaf drag, 146–48
Hookean materials, 184, 203
hoop stress. *See* stress
hopping, work storage, 80, 194, 292
horn: in bows, 181; strain energy storage, 188
horse: cost of transport, 311; ischial trabeculae, 214; metabolic scope, 53; results of fall, 7
hull speeds, 67–68, 103
humans: acceleration, 291–92; bends, 90; breaking bones, 243; collarbone, 227–28; cost of transport, 311; density, 66; exhalation temperatures, 129; falling, 6, 7; fetal metabolism, 54; gait transition, 68–69; kneecap, 267–68; large size, 5; metabolic scope, 53; ponderal index, 65–66; safety factors, 245; skeletal mass, 50; tongue, 270; under water, 86–87; walking on moon, 22; walking on water, 100

human technology, 11–13, 86–87, 149, 152, 155, 181, 197, 205, 218, 248–53, 264–65, 273, 275, 283–85, 315
hummingbirds: torpor, 52–53; wings, 156
hyaline membrane disease, 235
Hydra (coelenterate), nematocyst acceleration, 290–92
hydration, for motility, 254–55
hydraulic linkages, 272–74
hydrophilic surfaces: adhesion, 285; bubbles, 237; clean glass, 238; fabric "breathing," 239; in xylem, 237–38
hydrophobic surfaces: bubbles, 237–38; insect hairs, 240; plastics, 238; water repellent fabrics, 239
hydroskeletons, 219–24; constancy of loading, 243; as supportive systems, 253. *See also* muscular hydrostats
hysteresis. *See* resilience

I-beam, 208
icosahedrons, 55
imbibition, 255
induced drag, 154
induced power, 154
inebriation, 160–61, 163, 64
inertial forces: in Froude number, 67; in Reynolds number, 115
information, 168–69
insects: antennae, 111–12; caddisfly larvae, 135–36; chitin, 179; cuticle, 191; dragonflies, 268; flight, 28, 80, 268; flight muscle, 266–68; fruit-fly, 153, 155, 252; hollow skeleton, 205–206; leg exoskeleton, 250; plastron respiration, 101, 240; resilin, 80, 180; scaling of sounds, 44; silk, 178, 195; silkmoths, 111–12, 178, 195; size range, 39; suspension feeding, 113; tracheae, 158, 166, 174; walking on water, 99–100; wettability, 99–101; wingbeat, 266–68; wing flatness, 252; wing structure, 208
interface, air-water, 95–104, 105, 236–41
internal cohesion: of liquids, 92–94; in trees, 94–95
internally pressurized systems. *See* hydroskeletons, muscular hydrostats

intervertebral discs: adhesion, 285; viscoelasticity, 199
intestine, surface area, 42
intrinsic speed of muscle, 262
inulin, diffusion coefficient, 164
inviscid fluid, 130; and Bernoulli's principle, 131
iron, density, 27
isometry, 40–41; shell growth, mollusks, 56

jellyfish: buoyancy, 88; jelly (mesoglea), 181; nematocysts, 43
Jesus number, 70, 102
jetting, squid, 139, 224, 273
jib: collarbone as, 227–28; kneecap as, 267–68
joints: of frameworks, 215; of sea anemone, 212
joule, 19, 25, 62
jumping: cat, 76; click beetle, 291; flea, 291; galago, 291; locust, 269, 291; maximum height, 7, 289; spider, 274, 291

kangaroos: gait transition, 69; work storage in tendons, 80, 194, 292
Kelvin degree, 300, 308, 320, 326
keratin, 181; extensibility, 186; strain energy storage, 188; strength, 185
kidneys, hydraulic filter, 273
kinematic viscosity, 108
kneecap, as jib, 268–69

lamb, nuchal ligament, 180
laminar flow: in pipes, 125–29; turbulent and, 113–14
Laplace's law, 229–41
leaves: as cantilevers, 208–209; cell wall pores, 241; drag and flexibility, 146–48; evaporative pump, 175–76; flatness, 252; petiole, 206, 208–209; stomata, 258–59; surface area, 41; tearing, 198; thermoregulation, 302
length, 17, 19; characteristic, 46, 115; nominal measure, 45
levers, 32–33, 263–73; forearm, 33; hydraulic systems, 273; mechanical advantage, 64; muscular hydrostats, 270–73

water: capillarity, 98–99; density, 108; diffusion coefficients in, 164; heat capacity, 301; incompressibility, 96, 218–24; structural material, 13; surface tension, 95–98; surface waves, 101–104; tensile strength, 92–94; walking on, 70, 99–101

water strider, 100

watt, 19, 26, 326

wavelength, 101–102, 342

waves, 101–104; bow, 67; capillary, 101–102; gravity, 101–102; stern, 67; whirligig beetles, 103–104

weight: as force, 21; and mass, 8, 21–22; on moon, 22; tree, 278–80

wettability, 97–99

whales: density of tympanic bulla, 27; size, 5, 38, 41; skeletal mass, 49, 51

Wharton's jelly, 181

wheels, 274–76

whirligig beetles: echolocation, 104; surface waves, 103–104

wind: on leaves, 146–48; on tree, 243–44, 278–79. See also drag, fluid flow, etc.

windlass, 32–33

wind-throw, 278–80

wings. See airfoils, flight, lift

wood: anisotropy, 182, 193; composition, 181–82; compressive strength, 186, 193; density, 27; energy content, 305; reaction, 246; strain energy storage,

188; strength, 185; vs. timber, 193; toughness, 191

wool. See keratin

work, 7, 19, 25–26, 342; and efficiency, 64; and heat in gas, 303–304; to get lift, 153; in locomotion, 275, 309–312; of muscle, 51; negative, 309; in surface tension, 96–97

work of extension (strain energy storage): collagen, 193–94; in jumping, 292–93; kelp, 195; and safety factor, 246; spider silk, 195; from stress-strain graph, 79–80, 183, 187–88; units, 187

work of fracture. See toughness

worms: crawling, 273; use of water, 13. See also annelids, flatworms, nemerteans, roundworms

xylem: hydrophilic surface, 237; negative pressure in, 94–95, 231; tensile stress, 84, 94–95; and water transport, 94

Young's modulus of elasticity, 183–84; bone, 193, 212; chitin of plastron, 240; in flexural stiffness, 202; guard cells, 258; and habitat, 247; and length, 217; teeth, 193; wood, 193

Zoothamnium (protozoan), spasmoneme, 123